WATER RESOURCES IN THE MEDITERRANEAN REGION

WATER RESOURCES IN THE MEDITERRANEAN REGION

Edited by

MEHREZ ZRIBI
Director of Research, National Centre for Scientific Research (CNRS), France

LUCA BROCCA
Director of Research, National Research Council (CNR), Italy

YVES TRAMBLAY
Researcher, Institute of Research for Development (IRD), France

FRANÇOIS MOLLE
Director of Research, Institute of Research for Development (IRD), France

ELSEVIER

Elsevier
Radarweg 29, PO Box 211, 1000 AE Amsterdam, Netherlands
The Boulevard, Langford Lane, Kidlington, Oxford OX5 1GB, United Kingdom
50 Hampshire Street, 5th Floor, Cambridge, MA 02139, United States

Library of Congress Cataloging-in-Publication Data
A catalog record for this book is available from the Library of Congress

British Library Cataloguing-in-Publication Data
A catalogue record for this book is available from the British Library

ISBN: 978-0-12-818086-0

For information on all Elsevier publications
visit our website at https://www.elsevier.com/books-and-journals

Publisher: Candice Janco
Acquisitions Editor: Louisa Munro
Editorial Project Manager: Emerald Li
Production Project Manager: Paul Prasad Chandramohan
Cover Designer: Matthew Limbert

Typeset by SPi Global, India

Contents

List of contributors

Nassim Ait-Mouheb
INRAE, UMR GEAU, University of Montpellier, Montpellier cedex 5, France

Clément Albergel
CNRM, University of Toulouse, Meteo-France, CNRS, Toulouse, France

Antoine Allam
CREEN, ESIB, Univ. Saint-Joseph, Beirut, Lebanon; UMR LISAH, Univ. Montpellier, INRA, Montpellier, France

Jalel Aouissi
Institut National Agronomique de Tunisie, Laboratoire GREEN-TEAM, Université de Carthage, Tunis, Tunisia

Nicolas Baghdadi
IRSTEA, University of Montpellier, TETIS, Montpellier, France

Claude Bocquillon
CREEN, ESIB, Univ. Saint-Joseph, Beirut, Lebanon

Gilles Boulet
CESBIO (UPS/CNES/CNRS/INRAE/IRD), Toulouse, France

Luca Brocca
Research Institute for Geo-Hydrological Protection, National Research Council of Italy, Perugia, Italy

Jean-Christophe Calvet
CNRM, University of Toulouse, Meteo-France, CNRS, Toulouse, France

J. Julio Camarero
Instituto Pirenaico de Ecología, Spanish National Research Council (IPE-CSIC), Zaragoza, Spain

Stefania Camici
CNR, Istituto di Ricerca per la Protezione Idrogeologica - Via della Madonna Alta, Perugia, Italy

Luca Ciabatta
CNR, Istituto di Ricerca per la Protezione Idrogeologica - Via della Madonna Alta, Perugia, Italy

Fernando Domínguez-Castro
Instituto Pirenaico de Ecología, Spanish National Research Council (IPE-CSIC), Zaragoza, Spain

Ahmed El Kenawy
Department of Geography, Mansoura University, Mansoura, Egypt; Department of Geography, Sultan Qaboos University, Muscat, Oman

Younes Fakir
TREMA/GEOHYD, Cadi Ayyad University, Marrakech; CRSA (Center for Remote Sensing Application), UM6P, Benguerir, Morocco

Elie Gaget
Tour du Valat Research Institute for the Conservation of Mediterranean Wetlands, Arles, France

Thomas Galewski
Tour du Valat Research Institute for the Conservation of Mediterranean Wetlands, Arles, France

Simon Gascoin
CESBIO, Université de Toulouse, CNES/CNRS/INRA/IRD/UPS, Toulouse, France

Ilse Geijzendorffer
Tour du Valat Research Institute for the Conservation of Mediterranean Wetlands, Arles, France

Anis Guelmami
Tour du Valat Research Institute for the Conservation of Mediterranean Wetlands, Arles, France

Tarik Hartani
Centre Universitaire de Tipaza, Tipaza, Algeria

Mariano Moreno de las Heras
Institute of Environmental Assessment and Water Research, CSIC, Barcelona; Desertification Research Centre, CSIC-UV-GV, Valencia, Spain

Lionel Jarlan
CESBIO (UPS/CNES/CNRS/INRAE/IRD), Toulouse, France

Marcel Kuper
Cirad, UMR G-Eau, Univ Montpellier, Montpellier, France; IAV Hassan II, Rabat, Morocco

Noemí Lana-Renault
Universidad de La Rioja, Logroño, Spain

Michel Le Page
CESBIO (UPS/CNES/CNRS/INRAE/IRD), Toulouse, France

Delphine Leroux
CNRM, University of Toulouse, Meteo-France, CNRS, Toulouse, France

Nacho López-Moreno
Pyrenean Institute of Ecology, CSIC, Zaragoza, Spain

Jorge Lorenzo-Lacruz
University of La Rioja, Logroño, Spain

Anna Cinzia Marra
Institute of Atmospheric Sciences and Climate, National Research Council (ISAC/CNR), Rome, Italy

Christian Massari
CNR, Istituto di Ricerca per la Protezione Idrogeologica - Via della Madonna Alta, Perugia, Italy

Javier Mateo-Sagasta
International Water Management Institute (IWMI), Colombo, Sri Lanka

Pierre-Louis Mayaux
CIRAD, UMR GEAU, University of Montpellier, Montpellier cedex 5, France

Bruno Molle
INRAE, UMR GEAU, University of Montpellier, Montpellier cedex 5, France

François Molle
Institut de Recherche pour le Développement (IRD), UMR G-Eau, Univ Montpellier, Montpellier, France

Enrique Morán-Tejeda
Department of Geography, University of the Balearic Islands, Palma de Mallorca, Spain

Roger Moussa
UMR LISAH, Univ. Montpellier, INRA, Montpellier, France

Simon Munier
CNRM, University of Toulouse, Meteo-France, CNRS, Toulouse, France

Wajdi Najem
CREEN, ESIB, Univ. Saint-Joseph, Beirut, Lebanon

Hector Nieto
COMPLUTIG, Alcala de Henares, Spain

Claudia Notarnicola
Institute for Earth Observation, EURAC, Bozen-Bolzano, Italy

Albert Olioso
UMR EMMAH (INRAE, Avignon Université), Avignon, France

Giulia Panegrossi
Institute of Atmospheric Sciences and Climate, National Research Council (ISAC/CNR), Rome, Italy

Daniele Penna
Department of Agriculture, Food, Environment and Forestry, University of Florence, Florence, Italy

Christian Perennou
Tour du Valat Research Institute for the Conservation of Mediterranean Wetlands, Arles, France

Rafael Pimentel
Fluvial Dynamics and Hydrology Group, University of Cordoba-Andalusian Institute for Earth System Research, Córdoba, Spain

María J. Polo
Fluvial Dynamics and Hydrology Group, University of Cordoba-Andalusian Institute for Earth System Research, Córdoba, Spain

Carles Sanchis-Ibor
Centro Valenciano de Estudios del Riego (CVER), Universitat Politècnica de València, València, Spain

Yves Tramblay
HydroSciences Montpellier, Montpellier, France

Sergio M. Vicente-Serrano
Instituto Pirenaico de Ecología, Spanish National Research Council (IPE-CSIC), Zaragoza, Spain

Mehrez Zribi
CESBIO (UPS/CNRS/IRD/CNES/INRAE), Toulouse, France

Foreword

Water resources in the Mediterranean region: The challenge of the 21st century

The Mediterranean basin is a unique place of biological, cultural, economic, political, and religious diversity, in perpetual evolution. This incredible melting pot, unparalleled in the world, faces a new challenge: that of having to adapt quickly to the combined effects of global warming, demographic pressure, and the excessive use of natural resources.

While the temperature of the planet has increased by 1°C "only," global warming has reached 1.4°C in the Mediterranean basin since the start of the industrial era, with measurable consequences such as the increase in sea level and acidification, or the notable drop in precipitation in the South and East. The work of the IPCC anticipates for the coming decades an amplification of these phenomena which, in the Mediterranean more than elsewhere, directly threaten biodiversity and the living conditions of populations.

Among all the impacts of these changes, those that will affect the hydrological cycle and ultimately the water resources available will be at the center of the preoccupations of citizens and decision makers, concerned about their living conditions and the development of their societies. Most of the changes are already a reality for the Mediterranean populations through the intensification of droughts, the decrease in water reserves, and its quality or the increase in extreme rain events. And the trend will not be reversed in the short term.

In this context, climate and environmental researchers have a crucial role to play not only in making their knowledge of the state of the Mediterranean basin available but also in alerting, on the basis of their scientific results, to the risks associated with these future changes. It is with this objective that the CNRS and other French research organizations (CEA, IFREMER, INRAE, IRD, Météo-France) organized themselves in 2010 by federating within a large 10-year interdisciplinary research program on the Mediterranean—MISTRALS—and by promoting collaboration with research organizations from other Mediterranean countries.

For 10 years, more than 1000 scientists from 23 countries have analyzed air pollution particles using planes and satellites, modeled using supercomputers droughts, floods and the evolution of precipitation over the next decades, embarked on ships to measure currents and contaminants in the seawater, measured the impact of agriculture and urbanization on soil and groundwater pollution, or described the evolution of biodiversity in coastal waters or on the islands.

The impressive amount of knowledge resulting from their work and published in 1500 scientific articles has led to major advances in the understanding of the physical, chemical, and biological processes that affect the environment in the Mediterranean, and to respond to societal issues of sustainable water and soil management, preservation of air and water quality, prevention of natural risks (floods, storms, droughts, etc.). MISTRALS has also enabled the emergence of international networks of scientists, such as MedECC, to communicate results to society.

This book focuses on water resources and perfectly illustrates this approach and this commitment to research for a sustainable development in the Mediterranean. The authors do not have all the solutions that will enable us to adapt our lifestyles and societies to the current hydrological changes, but they clearly set out their scientific results and submit for our consideration the consequences of the foreseeable changes in water resources in the Mediterranean basin. It is essential that each one of us—citizen, entrepreneur, politician—quickly seize these results to invent solutions for the future of the Mediterranean.

Cyril Moulin
Director of the MISTRALS program, CNRS/INSU

Introduction

Mehrez Zribi[a], Luca Brocca[b], François Molle[c], Yves Tramblay[d]
[a]CESBIO (UPS/CNRS/IRD/CNES/INRAE), Toulouse, France
[b]Research Institute for Geo-Hydrological Protection, National Research Council of Italy, Perugia, Italy
[c]Institut de Recherche pour le Développement (IRD), UMR G-Eau, Univ Montpellier, Montpellier, France
[d]HydroSciences Montpellier, Montpellier, France

As demonstrated by numerous water-related crises, international meetings, and a growing scientific literature in all disciplines, water is fast becoming a strategic issue in most regions of the globe, mainly because of the degradation of the resources in terms of quantity and/or quality, the ever-increasing demand, and the social/political challenge of "sharing water scarcity."

In this context, the Mediterranean region is often considered a paradigmatic example of the extreme tension around water resources: 180 million people live in countries with less than 1000 m^3/year per capita of renewable water resources (425 in Tunisia and 750 in Morocco), while 80 million are expected to have less than 500 m^3/year in 2025. The overexploitation of water resources is already observed in most Mediterranean countries and the situation is bound to deteriorate further. Together with food security, water management is therefore the pivotal issue in the Mediterranean in the twenty-first century in the context of climate change. Overall water demand increases unremittingly, particularly in southern and eastern Mediterranean countries (SEM), while access to water for the poorest and/or more isolated populations is still limited. The supply of domestic water, considered the priority by all states, and irrigation are the two main demands to be met (tourism is included in domestic water demand). Predicting the future state of water resources in the Mediterranean requires consideration of many social, technological, and environmental processes that exist simultaneously and interact.

The Mediterranean climate is characterized by its variability and the magnitude and impact of extreme events, such as droughts and floods. These events impact natural processes (e.g., aquifer recharge, erosion) and human activities. Such adverse conditions led to the development of collective rules to share scarce resources at the community level and to the building of hydraulic infrastructure. Hence water has always been central to both development and the exercise of state power. Its management reveals and consolidates power relationships between scale, geographical area, social groups, and economic sectoral interests (irrigation, domestic use, tourism, industry, etc.). The Mediterranean's water resources are not the victim of an ineluctable and growing imbalance between demand and supply; their scarcity should be viewed and treated as a socio-political construct, where coarse Malthusian views are rejected in favor of an analysis of the spatial and social distribution of scarcity and its discursive manufacturing.

Knowledge of fluxes

The quantification of the water and energy fluxes over the Mediterranean Basin, as well as in other regions worldwide, is a challenging task. Ground observations and remote sensing are increasingly combined to improve our understanding of the different components of the water cycle, with a specific focus on the estimation of precipitation, evapotranspiration, and freshwater fluxes, i.e., river discharge. However, assessments of the spatial and temporal variability of water and energy fluxes are still affected by large uncertainties. Evapotranspiration is particularly difficult to measure and prone to substantial error. Establishing freshwater and snowmelt fluxes is hampered by the complex morphology and topography of the Mediterranean region. The estimation of extreme events in precipitation and river discharge (e.g., for flood and drought assessment) is challenging due to the nature of such events. These problems are exacerbated by the increasing human impact on water and energy fluxes, particularly the upward trend in irrigation water demand. This book outlines the main advances and limitations in estimating, monitoring, and controlling water and energy fluxes with specific focus on evapotranspiration (cf. Chapter 2), mountain hydrology (cf. Chapter 3), irrigation (cf. Chapter 4), remote sensing (cf. Chapter 6), flood (cf. Chapter 8), and drought (cf. Chapter 9). The current knowledge of water and energy fluxes is described in Chapter 1.

Remote sensing for water resources

Across the Mediterranean the networks for hydro-meteorological measurements are often scattered and poorly adapted to the real-time demand and needs of water-resource managers. Thus, remote sensing has shown great potential to complement in situ measurement networks when assessing the components of the hydrological cycle. These include precipitation, evaporation, water content in the soil, snow, surface and subsurface water storage, river, and lake water levels. To this must be added the possibility of generating accurate descriptions of land use and mapping the development of vegetation cover. Remote sensors contribute substantially to providing essential information for the management of water resources as well as the monitoring of extreme events and their impacts. The wide spatial coverage of these data should allow a better assessment of risk in terms of regional water and food security. Historically, several satellite products with low spatial resolution (greater than 1 km) have been made available in an operational mode. They are particularly used by meteorological services to improve the accuracy of weather-prediction models. They are able to produce data describing the properties of the canopy as well as the water content of the soil (cf. Chapter 6). These data are an essential tool in the monitoring of extreme events, particularly drought, through an analysis of statistical anomalies over increasing time series (AVHRR, ASCAT, SMOS, MODIS, etc.) (cf. Chapter 8 and 9) even exceeding thirty years. Given global changes (climatic and anthropogenic), these series are also very useful in identifying trends in different components of the water cycle, as well as helping to predict changes in the resource (cf. Chapter 7). With the arrival of various data with high spatial and temporal resolution in recent years we have witnessed dramatic developments in remote-sensing applications for water resource management, including the European Union's Copernicus Programme and future programs such as the SWOT interferometric altimeter. These data will gradually be able to meet the needs of local managers for different components of the hydrological cycle.

Governing water resources

Over the past five decades, water policies were based mainly on increasing supply through a greater mobilization of water resources via large-scale infrastructure (especially dams and interbasin transfers) and (tube)wells. This approach fueled the expansion of irrigation and the intensification and diversification of agricultural systems, agriculture representing about 80% of the total water demand in North Africa. Having exploited their resources to the limit most countries can no longer pursue this strategy and face a stark imbalance between declining exploitable water resources and ever-increasing demand. It is nonetheless tempting for states to continue developing resources that are increasingly scarce and costly to develop: this comes at higher (public) financial and (often) environmental costs but avoids curtailing existing uses based on the promise that more water can be produced (cf. Chapter 4). In some countries, it is still possible to increase production, notably through interbasin transfers (e.g., North-South transfer projects in Morocco and Tunisia) or desalination (using solar energy, as in Morocco).

The reuse of wastewater is often seen as a promising source of supply augmentation (cf. Chapter 5). However, in many countries (e.g., Egypt, Lebanon, northern Africa) untreated wastewater is already being used by farmers and the gain from treating it is mostly in terms of human and environmental health. While some countries (initially Tunisia and later Israel, for example) have treated wastewater on a large scale, the practice is only incipient in southern Europe where it has potential.

At a time of global change (climatic, demographic, economic, etc.) and erratic, ever-dwindling water resources, the challenges posed by food security in the Mediterranean region are enormous, with the need to intensify cropping and increase productivity while preserving resources. The agricultural sector, which accounts for 70%–80% of total water withdrawals, presents several important levers for action since there is still significant potential to both act on demand and reduce water losses while increasing water productivity. It has been shown that better management can save 10%–50% of applied irrigation water, depending on the irrigation technique used and the type of soil (cf. Chapter 4). Micro-irrigation has been seen by most governments as a "technological fix" that would "save water" while boosting production. Unfortunately, more often than not reducing the water applied has also reduced return flows that were already appropriated, while better irrigation scheduling and intensification have tended to enhance evapotranspiration (i.e., water consumption) rather than to decrease it.

Eventually, states are confronted with the imperative to enhance their management of available resources and make difficult trade-offs between the various conflicting sectors, needs, and preexisting water rights. In the particular case of groundwater resources, this means curtailing abstraction and banning drilling, something no state has been able to achieve satisfactorily. Acknowledging the finite and interconnected nature of water resources also includes ensuring or returning water to the environment. As the continuous degradation of Mediterranean wetlands demonstrates (cf. Chapter 11), this remain a tall order on all shores of the Mediterranean.

Models for water resources management

Integrated models enable the development and sharing of a coherent and synoptic view of the supply and demand of regional water resources by taking into account the evolution of their different components (climate, rivers, aquifers, reservoirs) and allocation to various uses (agriculture, domestic water, industry, tourism, environment, etc.). This tool allows physical processes (such as evapotranspiration on agricultural areas) to be combined with anthropogenic dynamics (irrigation, hydraulic infrastructure, allocation regulations, etc.). Thus they are a simulation and planning, as well as a communication, tool. However, several challenges need to be addressed. The level of complexity must be adjusted so that the tool is suitable for all three purposes; the system must be fed with reliable data; and the model must be calibrated. The integrated model can then be used for future projections by taking into account both anthropogenic and climatic reference scenarios. Hydraulic, economic, political, and social measures and policies are translated in terms of their impact on the demand for and allocation of water resources, and their overall impact on the system can be evaluated (cf. Chapter 7).

Climate scenarios

The Mediterranean region is being strongly affected by climate change: average temperatures are now around 1.5°C above the values observed before the industrial revolution (1880–99) compared to global warming of +1.1°C. The region is therefore considered a hotspot of climate change where, without mitigation, temperatures are likely to continue to increase sharply. The increased temperatures are most noticeable in summer, with more frequent and intense heatwaves. In terms of precipitation, the region is already characterized by a high degree of spatial and temporal variability, and all climate scenarios point to a decrease in precipitation with a consequent increase in the length of drought events, as has already been observed since 1950. The combination of an increase in the atmospheric evaporative demand linked to a rise in temperature and a decrease in precipitation sets the conditions for climate drying in the region (cf. Chapter 9). Paradoxically, heavy precipitation events are projected to increase for several Mediterranean countries due to climate change, possibly leading to more severe floods (cf. Chapter 8), particularly in urban areas. Alongside these climatic drivers, the region has seen an increase in water demand due to high population growth, particularly in southern and eastern countries where supply is already limited. Irrigation requirements are expected to increase by the end of the twenty-first century due to population growth and rising demand. In addition, intensive groundwater harvesting has caused groundwater levels to drop in many regions, and groundwater quality is deteriorating as a result of over-exploitation, pollution, urbanization, tourism development (cf. Chapter 10), and seawater intrusion due to rising water levels and aquifer drawdowns. As a consequence, the Mediterranean is expected to experience significantly decreased freshwater availability, and meeting the growing demand for drinking water and irrigation and industrial use will become increasingly problematic.

The objective of this book is to propose a comprehensive view of water resources in the Mediterranean at a time of significant climatic and anthropogenic change.

The first part of the book presents the main processes governing the functioning of water resource systems, with three chapters explaining the hydrological cycle, discussing evapotranspiration processes in semiarid regions and describing mountain hydrology.

The second part of the book focuses on the anthropogenic dimension of water resource management. The four chapters address the issues of irrigation and water governance, the use of wastewater in agriculture, the potential of remote sensing and, finally, the use of integrated modelling for water resource management.

The third part deals with extreme hydro-climatic phenomena, as well as environmental transformation at a time of anthropogenic and climatic change. Four chapters successively cover the issues of floods in the Mediterranean, drought events, land-use change and, finally, the status and evolution of wetlands.

This book is intended for researchers and stakeholders involved in the evaluation, management, and governance of water resources, and it has been elaborated by scientists who are internationally renowned in their respective fields. It will also be of interest to those involved in agricultural and water management, particularly in the Mediterranean Basin and in regions characterized by similar climatic and land-cover conditions.

We are grateful to all those who contributed to the development of this work. First, we thank the scientists, the authors of the chapters, and the experts at the Scientific Committee for their review of the chapters and their corrections. This project was conducted with the support of the French National Centre for Scientific Research (CNRS), the French Institute of Research for Development (IRD), and the National Research Council (CNR).

CHAPTER 1

Hydrological cycle, Mediterranean basins hydrology

Antoine Allam[a,b], Roger Moussa[b], Wajdi Najem[a], Claude Bocquillon[a]
[a]CREEN, ESIB, Univ. Saint-Joseph, Beirut, Lebanon
[b]UMR LISAH, Univ. Montpellier, INRA, Montpellier, France

1.1 Introduction

The Mediterranean climate is characterized by seasonal rainfall patterns, high spatio-temporal variability of precipitation, summer drought, and intense rainfall events. These climatic factors are generating several water challenges across the Mediterranean. Water resources become scarcer due to climate change and the increasing demands on water by various economic sectors [1]. Moreover, the irregular spatial distribution of precipitation leads to disproportional water availability across territories with long periods of drought. Furthermore, high intensity but short rainfall events cause intense flooding. Mediterranean catchments are therefore characterized by three main features: limited water resources, dry summers, and high-intensity rainfall events that generate flash floods.

Many attempts have been made to provide an overview on the hydrology of the Mediterranean region. Reviews have already detailed specific aspects of the hydrology of Mediterranean catchments, such as rainfall interception [2], the impact of the Mediterranean forest on catchment responses [3], the dryland hydrology [4], the impact of human activities on fluvial systems [5], erosion processes [6, 7], the hydrology of mountainous catchments [8], and more recently the review and the metaanalysis of hydrological response characteristics of Mediterranean catchments at different time scales [9].

This chapter aims to present an overview of the hydrological response characteristics of Mediterranean catchments. It starts by defining the study area, then the physical and hydro-climatic characteristics, followed by a synthesis of studies related to annual water balance, flood events, and droughts.

1.2 Mediterranean catchments

1.2.1 Boundary of the Mediterranean region

From the Latin word *Mediterranĕus* meaning "middle land," the Mediterranean refers to the sea and bordering region located in the middle of the Ecumene between the European, African, and Asiatic continents. With Köppen's classification [10], the definition designates

Water Resources in the Mediterranean Region
https://doi.org/10.1016/B978-0-12-818086-0.00001-7

a moderate climate with dry summers and extended geographically beyond the limits of the Mediterranean Sea. The question that arises is: how was the Mediterranean boundary defined? Several alternatives are considered in this case based on the practiced discipline, but the hydrological boundary was adopted for this study, as shown in Fig. 1.1, and in the following, we study the catchments draining toward the Mediterranean Sea.

- The climatic boundary could be defined according to Köppen's classification where a set of regions share similar temperature and precipitation characteristics and are known for their warm and dry summers and cold and humid winters, where either the precipitation in the driest month in summer is below 40 mm or below the third of the precipitation in the wettest month in winter (Cs) and the air temperature of the warmest month is above 22°C (Csa) or the number of months with air temperature above 10°C exceeds 4 months (Csb). It is limited by the African desert to the South and the temperate European countries to the North. This boundary might change according to the definition of this similarity. Some regions share a similar Mediterranean climate although they are located far outside the Ecumene, such as Chile, California, Australia, and South Africa.

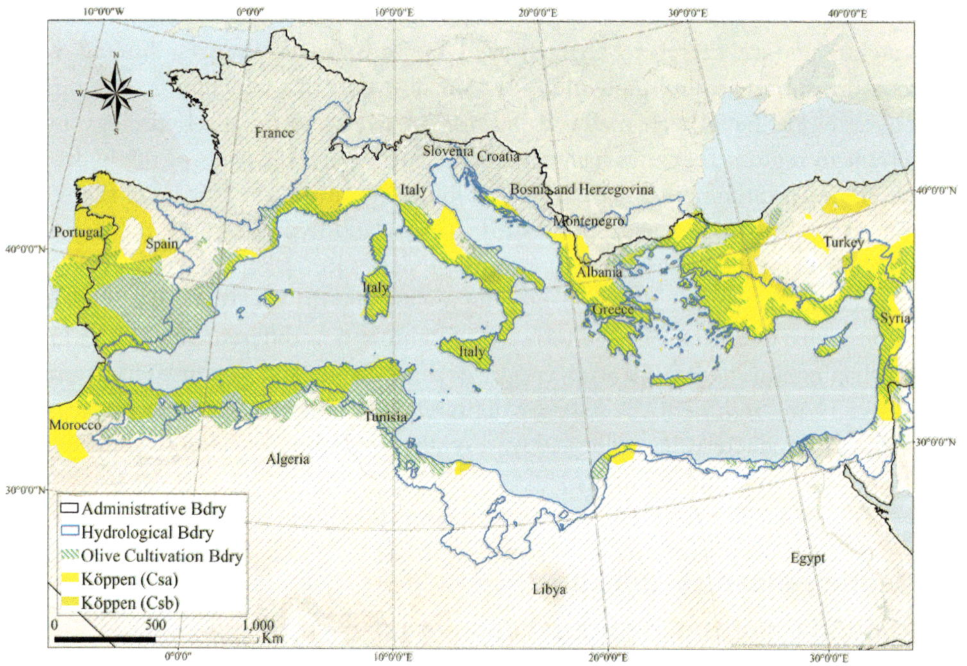

Fig. 1.1 Four Mediterranean region boundaries (Bdry) from Ref. [11]; first administrative, second topographic based on Ref. [12], third olive cultivation based on Ref. [13]; and fourth climatic based on Ref. [14].

- The hydrological boundary could be defined by the set of catchments draining toward the Mediterranean Sea [12]. This definition neglects some of Mediterranean climate regions like Portugal and part of Spain, and favors geographically adjacent regions like Egypt and Libya.
- The agricultural-bioclimatic boundary consists of the set of regions sharing the same types of vegetation considered as indicators of the Mediterranean region such as olives [13]. This definition is linked to human activity with the same nuances as the climatic limit.
- The administrative boundary of countries adjacent to the Mediterranean Sea has a problematic definition independent of any natural base [15]. These boundaries include several climatic classes and cover larger areas than the topographical limits.

1.2.2 Mediterranean catchments draining toward the Mediterranean Sea

The delimitation of these catchments was imported from the European Commission and Joint Research Centre (JRC) [16] and HydroSHEDS of the World Wildlife Fund [17]. According to these databases, the total number of catchments exceeding 1 km^2 and having a Mediterranean Sea mouth outlet is 3681, covering a total area of 1,781,645 km^2. It should be noted that the Nile and Black Sea were omitted—the Nile for its extent 3500 km to the south of the Mediterranean, and the Black Sea because it is fed by catchments located above latitude 45°. Catchments area distribution is shown in Table 1.1, where 1270 middle-range catchments, between 100 and 3000 km^2, constitute 35% of the total number and cover 28% of the total area.

To study regional tendencies in the Mediterranean zone, the study region was divided into three zones: (i) the north-western Mediterranean (NWM), encompassing Mediterranean Albania, Croatia, France, Italy, Montenegro, Portugal, Slovenia, and Spain; (ii) the eastern Mediterranean (EM), encompassing Cyprus, Egypt, Greece, Israel, Lebanon, Palestinian territories, Syria, and Turkey; and (iii) the southern Mediterranean (SM), encompassing Algeria, Egypt, Libya, Morocco, and Tunisia.

1.3 Physical characteristics

The physiographic characteristics of the Mediterranean region were derived from global data sets: topography, geology, landcover, lithology, soil, and karst (Fig. 1.2).

Table 1.1 Catchments distribution per area and ratio to total area.

Area range	Catchments	Ratio	Total area (km^2)	Ratio area
$1\,km^2 < A < 100\,km^2$	2333	63%	80,157	4%
$100\,km^2 < A < 3000\,km^2$	1270	35%	498,614	28%
$A > 3000\,km^2$	78	2%	1,202,874	68%

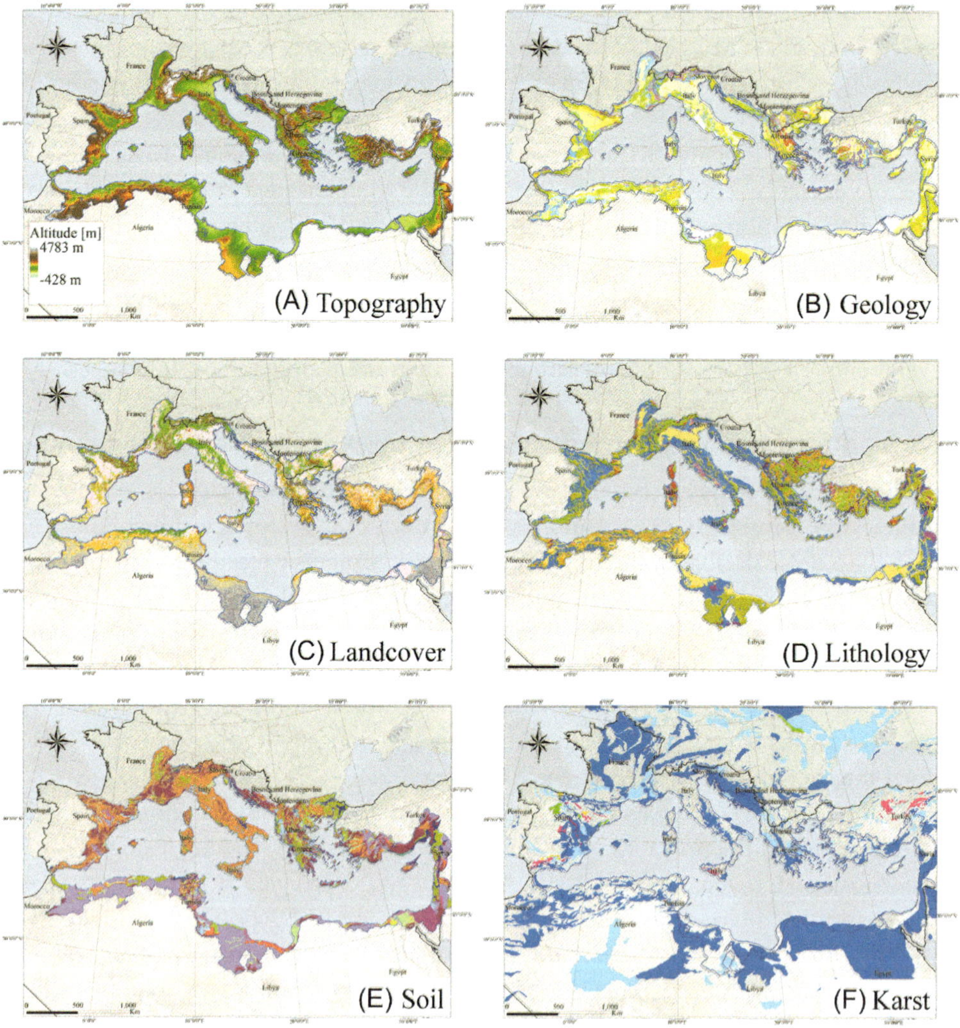

Fig. 1.2 Physical characteristics of the Mediterranean region within the hydrological boundary, derived from global data set: (A) topography, (B) geology, (C) landcover, (D) lithology, (E) soil, and (F) karst. Maps' color-coding details are available at the end of the chapter in Tables A1–A5.

1.3.1 Topography

A digital elevation map of the Shuttle Radar Topography Mission (SRTM) was used ([18]; Fig. 1.2A). Few catchments had below sea level mean altitudes, and these were mainly in Venice and some coastal areas in France usually submerged by seawater. About 50% of the Mediterranean catchments have a mean altitude below 200 m and cover 28%

of the area, highlighting the wide plains located in the region, while only 2% of the catchments have a mean altitude above 1000 m but cover 19% of the area.

The topographical maps show that the Mediterranean Sea is surrounded by high mountains except for Egypt and Libya: the Pyrenees in southwestern Europe, the Alps in southern France and north-western Italy, the Apennines along the Italian coast, the Dinaric Alps along the coast of the Adriatic Sea, the mountains of Greece, the Taurus Mountains in Turkey, Mount Lebanon in the eastern Mediterranean, and the Atlas Mountains in northern Africa. These mountains were shaped mostly during the Oligocene and Miocene epochs, when the African-Arabian continent collided with the Eurasian continent [19]. The proximity of the mountain ranges to the sea explains why a majority of the Mediterranean catchments are medium-sized and sloping. Lower hills, such as the Cevennes region in France, and plains also exist along the coasts and in some interior regions; thus, the landscape is quite heterogeneous.

1.3.2 Geology

The Mediterranean geologic age layers were imported from the 1:5 Million International Geological Map of Europe and Adjacent Areas (IGME 5000), which shows the prequaternarian geology of Europe and adjacent countries onshore and offshore (Fig. 1.2B). In addition to the geology attributed by age, petrography, and genesis, magnetic anomalies, tectonic structures, continental edges, metamorphism, and crusts are shown. The map was developed under the leadership of Bundesanstalt für Geowissenschaften und Rohstoffe (BGR) (Federal Institute for Geosciences and Natural Resources) and the auspices of the World Map Commission (CGMW) in cooperation with geological surveys of 48 countries and more than 20 scientific institutes [20]. Most of the underlying geology comprises limestone with sandstone, sedimentary deposits, and metamorphic granites [21]. The prevalence of limestone rocks means that karstic catchments are very common in the region. The layers covering more than 1% of the total Mediterranean area are shown in Table A1.

1.3.3 Landcover

Landcover was expressed according to Global Land Cover GLC 2000 major types [22]. GLC 2000 was adopted in order to have coverage of south Mediterranean countries and a temporal coverage of the same era as the hydroclimatic data. The Mediterranean landcover map in Fig.1.2C and Table A2 show that in the top rank are the Cultivated and Managed Areas that dominate the Mediterranean catchments with 31% of the total area. All types of shrub cover come in second place with a total of 23%, all tree cover types in third with 20%, and then bare areas at 13%, mainly in the southern countries.

Vegetation is mainly dominated by evergreen trees and shrubs (the famous Mediterranean "Maquia" or "Garrigue"). In mountainous areas with wetter conditions, several deciduous tree species prevail. Conversely, the driest areas are dominated by steppe [23]. However, Mediterranean catchments exhibit a particular relief-driven organization in the distribution of land types. Hence, forests grow directly under the rocky summits of Mediterranean mountains; agricultural terraces are found downslope. Further down, the Mediterranean Garrigue is present. Finally, agriculture dominates the coastal plains. This pattern is partly attributed to past human activities, but it is now disturbed by the intense urbanization on Mediterranean coastal plains and the abandonment of the agricultural terraces for Mediterranean forests, particularly in Europe (e.g., [1, 24–26]). These changes undoubtedly have an impact on the hydrological responses of Mediterranean catchments.

1.3.4 Lithology

Lithology describes the physical properties of rocks which play a key role in hydrological processes. Understanding these processes at the regional scale requires a high-resolution description of lithology. The lithologic structure of Mediterranean catchments was deduced from the new Global Lithological Map database (GLiM) ([27]; Fig. 1.2D and Table A3). The Mediterranean is covered by 85% of sedimentary rocks, mainly carbonate sedimentary rocks (33%).

1.3.5 Soil

A Mediterranean soil map was derived from the Harmonized World Soil Database (HWSD; [28]; Fig. 1.2E and Table A4). Cambisols (26%), Calcisols (22%), Leptosols (20%), and Luvisols (10%) are the soil types most commonly present in the Mediterranean. Cambisols are mainly present in European countries, covering 26% of the total area; they contribute to good agricultural land and are considered among the most productive soils on earth. Calcisols are soils with substantial accumulation of secondary lime, mostly present in North African countries and to the East; they are known for their drought tolerance and were used for cotton cultivation in the Mediterranean. Leptosols are shallow soils; they represent mineral soils whose formation was conditioned by the topography/physiography of the terrain. They are present all over the Mediterranean mountains and cover 20% of them in total. Luvisols are fertile soils and rich in clay; they were conditioned by subhumid temperate regions climate hence their Mediterranean specificity [29].

1.3.6 Karst

The Mediterranean karst map (Fig. 1.2F and Table A5) shows that overall, 41% of the Mediterranean contains what is qualified as karstifiable rocks and Continuous Carbonate Rocks. The map was derived from the World Karst Aquifers Mapping project (WOKAM) [30].

1.4 Hydro-climatic characteristics

This section presents the characteristics of the main variables controlling the hydrological components of the annual water balance: precipitation, snow cover, temperature, and the runoff coefficient, the ratio of runoff depth to precipitation depth, all illustrated in Fig. 1.3.

Mean annual precipitation (Fig. 1.3A) ranging between 5 and 3000 mm and mean annual temperature (Fig. 1.3B) ranging between −14°C (Mont Blanc, Alps, France) and +26°C (Karak, Jordan) were generated from WorldClim-2 new 1-km spatial resolution climate surface data, which consist of long-term average monthly temperature and precipitation, solar radiation, vapor pressure, and wind speed data, aggregated across a target temporal range for the years 1970–2000, using data from 9000 to 60,000 weather stations [31].

The snow cover map (Fig. 1.3C) expresses the average duration, as a ratio of the year, of the Mediterranean regions covered by snow. It is based on the MODIS/Terra Snow Cover Daily L3 Global 0.05Deg CMG, Version 6 [32]. This data set was generated from the Normalized Difference Snow Index (NDSI) snow cover in the MOD10A1 data set from 2000 to 2013. The MOD10A1 data set contains daily gridded snow cover and albedo observations at 500 m, mapped into 0.05° (approx. 5 km) CMG cells, derived from radiance data acquired by the Moderate Resolution Imaging Spectroradiometer (MODIS) on board the Terra satellite. The map in Fig. 1.3C shows that snow covers, for at least 1 month per year, 6% of the Mediterranean catchments and 12.2% of its area, mainly in Northern countries in addition to Mount Lebanon and Turkish mountains in the East and the Moroccan Atlas Mountains in the South West.

Finally, Fig. 1.3D shows the mean annual runoff coefficient for 50 Mediterranean catchments across 15 countries in different geologic, soil, and landcover settings. We observe a very wide variability of this runoff coefficient—between 1% and 80%. The water balance components for the 50 catchments were calculated for an average period of 28 years between 1950 and 2017, and added in Table B1 of Appendix B. The table includes: the mean annual precipitation (MAP), which ranges between 230 mm for Kert,

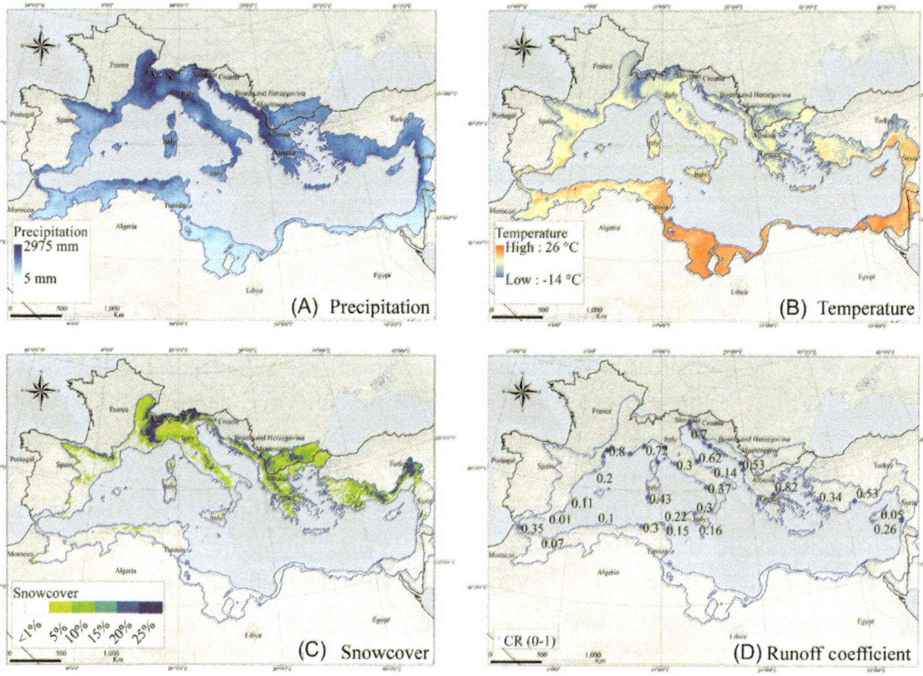

Fig. 1.3 Hydroclimatic characteristics of the Mediterranean region within the hydrological boundary: (A) mean annual precipitation ranging between 5 and 3000 mm; (B) mean annual temperature ranging between −14°C and +26°C [31]; (C) Mediterranean Snow Cover Duration Map (ratio of the year) from the MODIS/Terra Snow Cover Daily [32]; (D) average runoff coefficient for few Mediterranean catchments.

Morocco and 3383 mm for Moraca, Montenegro; the mean annual potential evapotranspiration (MPET) based on Turc's formula and ranging between 566 mm for Cervaro, Italy and 988 mm for Andarax, Spain; the mean annual runoff (MAQ) ranging between 2 mm for Andarax and 1778 mm for Moraca; and finally the runoff coefficient (CR), which ranges between 0.01 for Andarax and 0.82 for Sperchios in Greece.

1.4.1 Precipitation seasonality

Precipitation seasonality, the main characteristic of the Mediterranean region, could be expressed by a seasonality index I_S. It corresponds to the precipitation difference between humid and dry seasons. Knowing the monthly rain at each station, the calculation of this index is as follows, where P_i is the monthly precipitation of the month i in mm, i being the month rank starting from the first month of the hydrological year—September for the Mediterranean region.

$$I_s = 1 - \frac{\min C_i}{\max C_i} \text{ with } C_i = \sum_{i=1}^{i+2} P_i$$

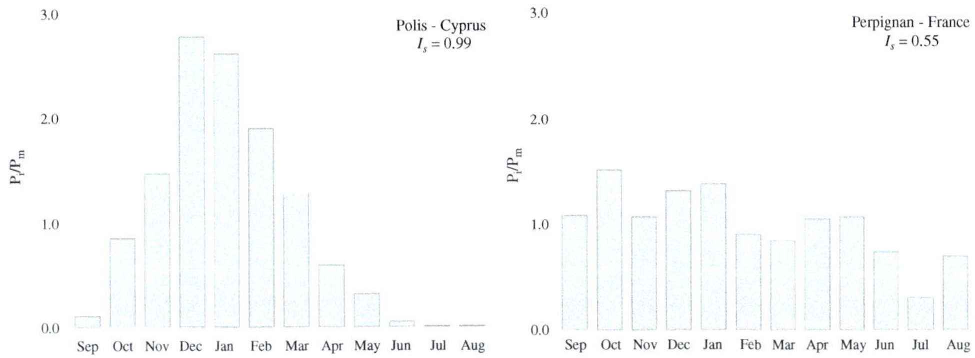

Fig. 1.4 Monthly distribution of the precipitation at Polis-Cyprus (left) and Perpignan-France (right).

The charts in Fig. 1.4 indicate the monthly distribution over the hydrological year of the nondimensional precipitation ratio (P_i/P_m) (where P_i is the monthly precipitation divided by P_m the monthly average of annual precipitation) corresponding to Polis station in Cyprus with a strong $I_S = 0.99$ and centered distribution indicating the seasonal precipitation in contrary to Perpignan in France with a low $I_S = 0.55$ and flattened distribution indicating a regular distribution of rain throughout the year [33]. The Mediterranean would be qualified by high seasonality, an interesting criterion for the study of interannual variability.

1.4.2 Stochastic characteristics of Mediterranean wet and dry periods

Several authors pointed out the stochastic behavior of wet and dry periods alternation in the Mediterranean to describe the climatic variability synthetically. The single-site stochastic precipitation models that include precipitation occurrence and intensity were useful to generate long daily rainfall series and model hydrological responses to predict flood and erosion impacts. Efremides [34] proposed a one-part stochastic rainfall model to simulate the rainfall process in the Greek islands of Rhodes and Chios. The Markovian process was found suitable for modeling daily precipitation data in Andalusia, Spain [35], and the Middle East [36, 37]. Najem [38] has demonstrated that precipitation in Lebanon, which is a unidimensional time-series observed and recorded over time at a certain location, follows a stochastic model. This stochastic model was verified [33], and its parameters were identified on a set of stations around the Mediterranean which helped in finding climate variability indicators. The model is composed of two mechanisms with four parameters (see Fig. 1.5):

- The first mechanism is the alternation of rain and non-rain events, which is a first-rate Markovian process with T1 the average duration of nonrain episodes, T2 the average duration of rainy episodes.
- The second mechanism is the rain impulses which have a nonzero random variables structure which follows a Markovian process of order 0. With *N* the mean duration between two successive showers within the same rain event and G the rain intensity. In other words, the intensity of today's rain is independent of the intensity of the previous day.

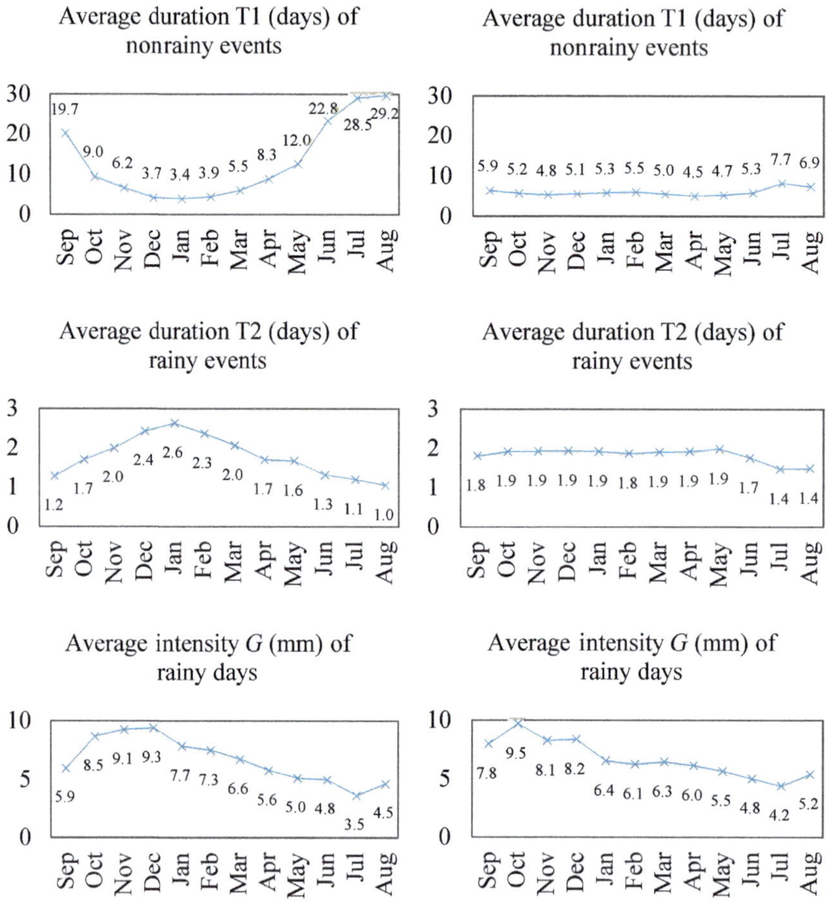

Fig. 1.5 Stochastic parameters for precipitation time series in Polis-Cyprus (left) and Perpignan-France (right).

1.5 Annual water balance of Mediterranean catchments

At the Mediterranean scale, the total inflow from precipitation is estimated at 1070 km^3, total runoff potential at 492 km^3, and total evaporation at 685 km^3 [39]. Merheb et al. [9] synthesized 58 annual water balance studies on Mediterranean catchments, which highlighted the wide variability of mean annual precipitation (MAP), evapotranspiration (ET0), and mean annual runoff (MAQ) across the three Mediterranean subregions, North Western (NWM), Eastern (EM), and Southern (SM). The median values show decreasing MAP, MAQ, and MAQ/MAP, and an increasing ET0 from NWM to EM and SM.

Merheb et al. [9] have identified a high correlation rate ($r^2 = 0.69$) between the MAP and MAQ of all the studied catchments (Fig. 1.6); however, catchments distribution analysis revealed that physical characteristics contribute to MAQ variability. While low-yielding catchments are mostly located in the SM and high-yielding catchments are located in the NWM, some catchments receiving large MAP are yielding very low runoff due to highly permeable environments. Few catchments' MAQ are exceeding MAP (especially in the NWM and EM); this could be explained by underestimated snow contribution in mountainous catchments and karst presence greatly increasing the effective catchment area. In Fig. 1.6, we can see that the trend for the EM and the NWM catchments are quite similar with an intercept close to 500–550 mm and comparable slopes, while the trend of SM catchments is different, with an intercept close to 250–300 mm and a lower slope.

A similar analysis (not shown) that accounted for catchment area and elevation was undertaken. The catchment area does not appear to influence its response. The impact of elevation, however, could not be neglected because it was found that both the MAP and catchment runoff increase with elevation.

Fig. 1.7 illustrates the different water balance components interrelations and highlights the spatial, climatic, and physical variability of Mediterranean catchments. It shows a Budyko-type plot representing the mean annual runoff coefficient MAQ/

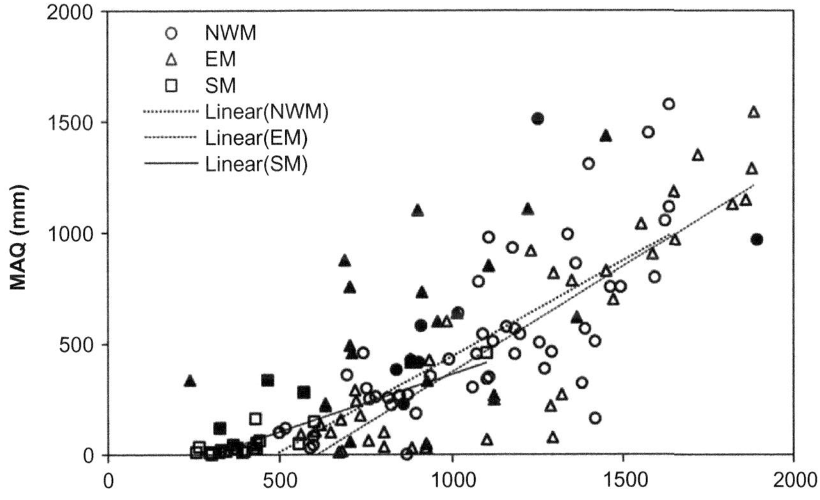

Fig. 1.6 Relationship between mean annual runoff (MAQ) and mean annual precipitation (MAP) for the three studied subregions (NWM, EM, and SM). Filled symbols indicate karstic catchments in each subregion. *From M. Merheb, R. Moussa, C. Abdallah, F. Colin, C. Perrin, N. Baghdadi, Hydrological response characteristics of Mediterranean catchments at different time scales: a meta-analysis, Hydrol. Sci. J. 61 (14) (2016) 2520–2539.*

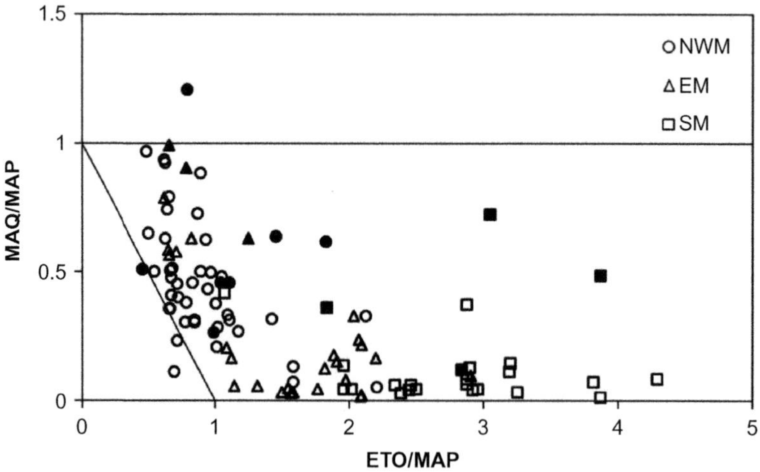

Fig. 1.7 Plot of mass balance data from the study catchments on a Budyko diagram: mean annual runoff coefficient MAQ/MAP function of the aridity index ET0/MAP. MAP: mean annual precipitation; MAQ: mean annual runoff; ET0: mean annual reference evapotranspiration for the three subregions NWM, EM, and SM. Filled symbols indicate karstic catchments in each subregion. *From M. Merheb, R. Moussa, C. Abdallah, F. Colin, C. Perrin, N. Baghdadi, Hydrological response characteristics of Mediterranean catchments at different time scales: a meta-analysis, Hydrol. Sci. J. 61 (14) (2016) 2520–2539.*

MAP versus the aridity index AI = ET0/MAP. The aridity index expresses the catchment prevailing conditions, where ET0/MAP < 1 means wet conditions and ET0/MAP > 1 means dry conditions, with most Mediterranean catchments (60%) being considered water stressed. The lines MAQ/MAP = 1 and MAP = MAQ + ET0 represent the water and energy limits. While all catchments are expected to fall within the limits, four catchments fell outside the water or energy limits. It appeared that these catchments are karstic, which may explain the presence of underground water gain or loss processes.

The spatial distribution analysis shows a clustering of NWM catchments, with the lowest aridity and a vertical spread of MAQ/MAP indicating a wide range of runoff. SM had the lowest yield, while EM catchments show a large spread on the chart, marking the heterogeneity between physioclimatic features and hydrological responses. This heterogeneity could be explained by the complex geomorphologic features, mountainous landforms, and dominating karstic presence.

To summarize, Mediterranean catchments exhibit high variability in terms of both climatic characteristics and catchment hydrological responses at the annual scale. The latter can vary greatly, even for the same amount of rainfall input, which may seriously challenge any modeling approach. Nevertheless, some regional patterns exist, and catchments in each of the abovementioned subregions appear to have somewhat similar characteristics.

1.6 Flood events

Floods and flash floods constitute one main aspect of Mediterranean hydrology and are of growing concern due to their increasing frequencies and damage produced (Chapter 8). Hydrological characteristics, catchment area, event duration, rainfall depth, and unit peak discharge were collected from 191 flood events in the Mediterranean region and analyzed by Ref. [9]. The study resulted with three charts characterizing Mediterranean floods, as represented in Fig. 1.8.

Fig. 1.8A represents a log-log diagram of the unit peak discharges versus catchment areas with two envelope curves developed by Tarolli et al. [40] for NWM (France, Italy, and Spain) and EM (Israel) flash floods. The chart shows an inverse correlation of the unit peak discharges with catchment area where unit peak discharges decreases when catchment area increases. This correlation may reflect the spatial variability in the rainfall events at the local scale. The chart also reflects a spatial variability at the regional scale where the highest peak discharges are recorded in the NWM, gradually decreasing for the SM and then the EM.

Fig. 1.8B represents unit peak discharges versus rainfall depth and shows a clear rainfall threshold of 100 mm. For events between 100 and 3000 mm, the unit peak discharge seems to increase continuously with the amount of event rainfall, while no correlation could be found for rainfall depths below 100 mm.

Fig. 1.8C represents unit peak discharges versus event duration. This chart shows that for short event duration, the unit peak discharges vary greatly, but for longer event duration, the unit peak discharge appears to increase with increasing event duration. This relationship could be explained through soil characteristics where longer events seem to modify soil moisture conditions and runoff processes to resemble more closely those of humid conditions. Once again, the physical characteristics are playing a role in hydrological variability in the Mediterranean, but temporally this time.

Event runoff ratios versus Mediterranean catchments area, rainfall depth, and event duration vary over local and regional scales. This can be explained by the temporal and spatial variability of climatic and physical features. The karstic nature of the catchment increases the effective catchment areas, whereas longer and heavier rainfall events modify soil characteristics and trigger high runoff ratios and peak discharges, though with thresholds.

1.7 Droughts

Climatic trends in the region show an overall decrease in the available water resources due to a reduction in the annual rainfall and an increase in the annual temperature and reference evapotranspiration [41–46]. However, important seasonal discrepancies

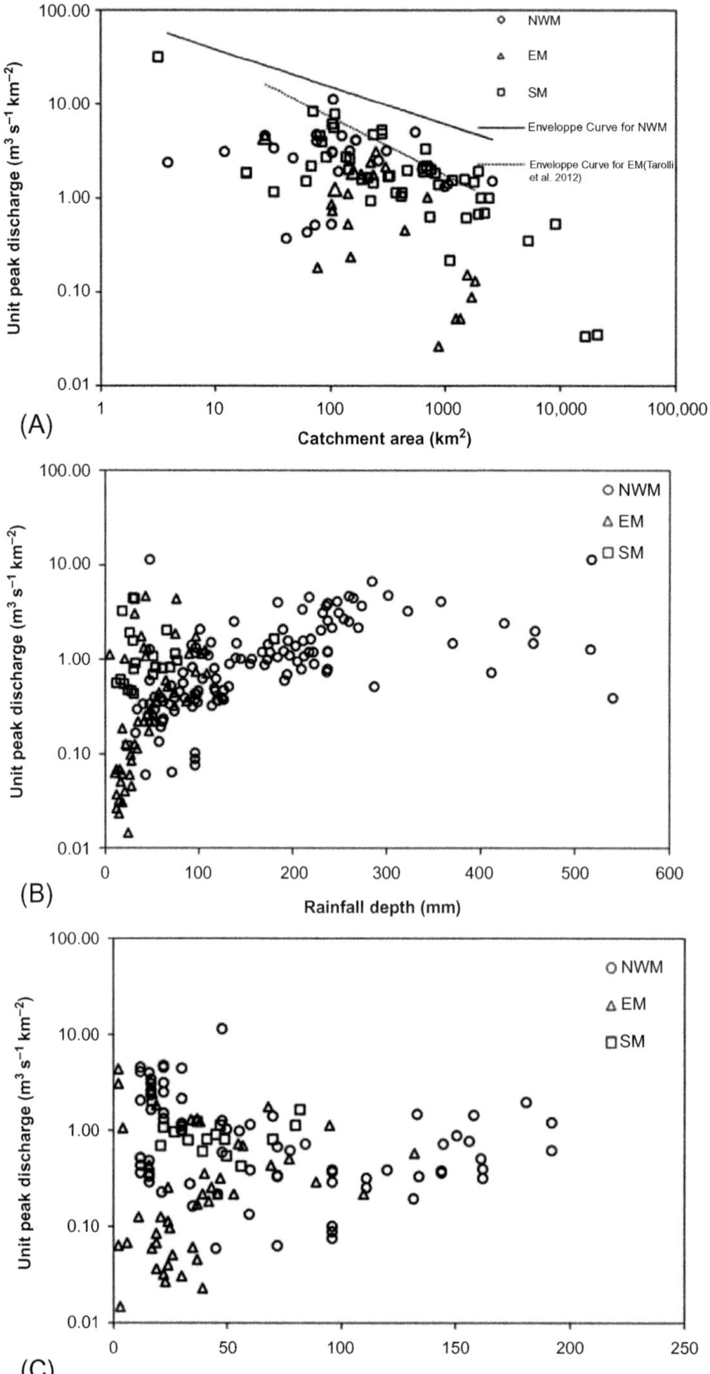

Fig. 1.8 Relationship between unit peak discharge and (A) catchment area, (B) rainfall depth, and (C) event duration. *From M. Merheb, R. Moussa, C. Abdallah, F. Colin, C. Perrin, N. Baghdadi, Hydrological response characteristics of Mediterranean catchments at different time scales: a meta-analysis, Hydrol. Sci. J. 61 (14) (2016) 2520–2539.*

exist in the evolution of different components of the water balance equation across areas of the Mediterranean. All studies show a high spatial and temporal variability of drought events across the Mediterranean, in the EM, the SM, parts of the Iberian Peninsula, the main islands, and south-eastern Italy (Chapter 9).

Climate change impact projections for the 21st century in the Mediterranean converge toward a drier climate by the end of the century (e.g., [47–51]). The water management scenarios in different catchments show a large decrease in water availability, particularly during dry periods, due to the pressure of a large anthropogenic increase of population in these catchments. However, important disparities can exist within the same catchment.

1.8 Discussion, conclusion, and perspectives

This chapter presented an overview of the hydrological response characteristics of Mediterranean catchments. Cartographic data are presented and a spatial analysis conducted. First, various definitions of the boundaries of the Mediterranean region according to climatic, hydrological, agro-bioclimatic, and administrative indexes were discussed, then physiographic characteristics were derived from global data sets of topography, geology, landcover, lithology, soil, and karst. Second, the analysis of the two main hydro-climatic variables, precipitation and temperature, controlling the annual water balance components showed the main characteristics of the Mediterranean climate characterized by seasonality, alternating very dry periods and high rainfall intensities.

We then presented a synthesis of the annual water balance, flood events, and drought studies in the Mediterranean and discussed the spatial variability and the physiographic and climatic contribution to the hydrological response. The annual water balance of catchments located in the SM has low-yielding catchments while high-yielding catchments are located in the NWM. At the flood event scale, the highest peak discharges are located in the NWM, gradually decreasing for the SM and EM. Finally, drought periods increase in the EM, the SM, and islands.

The demographic growth of the countries around the Mediterranean, especially in the south and east, accompanied by increasing pressure on water resources, is a major challenge not only for future research in water sciences but also for stakeholders dealing with the management of water resources, flood risk, and droughts.

Appendix A Color legend and characteristics of Fig. 1.2

See Table A1–A5.

Table A1 Mediterranean geologic ages for layers covering more than 1% of the total area with original color ID legend of Fig. 1.2B.

Color ID	Geologic age name	Age (MYears)	Area (km²)	Percentage
	Miocene	23.8–5.32	198,957	11.6%
	Late Cretaceous	98.9–65.5	197,167	11.4%
	Pliocene	5.32–1.81	152,630	8.9%
	Neogene	23.8–1.81	117,227	6.8%
	Quaternary	1.81–0	101,116	5.9%
	Early Cretaceous	142–98.9	70,081	4.1%
	Eocene	55–33.7	67,089	3.9%
	Pliocene—Quaternary	5.32–0	52,118	3.0%
	Paleocene	65.5–55	41,537	2.4%
	Paleozoic	545–250	41,051	2.4%
	Jurassic	205.1–142	40,593	2.4%
	Oligocene	33.7–23.8	38,675	2.2%
	Eocene—Oligocene	55–23.8	36,556	2.1%
	Late Jurassic	159.4–142	34,740	2.0%
	Paleocene—Eocene	65.5–33.7	29,629	1.7%
	Cretaceous	142–65.5	26,651	1.6%
	Triassic	250–205.1	26,348	1.5%
	Jurassic—Cretaceous	205.1–65.5	24,322	1.4%
	Oligocene—Miocene	33.7–5.32	23,102	1.3%
	Middle Jurassic	180.1–159.4	20,799	1.2%
	Permian	292–250	19,974	1.2%
	Mesozoic	250–65.5	19,096	1.1%
	Pliocene—Pleistocene	5.32–0.01	18,431	1.1%

Table A2 Mediterranean landcover major types according to GLC 2000 and color legend of Fig. 1.2C.

Color ID	Global landcover type	Code	Area (km²)	Percentage area
	Cultivated and managed areas	CMA	553,874	31%
	Shrub cover, closed-open, deciduous	SC_CO_D	278,599	16%
	Bare areas	BA	235,192	13%
	Sparse herbaceous or sparse shrub cover	SH or SHC	216,329	12%
	Tree cover, broadleaved, deciduous, closed	TC_BD_C	166,171	9%
	Tree cover, needle-leaved, evergreen	TC_NL_E	127,713	7%
	Tree cover, mixed leaf type	TC_MLT	70,192	4%
	Herbaceous cover, closed-open	HC_CO	37,566	2%
	Shrub cover, closed-open, evergreen	SC_CO_E	27,045	2%

Table A3 Mediterranean lithology types with ratio to total area, color-coded as original; see Fig. 1.2D.

Color ID	Lithology type	Code	Area (km^2)	Percentage area
	Carbonate sedimentary rocks	sc	576,222	33%
	Mixed sedimentary rocks	sm	308,533	18%
	Unconsolidated sediments	su	301,987	17%
	Siliciclastic sedimentary rocks	ss	288,115	17%
	Metamorphic rocks	mt	120,331	7%
	Plutonic rocks	pa/pb/pi	64,290	4%
	Volcanic rocks	va/vb/vi	53,636	3%

Table A4 Mediterranean soil types table as per the Harmonized World Soil Database and the color legend of Fig. 1.2E.

ID	Soil type	Code	Area (km^2)	Percentage area
	CaMbisols	CM	470,896	26.4%
	CaLcisols	CL	396,947	22.3%
	LePtosols	LP	353,463	19.8%
	LuVisols	LV	182,671	10.3%
	FLuvisols	FL	114,409	6.4%
	ReGosols	RG	63,978	3.6%
	VeRtisols	VR	48,205	2.7%
	GYpsisols	GY	31,462	1.8%
	Not identified	NI	37,615	2.1%
	Other	Other	82,026	4.6%

Table A5 Mediterranean karstifiable rocks and color legend of Fig. 1.2F.

Color ID	Karstifiable rocks	Area (km^2)	Percentage area
	Continuous carbonate rocks	582,179	32.68%
	Discontinuous carbonate rocks	122,793	6.89%
	Continuous evaporite rocks	13,038	0.73%
	Discontinuous evaporite rocks	2661	0.15%
	Mixed carbonate and evaporite rocks	1677	0.09%

Appendix B Catchments hydrological characteristics table of Fig. 1.3D

See Table B1.

Table B1 Water balance components of 50 Mediterranean catchments; mean annual precipitation (MAP), mean annual potential evapotranspiration (MPET) estimated according to Turc's formula, mean annual runoff (MAQ) and runoff coefficient (CR); see Fig. 1.3D.

ID	Country	Name	AREA (km^2)	MAP (mm)	MPET (mm)	MAQ (mm)	CR
1	AG	MELAH[a]	6334	587	881	174	0.30
2	AG	MAZAFRAN	2039	574	911	57	0.10
3	AL	ERZENIT	760	1202	841	553	0.46
4	CY	SERRAKHIS[a]	737	1006	950	188	0.19
5	CY	VASILIKOS	162	629	941	32	0.05
6	CY	EZOUSAS	225	669	934	58	0.09
7	CY	DHIARIZOS	260	675	927	173	0.26
8	ES	CENIA	200	529	830	281	0.53
9	ES	AMADORIO	205	359	888	38	0.11
10	ES	FLUVIA	974	802	819	221	0.28
11	ES	LA MUGA	854	714	797	77	0.11
12	ES	ELTER	3010	701	819	129	0.18
13	ES	EL BESÒS	1036	644	807	126	0.20
14	ES	ANDARAX	490	242	988	2	0.01
15	FR	FIUM-ALTO	127	749	850	390	0.52
16	FR	LEZ	194	629	840	382	0.61
17	FR	LOUP	264	755	835	541	0.72
18	FR	HERAULT	2625	629	832	502	0.80
19	FR	AUDE[a]	5327	557	775	342	0.61
20	FR	GAPEAU	548	536	852	252	0.47
21	FR	TECH	729	554	848	351	0.63
22	FR	TET	1371	557	848	244	0.44
23	GR	SPERCHIOS	1830	701	890	575	0.82
24	HR	MIRNA	579	1164	727	384	0.33
25	HR	KRKA	2103	1074	803	771	0.72
26	IS	ALEXANDER	953	603	888	11	0.02
27	IS	SOREQ	492	543	949	18	0.03
28	IT	FIORA	825	959	699	287	0.30
29	IT	FLUMINIMAGGIORE	83	747	867	320	0.43
30	IT	SERCHIO	1525	1831	571	997	0.54
31	IT	ARGENTINA	209	982	871	703	0.72
32	IT	CECINA	640	857	814	262	0.31
33	IT	PESCARA	3153	783	811	487	0.62
34	IT	ALENTO	415	1253	875	462	0.37
35	IT	SAN LEONARDO	522	660	820	195	0.30
36	IT	F. IMERA MERIDIONALE	2013	633	766	102	0.16

Table B1 Water balance components of 50 Mediterranean catchments; mean annual precipitation (MAP), mean annual potential evapotranspiration (MPET) estimated according to Turc's formula, mean annual runoff (MAQ) and runoff coefficient (CR); see D—cont'd

ID	Country	Name	AREA (km^2)	MAP (mm)	MPET (mm)	MAQ (mm)	CR
37	IT	CERVARO	657	713	566	103	0.14
38	LB	NAHR BEYROUTH	217	1032	935	454	0.44
39	LB	NAHR EL KALB	249	1440	812	904	0.63
40	LB	NAHR IBRAHIM	326	1450	812	1005	0.69
41	LB	NAHR LITANI	2172	637	825	289	0.45
42	ME	MORACA[a]	3257	3383	824	1778	0.53
43	MO	KERT[a]	2622	230	872	16	0.07
44	MO	EMSA	110	572	887	203	0.35
45	SI	RIŽANA	205	941	801	606	0.64
46	SI	DRAGONJA	147	956	801	216	0.23
47	SI	ISONZO[a]	3334	2254	672	1569	0.70
48	TR	KOEPRUE[a]	3003	632	703	336	0.53
49	TR	CINE CAYI[a]	2754	632	703	217	0.34
50	TS	MILIANE[a]	2283	413	935	64	0.15

[a]Subcatchments.

References

[1] E. Morán-Tejeda, A. Ceballos-Barbancho, J.M. Llorente-Pinto, Hydrological response of Mediterranean headwaters to climate oscillations and land-cover changes: the mountains of Duero River basin (Central Spain), Glob. Planet. Chang. 72 (1–2) (2010) 39–49.

[2] P. Llorens, F. Domingo, Rainfall partitioning by vegetation under Mediterranean conditions. A review of studies in Europe, J. Hydrol. 335 (2007) 37–54.

[3] C. Cosandey, V. Andréassian, C. Martin, J.-F. Didon-Lescot, J. Lavabre, N. Folton, D. Richard, The hydrological impact of the Mediterranean forest: a review of French research, J. Hydrol. 301 (1–4) (2005) 235–249.

[4] C. Cudennec, C. Leduc, D. Koutsoyiannis, Dryland hydrology in Mediterranean regions—a review, Hydrol. Sci. J. 52 (6) (2007) 1077–1087.

[5] J. Hooke, Human impacts on fluvial systems in the Mediterranean region, Geomorphology 79 (3–4) (2006) 311–335.

[6] R. Shakesby, Post-wildfire soil erosion in the Mediterranean: review and future research directions, Earth Sci. Rev. 105 (3–4) (2011) 71–100.

[7] J.M. García-Ruiz, E. Nadal-Romero, N. Lana-Renault, S. Beguería, Erosion in Mediterranean landscapes: changes and future challenges, Geomorphology 198 (2013) 20–36.

[8] J. Latron, P. Llorens, F. Gallart, The hydrology of Mediterranean mountain areas, Geogr. Compass 3 (6) (2009) 2045–2064.

[9] M. Merheb, R. Moussa, C. Abdallah, F. Colin, C. Perrin, N. Baghdadi, Hydrological response characteristics of Mediterranean catchments at different time scales: a meta-analysis, Hydrol. Sci. J. 61 (14) (2016) 2520–2539.

[10] W. Köppen, Das geographische System der Klimate, Borntraeger, 1936.

[11] A. Allam, R. Moussa, W. Najem, C. Bocquillon, Mediterranean specific climate classification and future evolution under RCP scenarios, Hydrol. Earth Syst. Sci. Discuss. (2019) 1–25.

[12] M. Milano, D. Ruelland, S. Fernandez, A. Dezetter, J. Fabre, E. Servat, G. Thivet, J.-M. Fritsch, S. Ardoin-Bardin, G. Thivet, Current state of Mediterranean water resources and future trends under climatic and anthropogenic changes, Hydrol. Sci. J. 58 (3) (2013) 498–518.

[13] J.A.O. Moreno, Modelización del ciclo fenológico reproductor del olivo (*Olea europaea* L.), Universidad de Córdoba, 2014.

[14] M.C. Peel, B.L. Finlayson, T.A. McMahon, Updated world map of the Köppen-Geiger climate classification, Hydrol. Earth Syst. Sci. Discuss. 4 (2) (2007) 439–473.

[15] J.B. Thornes, J. Wainwright, Environmental issues in the Mediterranean: processes and perspectives from the past and present, Routledge, 2004.

[16] A.L. de Jager, J.V. Vogt, Development and demonstration of a structured hydrological feature coding system for Europe, Hydrol. Sci. J. 55 (5) (2010) 661–675.

[17] B. Lehner, G. Grill, Global river hydrography and network routing: baseline data and new approaches to study the world's large river systems, Hydrol. Process. 27 (15) (2013) 2171–2186.

[18] A. Jarvis, H.I. Reuter, A. Nelson, E. Guevara, Hole-Filled SRTM for the Globe Version 4. Available from the CGIAR-CSI SRTM 90m Database, vol. 15, 2008, pp. 25–54.

[19] K.J. Hsü, L. Montadert, D. Bernoulli, M.B. Cita, A. Erickson, R.E. Garrison, R. Kidd, F. Mèlierés, C. Müller, R. Wright, History of the Mediterranean salinity crisis, Nature 267 (5610) (1977) 399.

[20] WCGM, K. Asch, S. Bellenberg, The 1: 5 Million International Geological Map of Europe and Adjacent Areas (IGME 5000): Bundesanstalt für Geowissenschaften Und Rohstoffe, 2005.

[21] F. Di Castri, Mediterranean type shrublands of the world, Ecosyst. World 11 (1981) 1–43.

[22] E. Bartholome, A.S. Belward, GLC2000: a new approach to global land cover mapping from Earth observation data, Int. J. Remote Sens. 26 (9) (2005) 1959–1977.

[23] N. Bonada, V.H. Resh, Mediterranean-climate streams and rivers: geographically separated but ecologically comparable freshwater systems, Hydrobiologia 719 (1) (2013) 1–29.

[24] F. Gallart, P. Llorens, Observations on land cover changes and water resources in the headwaters of the Ebro catchment, Iberian Peninsula, Phys. Chem. Earth Parts A/B/C 29 (11–12) (2004) 769–773.

[25] N. Lana-Renault, J. Latron, D. Karssenberg, P. Serrano-Muela, D. Regüés, M. Bierkens, Differences in stream flow in relation to changes in land cover: a comparative study in two sub-Mediterranean mountain catchments, J. Hydrol. 411 (3–4) (2011) 366–378.

[26] A. Ceballos-Barbancho, E. Morán-Tejeda, M.Á. Luengo-Ugidos, J.M. Llorente-Pinto, Water resources and environmental change in a Mediterranean environment: the south-west sector of the Duero river basin (Spain), J. Hydrol. 351 (1–2) (2008) 126–138.

[27] J. Hartmann, H.H. Dürr, N. Moosdorf, M. Meybeck, S. Kempe, The geochemical composition of the terrestrial surface (without soils) and comparison with the upper continental crust, Int. J. Earth Sci. 101 (1) (2012) 365–376.

[28] N.H. Batjes, ISRIC-WISE derived soil properties on a 5 by 5 arc-minutes global grid (ver. 1.2), 2012.

[29] IUSS Working Group WRB. 2007. World reference base for soil resources 2006, first update 2007. World soil resources reports no. 103. FAO, Rome.

[30] Z. Chen, A.S. Auler, M. Bakalowicz, D. Drew, F. Griger, J. Hartmann, G. Jiang, N. Moosdorf, A. Richts, Z. Stevanovic, The world karst aquifer mapping project: concept, mapping procedure and map of Europe, Hydrogeol. J. 25 (3) (2017) 771–785.

[31] S.E. Fick, R.J. Hijmans, WorldClim 2: new 1-km spatial resolution climate surfaces for global land areas, Int. J. Climatol. 37 (12) (2017) 4302–4315.

[32] G.A. Riggs, D.K. Hall, M.O. Román, MODIS Snow Products Collection 6 User Guide, National Snow & Ice Data Center, 2015.

[33] A. Hreiche, Modélisation conceptuelle de la transformation pluie-débit dans le contexte méditerranéen, Université Montpellier II-Sciences et Techniques du Languedoc, France, 2003.

[34] D. Efremides, G. Tsakiris, Stochastic modelling of point rainfall in a Mediterranean island environment, Water Resour. Manag. 8 (3) (1994) 171–182.

[35] P. Ramírez-Cobo, X. Marzo, A.V. Olivares-Nadal, J.Á. Francoso, E. Carrizosa, M.F. Pita, The Markovian arrival process: a statistical model for daily precipitation amounts, J. Hydrol. 510 (2014) 459–471.

[36] M. Mhanna, W. Bauwens, Assessment of a single-site daily rainfall generator in the Middle East, in: Paper presented at the 2009 Second International Conference on Environmental and Computer Science, 2009.
[37] W. Najem, S. Catafago, Contribution à l'étude de la pluviométrie Libanaise, Thèse de Doctorat présentée à l'Université Montpellier II, France, 1976.
[38] W. Najem, A continuous point process model for daily rainfall, Stoch. Hydrol. Hydraul. 2 (3) (1988) 189–200.
[39] J. Thornes, F. Lopez-Bermudez, J. Woodward, Hydrology, River Regimes and Sediment Yield, Oxford University, The Physical Geography of the Mediterranean, 2009, pp. 229–253.
[40] P. Tarolli, M. Borga, E. Morin, G. Delrieu, Analysis of flash flood regimes in the North-Western and South-Eastern Mediterranean regions, Nat. Hazards Earth Syst. Sci. 12 (2012) 1255–1265.
[41] H.K. Kafle, H.J. Bruins, Climatic trends in Israel 1970–2002: warmer and increasing aridity inland, Clim. Chang. 96 (1–2) (2009) 63–77.
[42] K. Chaouche, L. Neppel, C. Dieulin, N. Pujol, B. Ladouche, E. Martin, D. Salas, Y. Caballero, Analyses of precipitation, temperature and evapotranspiration in a French Mediterranean region in the context of climate change, Comptes Rendus Geosci. 342 (3) (2010) 234–243.
[43] T. Mavromatis, D. Stathis, Response of the water balance in Greece to temperature and precipitation trends, Theor. Appl. Climatol. 104 (1–2) (2011) 13–24.
[44] S.M. Vicente-Serrano, C. Azorin-Molina, A. Sanchez-Lorenzo, J. Revuelto, J.I. López-Moreno, J.C. González-Hidalgo, E. Moran-Tejeda, F. Espejo, Reference evapotranspiration variability and trends in Spain, 1961–2011, Global Planet. Change 121 (2014) 26–40.
[45] E. Hertig, Y. Tramblay, Regional downscaling of Mediterranean droughts under past and future climatic conditions, Glob. Planet. Chang. 151 (2017) 36–48.
[46] G. Naumann, L. Alfieri, K. Wyser, L. Mentaschi, R. Betts, H. Carrao, J. Spinoni, J. Vogt, L. Feyen, Global changes in drought conditions under different levels of warming, Geophys. Res. Lett. 45 (7) (2018) 3285–3296.
[47] A. Abouabdillah, O. Oueslati, A.M. De Girolamo, A.L. Porto, Modeling the impact of climate change in a Mediterranean catchment (Merguellil, Tunisia), Fresenius Environ. Bull. 19 (10) (2010) 2334–2347.
[48] A. Capra, S. Consoli, B. Scicolone, Long-term climatic variability in Calabria and effects on drought and agrometeorological parameters, Water Resour. Manag. 27 (2) (2013) 601–617.
[49] M. Marquès, R.F. Bangash, V. Kumar, R. Sharp, M. Schuhmacher, The impact of climate change on water provision under a low flow regime: a case study of the ecosystems services in the Francoli river basin, J. Hazard. Mater. 263 (2013) 224–232.
[50] J. Ruffault, N.K. Martin-StPaul, S. Rambal, F. Mouillot, Differential regional responses in drought length, intensity and timing to recent climate changes in a Mediterranean forested ecosystem, Clim. Chang. 117 (1–2) (2013) 103–117.
[51] A.-E. Vrochidou, I. Tsanis, M. Grillakis, A. Koutroulis, The impact of climate change on hydrometeorological droughts at a basin scale, J. Hydrol. 476 (2013) 290–301.

Further reading

J.R. McNeill, The Mountains of the Mediterranean world, Cambridge University Press, 2002.

CHAPTER 2

Evapotranspiration in the Mediterranean region

Gilles Boulet[a], Lionel Jarlan[a], Albert Olioso[b], Hector Nieto[c]
[a]CESBIO (UPS/CNES/CNRS/INRAE/IRD), Toulouse, France
[b]UMR EMMAH (INRAE, Avignon Université), Avignon, France
[c]COMPLUTIG, Alcala de Henares, Spain

2.1 Context: Evapotranspiration estimation as a main component of sustainable water management in the Mediterranean region

2.1.1 ET as an estimate of the water demand for crop and natural ecosystems

Evapotranspiration (ET) is the sum of the evaporation (E) from the soil (from the bare soil adjacent to the plant shoots and/or the soil beneath the canopy, and also the intercepted rain or aspersion by the plant) and the transpiration (T) of the vegetation, mostly through the leaves.

ET is the main component of green water, which is the water stored in the soil root zone and consequently evaporated or extracted by plants; it is used as an operational proxy to compute agricultural water demand even though the plant water use corresponds chiefly to transpiration. Both the transpiration/evaporation partitioning of ET and the drainage rates are hard to assess routinely, which hampers a more precise monitoring of irrigation water, for instance. In addition, ET may also be seen as a net groundwater uptake for pumping estimate, and hence it can be used to detect and quantify unregulated water withdrawals. In the case of rainfed or gravity/aspersion irrigation, the entire surface is wet (up to what can be intercepted by the foliage, of course) and an evapotranspiration rate that is determined from a maximum evaporation and a maximum transpiration, that is, not limited by water availability, is a fairly good approximation of the plant water need. One commonly differentiates several types of ET depending on the moisture status or the vegetation cover [1]. Actual ET accounts for all environmental factors, and corresponds to the surface water vapor loss toward the atmosphere that actually takes place. Potential ET is the theoretical rate of ET in actual weather and vegetation conditions when soil moisture is not limiting, for example, the surface is saturated. This potential ET is a theoretical value that can be even higher than the actual maximum rate if the surface were really saturated, because in actual weather conditions the air can be drier and hotter than what would be observed if the soil was moist, due to

Water Resources in the Mediterranean Region
https://doi.org/10.1016/B978-0-12-818086-0.00002-9

surface-atmosphere feedbacks. Finally, reference ET is the potential ET for a reference crop, for example, a short well-watered green grass, and can be considered as the summary of the weather influence into an evapotranspiration rate corresponding to the atmospheric evaporative demand. Crop ET is the maximum ET rate for a given crop when plant water availability is sufficient to warrant maximum transpiration. Reference and crop ET rates are usually computed with the FAO-56 method ([2], Fig. 2.1). This approach explicitly (but coarsely) accounts for the irrigation type and frequency through the relative fraction of wetted area corresponding to each particular technique, for example, aspersion, drip, or gravity irrigation. It is a rather efficient tool to compute agricultural water demand and use. Crop evapotranspiration calculated with this approach is the closest estimate to what can be named "crop water demand". Water managers rely on water demand estimate before and during the cropping season for water allocation—for instance, from regulated wells or dams. On the other hand, water use can then be derived from actual ET. Actual ET is obtained by multiplying a potential ET or the FAO-56 crop ET by a reduction coefficient representing crop water stress.

2.1.2 Actual ET and management issues

(i) Rainfed eco- and agro-systems: assessing vulnerability, drought monitoring, and early warning systems

In the Mediterranean basin, droughts and dry spells are frequent and have an impact on growth and food production of rainfed ecosystems. Long dry-downs during winter and spring can affect yields when they occur at specific phenological stages of the crops.

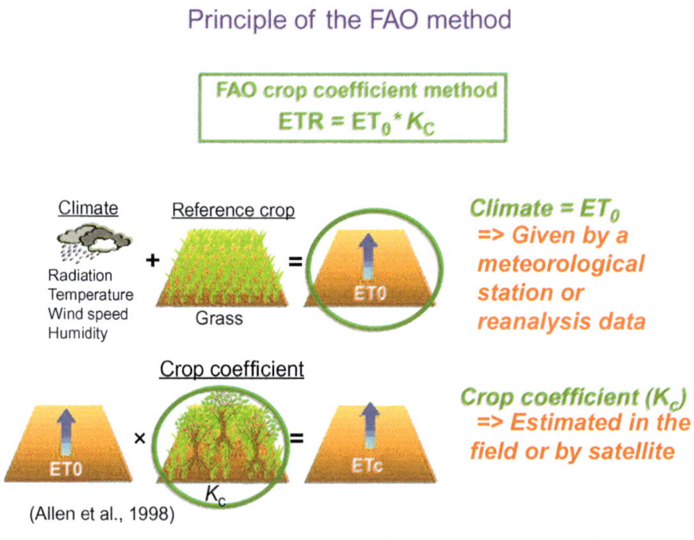

Fig. 2.1 Reference and crop evapotranspiration in the FAO-56 method for crop water demand assessment.

Sometimes, insufficient rainfall leads farmers to give up on harvesting the crops, leaving it as food for livestock, and/or to demand economic compensation from insurance agencies. Furthermore, successive summers with very little rain can trigger tree mortality. Rainfed tree crops are usually planted with an interspace calculated on the basis of rainfall climatology so that the rooting system of each tree can investigate a sufficient part of the unsaturated zone. Given the lower rainfall predictions from most climate change scenarios, this planting distance might not be adapted to future climates and will probably require additional irrigation, specifically at the end of the driest summers when root-zone soil water becomes almost depleted. For those areas, the main issue is thus to monitor water stress and to assess the resulting vulnerability of the vegetation. Combining both can lead to efficient early warning systems. Water stress is usually computed from the ratio of actual to potential ET. Estimation of actual ET is well constrained by rainfall estimates, contrarily to irrigated zones for which very little is known about the irrigation volumes, in particular at field scale. Therefore, water budget methods can perform well in calculating actual ET for those environments if soil hydric properties and rainfall are well-known.

(ii) Irrigated lands: challenges for the future, deficit irrigation, etc.

For irrigated lands, there is a large demand from most stakeholders for better (if not sustainable) management of scarce water resources. ET models, in diagnostic and prognostic modes, are key tools for, respectively, assessing the water uptake from surface reservoirs or pumped from the groundwater systems, or forecasting the water use at seasonal and interannual scales. In diagnostic modes, irrigation volumes are almost impossible to measure directly at a scale that is useful for resource managers (i.e., perimeter) or the allocation planers (who base their decisions on regional statistics). Irrigation must therefore be deduced through modeling of the agricultural practices. Since modeling involves significant uncertainty, additional information to the water budget closure is crucial. As we will see below, this additional information can be obtained by remote sensing.

2.1.3 Evapotranspiration as a key issue in the natural environment and groundwater

Evapotranspiration is one of the main factors controlling aquifer recharge in the Mediterranean region. A large part of the precipitation is returned to the atmosphere through evapotranspiration so that aquifer recharge is usually low. For example, classical values of precipitation in the Northern Mediterranean areas range between 500 mm/year and 1500 mm/year depending on the year, altitude, and the proximity to the sea, while reference evapotranspiration range between 1100 mm/year and 1300 mm/year. The level of evapotranspiration is strongly controlled by the water-holding capacity of the soil, which impacts the storage of water in the soil and the development of vegetation and then transpiration capacity. In particular, the Mediterranean region is often characterized by a karstic environment over carbonated formations. In these situations, water-holding capacity of the soil can be very low, down to a few tenths of mm, because of the amount of rocks and gravels. In consequence, the vegetation is poorly developed. Infiltration is fast to the aquifer and streams. In these situations, transpiration is low compared to what might be expected given the rainfall. Moreover, the presence of rocks close to the surface tends to limit the level of evaporation at the soil surface. Recharge of karstic hydrosystems can be large (ex. [3]).

The relations between plants and the aquifers in the Mediterranean karstic environment are very complex to investigate. The root system, in particular for trees, can develop to several meters or even tens of meters, through the fractured rocks, so that trees can extract waters from very deep layers in the ground. This is a crucial process for tree survival in such environment, where the amount of available water may decrease to very low levels in summer, inducing strong water stress ([4], Fig. 2.2).

The assessment of ET over natural vegetation in the Mediterranean area is also of interest for monitoring water stress in order to prevent and mitigate wildfires. Vegetation water content and stress are crucial variables for assessing ignition danger and spread risk. In addition, drought monitoring can be used as an indicator of forest health, as stressed vegetation is more likely to be affected by plagues and diseases, or for surveying forest degradation and in fine desertification processes. Finally, a holistic assessment of evapotranspiration and water fluxes is required for atmospheric applications, such as weather forecast, as well as for hydrology.

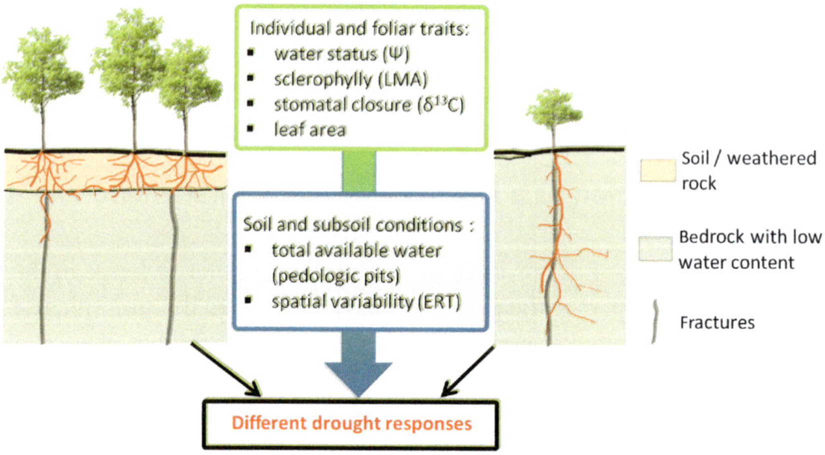

Fig. 2.2 Link between the various ecophysiological and edaphic drivers of drought. *(From S.D. Carrière, C. Doussan, J. Ruffault, F. Pimont, G. Simioni, I. Scotti, F. Courdier, H. Davi, N.K. Martin-StPaul, K. Chalikakis, J.M. Limousin, C.B. Cakpo, Impact of local soil and subsoil conditions on inter-individual variations in tree responses to drought: insights from electrical resistivity tomography, Sci. Total Environ. 698 (2020) 134247. https://doi.org/10.1016/j.scitotenv.2019.134247 (web archive link).)*

2.2 Evapotranspiration as a link between the water and energy cycles

2.2.1 The main drivers of ET: Climate, water, and plants

Climate is the first limiting factor of ET after a large rainfall or irrigation ("climate controlled" stage during a dry-down). Maximum evapotranspiration, in the absence of soil limiting factors such as insufficient soil moisture, is usually computed using radiation-based (Priestley-Taylor) or combination (Penman-Monteith) equations. It can be synthesized in the reference ET (ET_0, Fig. 2.3). The reference ET is the potential ET of a full canopy cover (usually grass or alfalfa) taken as a reference crop. It is generally computed with the Penman-Monteith ET formulation for a given meteorological forcing and specific surface properties corresponding to a well-watered surface.

In situations with no water stress, the main climate driver of ET is the incoming solar radiation, Rg. Several ET models are driven by the sole Rg. Air temperature is also important and impacts many processes, including the capacity of the atmosphere to store water vapor before saturation, which is called the air relative humidity or vapor pressure deficit. This deficit is often large in Mediterranean climates and favors the use of the Penman-Monteith formulation over the Priestley-Taylor formulation [5] to compute the "climate controlled" (or potential) rate of ET. Wind speed affects the turbulent transfer and thus the ability to replace through convection by eddies the air enriched by vapor at the surface with drier air from above.

Vegetation development is the second main driver of ET. Direct evaporation from a wet soil has a fast decreasing rate since surface soil moisture is quickly evaporated and less easily replaced by upward capillary or vapor transfer through the topsoil porous medium. When the soil surface is completely dry, the vapor transfer within the soil porous medium is always far below the potential evaporation rate. On the other hand, plants are able to extract water through their roots. They can therefore explore a large

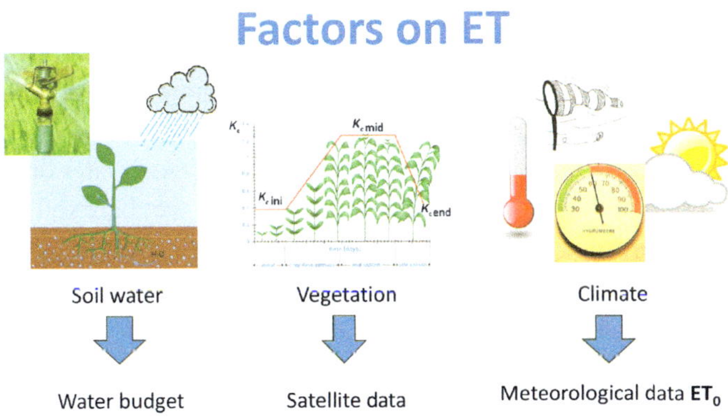

Fig. 2.3 The main drivers of ET.

portion of the unsaturated zone, and withdraw water from at least the top first meter of soil at maximum development, in contrast to the few upper centimeters from which most evaporation water is coming. ET therefore depends on rooting extent (vertically and horizontally) as well as the Leaf Area Index (LAI, the total upper leaf surface above a given square meter of soil) and its distribution (clumping effects, i.e., the concentration of vegetation in dense clumps interspaced by large patches of bare soil or herbaceous vegetation cover). For homogeneous vegetation (herbaceous cover mostly), E is usually negligible with respect to T when LAI values are larger than 3, because in those cases the vegetation fraction cover is often close to 100%. However, for annual crops, when considering the full annual cycle, cumulative evaporation may be larger than transpiration because of the long periods of bare soil between successive crops [6].

Finally, water availability (through rain and irrigation amounts, frequency, and intensity) is the third main driver for ET. When soil moisture availability is decreasing, E or T can enter the "soil controlled" stage with a rapid collapse of each rate. E is driven by shallow surface and T by the root zone soil moisture, but the falling rate accounts also for the ability of the soil to store or release water through its water retention and conductivity curves (and can therefore be seen as a fourth controlling factor).

2.2.2 ET in the water budget

ET can be computed from the water budget either simply as the residual term ($ET = P + I - D - R - \Delta S$, where P is precipitation, that is, rain and snowmelt, I is irrigation, D is drainage, R is runoff, and $-\Delta S$ is the change in the root zone water storage) or by computing each term of this budget and solving for the corresponding state variable representing the water content of the soil (volumetric soil moisture, water depth, matric potential). There is a range of water balance models of increasing complexity, from simple one- or two-reservoir models to detailed discrete mechanistic approaches (land surface models, or LSMs). The FAO-56 method occupies the lead position amongst the most simple and operational models. In this method, ET is computed as a combination of the reference evapotranspiration ET_0 (the climate driver), a basal crop coefficient (K_{cb}) representing the development of the vegetation (the vegetation driver), and two reduction factors (ranging between 0 and 1) dealing with the impact of insufficient soil moisture at the surface (K_e), for E, or in the root zone (K_s), for T (soil water driver), weighted by a factor representing the respective fraction cover of the wet bare soil (defined by the irrigation type) and the vegetation, respectively (Fig. 2.4):

$$ET = (K_e + K_{cb} * K_s) * ET_0 \tag{2.1}$$

For mountain environments, the interplay between the various vertical and lateral redistribution processes makes it difficult to close the water budget at basin scale and extract a meaningful ET value. Climate forcing in those environments is also very hard to distribute spatiotemporally.

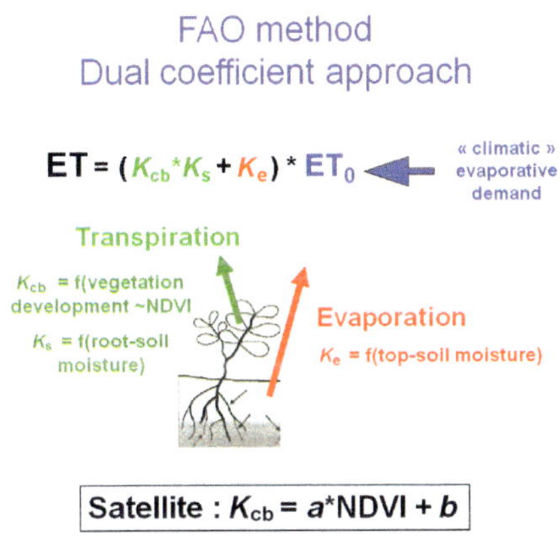

Fig. 2.4 The soil-plant-atmosphere interface according to the FAO-56 method and the use of NDVI.

The basal crop coefficient depends on the crop type and is usually obtained empirically from experimental data such as flux towers [7] or lysimeters. The rationale is to look at maxima in the ratio of observed ET to ET_0 and plot those values against the phenological status or an index representing the vegetation development. The number of such databases in the Mediterranean is rather limited with respect to the variety of crops and agricultural practices that exist throughout the region. K_{cb} depends mostly on the fraction cover, and reaches values of up to 1.2 in the FAO-56 method, but can remain far below 1 for sparse or row vegetation. K_{cb} is thus linearly related to the Normalized Differential Vegetation Index (NDVI), which can be derived from most remote sensing platforms in the optical domain (Sentinel 2, Landsat, Sentinel 3, MODIS, etc.). To do so, a linear regression is performed on the ET/ET_0 versus NDVI scatterplots for each tower once points above the linear scatter for low NDVI values (which correspond to E dominating ET) and the points below for larger NDVI values (which correspond to vegetation stress) are discarded.

Other methods using satellite information exist. Most of them are based on LSM driven by, at least, NDVI or Leaf Area Index estimated from remote sensing data, and sometimes (e.g., GLEAM, [8]) surface soil moisture information at coarse scale (0.25° typically) from passive microwave sensors (ASCAT, SMOS, etc., [9])

2.2.3 ET in the energy budget

ET is part of the energy balance as the total latent heat flux LE, that is, the energy transferred to the atmosphere through turbulent transfer of water vapor. As for the water budget pathway, LE can be retrieved either directly as the residual of the energy budget expressed for the whole surface ($LE = R_n - H - G$ where R_n is the net radiation, H is the sensible heat flux, and G is the soil heat flux) or by solving for the surface state variable, the total surface radiative temperature, in a multilayer energy budget model.

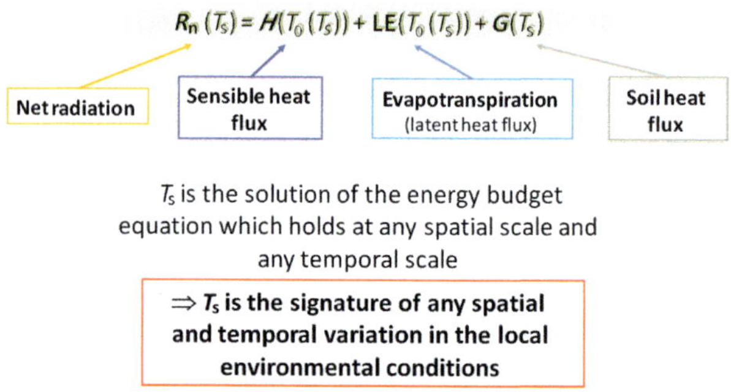

Fig. 2.5 The surface energy balance and its signature, the radiative surface temperature.

In both cases, the remotely sensed radiative surface temperature obtained from satellite measurements is used to constrain R_n, H, and G (Fig. 2.5). Again, there is a large variety of models to retrieve ET from information in the Thermal InfraRed domain [10]. Most of them do not separate explicitly E and T, but some [11, 12] solve simultaneously the energy budgets of the soil and the vegetation, and offer separate estimates of E and T. Since the radiative surface temperature is usually obtained once a day at the time of the satellite overpass, an instantaneous energy budget is solved at that time. Consequently, methods that rely on this budget expressed for a very short period during the day do not propagate the temperature in time to compute G (which is in that case parameterized as a fraction of soil or total net radiation). However, surface temperature data can also be assimilated in LSMs that compute the evolution of the soil temperature profile [13], being able to track, and even forecast, the subdaily ET trends.

We can summarize the estimation of evapotranspiration, within the energy balance framework, as the modeling of three processes: (1) the estimation of available energy at the surface (i.e., the net radiation minus the soil heat flux); (2) the evaluation of the effectivity of heat, momentum, and water vapor turbulent transport; and (3) the derivation of the reference source temperature for the convective transport of the air entrapped in the vegetation (i.e., the "aerodynamic temperature"—see below) from the surface radiative temperature.

Net radiation (R_n) is a key element in the energy budget of the land surface. It determines the available energy that the land can use for water evapotranspiration (latent heat flux, LE) and for heating up the overlying air layer (sensible heat flux, H). Since evapotranspiration is the combined process of soil evaporation and canopy transpiration, R_n can also be partitioned into soil and canopy net radiation.

Remote sensing has proven to be a sound alternative to derive R_n. Different approaches have been proposed to estimate surface albedo (which represents the proportion of solar radiation reflected back to the atmosphere), ranging from empirical relationships between ground-measured albedo and the different spectral bands of the satellite instrument [14] to more physically based methods relying on modeling the surface anisotropic effects [15, 16]. Indeed, one of the major challenges when estimating albedo with satellite remote sensing data is that such sensors typically measure the outgoing radiance spectrally and in a given direction, while the estimation of albedo needs to account for the outgoing radiance in all the directions of the hemisphere and on the overall solar spectrum and at any time of the day while the sun position is changing. Methods based on the modeling of those bidirectional effects have proven to be effective to overcome this challenge. However, Mira et al. [17] showed that the accuracy of remote sensed estimates of albedo is adequate for modeling R_n when retrieving the surface energy.

Aerodynamic resistances can be viewed as a simplification of the complex turbulent transport of heat, momentum, and water vapor, by similarity with Ohm's law for electric transport. These resistances therefore represent how efficiently a scalar (heat, momentum, or water vapor) is transported from one point to another following a gradient (i.e., vertical differences of temperature and/or vapor pressure). Several formulations and/or parameterizations have been proposed to describe these turbulent transport processes but in summary, they include variables related to surface aerodynamic roughness, in particular the vegetation height, wind speed as well as wind attenuation through the canopy, and atmospheric stability.

The resistance-energy balance models were originally formulated for computing H using the gradient between an effective surface temperature, called the "aerodynamic surface temperature" (T_0), and the air temperature at the reference height (2 m, a common figure for all agrometeorological stations or 10 m for the lowest boundary of the atmospheric temperature profile simulated by climate models). The drawback of this type of methodology is that the aerodynamic temperature cannot be directly estimated by remote sensing. Overall, remote sensing ET models differ from each other on how the difference between surface temperature T_s and T_0 is considered (Fig. 2.6).

Single-source or bulk transfer schemes for modeling sensible heat flux (H) often employ an additional resistance term (R_{ex}) to relate T_0 to T_s [18–20]. Appropriately calibrated, one-source models have shown satisfactory estimates of surface energy fluxes [21–23].

Due to the difficulty in parameterizing R_{ex} robustly and parsimoniously in the application of the one-source scheme for different landscapes, climates, and observational configurations, the two-source modeling approach was developed. Two-source models partition the surface energy fluxes and the radiometric temperature between nominal soil and canopy sources, and include a more physical representation of processes related to radiometric temperature and T_0. However, two-source models need to solve for both the canopy and soil temperatures. Partitioning of the radiometric temperature between those two component temperatures (i.e., solving two unknowns from a single piece of information) requires some assumptions on the water status of either the soil or the

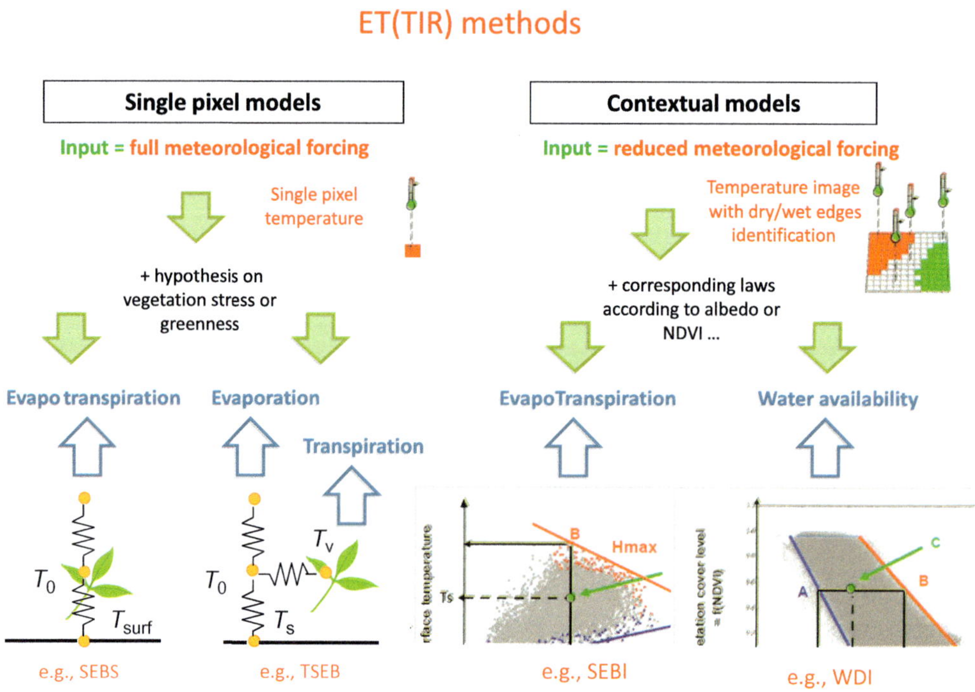

Fig. 2.6 The various energy budget methods to derive ET from TIR data.

canopy. To overcome this issue, an iterative process is carried out, in which in a first stage a prior canopy temperature is derived based on an estimate of potential transpiration, which is subsequently reduced until realistic temperatures and fluxes (such as negative fluxes meaning that condensation occurs, which is very unlikely during daytime) are obtained. Different models are proposed in order to initialize this iterative process, using either a Priestley-Tailor potential transpiration [12], an efficiency parameter from the imposed transpiration equation [11], or the Penman-Monteith transpiration [5]. These models are likely to be the most robust when applied to a wider range of conditions and environments, especially for sparse Mediterranean vegetation, but they require a larger set of remote sensing and ancillary input data, and are rather sensitive to input errors.

Contextual algorithms derive LE (or evaporative fraction, the LE/$(R_n - G)$ ratio) based on the distance of actual surface radiometric temperature between hot and cold temperature end-members corresponding to the driest and the wettest conditions, respectively. These hot and cold end-members are either derived empirically from the satellite image itself or retrieved theoretically by inversion of the Penman-Monteith equation. Models within this category differentiate themselves according to how these end-members are retrieved, such as S-SEBI [24], three-temperature method [25], SEB-4S [26], SEBAL [27], or METRIC [28]. Several assumptions and conditions must be met in these models, mainly as they assume that spatial variability of radiometric surface temperature is driven primarily by soil moisture variability. Therefore, the area where contextual models are applied should be under the same meteorological forcing (irradiance, air temperature, vapor pressure deficit, and wind speed). Furthermore, canopies should present similar heat transport efficiency (i.e., similar aerodynamic roughness) and hence have similar canopy architecture, and for those models that aim to retrieve the temperature end-members from the image itself, a wide range of moisture and vegetation density variability should be present in the scene. On the other hand, these methods are rather easy to use, require few ancillary and remote sensing inputs, and are less prone to transfer errors in surface radiometric temperature to the ET outputs.

2.3 Specificities of evapotranspiration processes in the Mediterranean

2.3.1 Plant scale: Complex canopies, sparse vegetation, and the competition for water and radiation

As is the case in many semiarid areas, natural vegetation is often sparse, with competition between species, especially woody and herbaceous vegetation, even though some cooperation strategies exist (cf. hydraulic lift); rainfed systems, with adapted cropping systems, do not have that problem; irrigated systems are often complex, with systems to increase water efficiency: intercropping, several crops in one hydrological year, etc.

Native Mediterranean plants, or plants adapted to semiarid lands, have developed specific mechanisms to control the loss of water while maximizing the CO_2 uptake (cuticle, stomatal regulation (e.g., *Crassulaceae* family)), as well as mechanisms to reduce their activity in the driest periods while favoring regrowth and maximizing carbon assimilation in more favorable conditions.

2.3.2 Soil processes: Vapor transfer, crusting, etc

Evaporation depends on the soil hydraulic properties of the soil surface. Given the high incoming radiation levels and the moderate rainfall, as well as the extent of bare soil surfaces, a large fraction of soil surfaces is drying very quickly toward residual soil moisture levels. In that case, evaporation takes place below the surface, and an "evaporation front," above which all transfers are in the vapor form, is rapidly going down deeper in the soil. Consequently, there is always a small but nonnegligible vapor flux even when the soil is very dry, with values up to 30–50 W/m^2 around midday [29]. Vapor and liquid transfer depend on the retention and hydraulic conductivity curves of the soil-porous medium. Very few LSMs properly take these into account in their formulation of the evaporation rate [30].

In addition, even if cumulative rainfall over the year can be very low, rainfall intensities can be high, and may directly impact the soil. This is especially true for sparse vegetation, since the soil surface receives the drops directly without any interception (and thus kinetic intensity depletion) by the leaves. A small fraction of clay content can induce crusting and impact the evaporation and runoff processes.

Another characteristic for ET in those environments is the high frequency of large stone fractions or impermeable surfaces such as those of karstic systems, with little evaporation capacities once the overland water has been evaporated.

2.3.3 Atmospheric processes: Oasis effects at a range of scales

Mediterranean landscapes, as well as many semiarid areas, are characterized by a large heterogeneity of surface conditions. Typically, dry hot surfaces, usually bare soils, can coexist with cooler and wetter areas, such as when irrigation is present. Advection of sensible heat from the hot/dry patches to the cool/wet ones can then occur, but is not accounted for in a correct way in the usual monodimensional (i.e., vertical) parameterization of surface models. When this "oasis effect" happens at a very local scale—for example, in the case of sparse or row vegetation—two-source models allow modeling of this transfer from the soil source to the vegetation source and enhance transpiration. At larger scales (e.g., perimeter scale) advection is rarely taken into account and remains a challenge in modeling complex land surface–atmosphere transfers accurately. Indeed, most methods (reference ET included) assume that there is no net influx of sensible or latent heat from one field to the other.

2.4 Regional evapotranspiration and its challenges

2.4.1 The increasing availability of remote sensing data and their use in computing ET

Spatially distributed information on surface properties and surface state variables can be retrieved from remote sensing data. Some atmospheric variables can also be derived with proper inverse radiative transfer modeling. Indeed, global incoming radiation (cf. SAF Land data) as well as air temperature and humidity can be estimated though radiative transfer model inversion from many geostationary and heliosynchronous platforms at a scale relevant to model forcing, with a higher resolution than the actual products obtained from reanalysis data (outputs of meteorological models, cf., e.g., ERA5 climate product).

The main biophysical variables classically derived from remote sensing (often available as downloadable products) are (see Fig. 2.7): vegetation fraction cover and leaf area index, either obtained from reflectance data using radiative transfer models, neural network approaches, or from empirical relations based on indices combining two reflectances

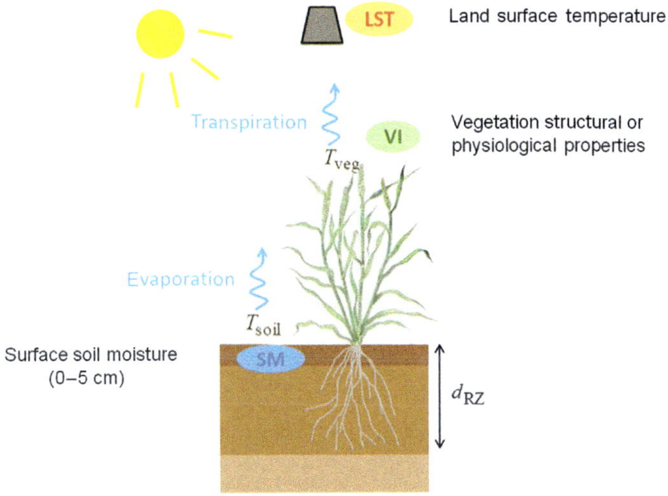

Fig. 2.7 The various biophysical variables that can be obtained from remote sensing.

(e.g., NDVI); energy budget variables, such as albedo, emissivity, and land surface temperature; and soil moisture from passive or active microwave sensors. Some other indices, such as active and passive fluorescence, the Photochemical Reflectance Index, or the Shortwave Infrared Water Stress Index, can also inform on soil moisture availability and the physiological activity of the plant, amongst it chiefly the photosynthesis process. However, those indices depend heavily on the structural properties of the vegetation. Some insight on soil textural properties and the surface clay content can also be derived from visible/middle infrared data [31]. Vegetation height can also be derived from lidar measurements. The Copernicus Sentinels offer unprecedented high-resolution data with a revisit frequency that enables tracking of most surface changes, including phenological stages (emergence, plant growth, senescence, harvest) and crop practices (plowing, irrigation at early stages).

Remote sensing variables can also drive directly, as forcing variables, the ET component of hydrological models; for instance, the SAMIR model relies on empirical linear relationships between K_{cb} and NDVI established on literature data and a large flux tower database collected since 2002 in the Mediterranean [7].

2.4.2 Water balance pathway and its pitfalls

A significant advantage of the water budget method is that it provides time- and space-continuous estimates of ET. However, it suffers from the difficulty of providing spatially meaningful values of the many input parameters that water balance models and LSM require. Notably, those governing the evaporation process from the topsoil, as well as the root zone moisture availability, are particularly difficult to infer (ex. [6]). For irrigated areas, water inputs are difficult to assess at the plot scale over large areas. The development of generic approaches to account for irrigation practices that are meaningful for large areas is difficult, as most practices vary from one farm to another. In addition, decisions on irrigation cannot always be reproduced by simple rules. Another issue is the description of poorly known processes such as deep drainage or overland flow, for which remote sensing data is of little help. At basin scale, the lack of knowledge on the subsurface flow and the interplay between the Wadi river flow and the flow of the saturated zone below does not enable to estimate ET as the residual term of the water budget for the whole catchment.

2.4.3 Energy balance pathway and its issues

The energy balance method is particularly useful in irrigated lands because it provides ET estimates for all areas for which times series of cloud-free images are available. For irrigated zones, the energy balance method is particularly interesting because, unlike the water budget method, it does not require an estimate of irrigation amount, neither an estimation of soil hydro-parameters. Moreover, it requires a smaller amount of input parameters compared to the water budget method, the most uncertain ones being the empirical formulation of the soil heat flux (all models), of the relationship between the aerodynamic and the surface temperature (single source models), or those affecting the maximum transpiration (minimum stomatal resistance, Priestley-Taylor coefficient) for the dual source approach.

On the other hand, evapotranspiration is obtained at the time of the satellite overpass for cloud-free images only [32, 33]; there is therefore not only an issue of producing a total daily estimate (temporal extrapolation issue), but also an issue of providing estimates in between two successive images or when clouds are present (temporal interpolation issue).

Moreover, since most Mediterranean landscapes are fragmented with agricultural fields sized around 1 ha, only high-resolution data (available up to now every 16 days at most for thermal infrared data) can meet the needs to evaluate ET at plot scale. This revisit frequency is of course incompatible with the temporal extension of most hydrological events (especially the time between two successive irrigations, the length of a typical dry-down, the period of observation to detect a water stress, etc.). However, daily estimates at kilometric scale can be produced and remain useful for, say, regional water budget assessment and groundwater resources management. Estimating the surface energy balance components is also difficult in zones with topography, even though methods exist to correct the images for slope and azimuth, because most hypotheses (in particular those driving turbulent and radiation exchanges) are no longer valid, and meteorological forcing does not exist at proper space and time scales.

2.5 Case studies

2.5.1 Rainfed olive yards: An emblematic agrosystem of the Mediterranean

Olives are of prime economic importance in the Mediterranean. The Tunisian olive represents 20% of the world olive yards' surface area. Rainfed olive trees have been planted according to a density computed from broad Tunisian climatic zones: 15 m interspacing in the north, 20 m in the center, and 25 m in the south. This ensures proper availability of water even during the driest months. However, this spacing might not be sufficient in the future if, because of climate change, rainfall patterns are modified and drought spells become more frequent. In order to monitor the impact of water stress on olive production, as well as to anticipate complementary water allocation and prevent tree mortality, one needs to understand the thermal and water functioning of this sparse agrosystem, as well as to reproduce this functioning through soil-plant-atmosphere-transfer modeling. To do so, an experiment was set up in the Kairouan area in central Tunisia. It consists of two sets of energy budget aerial component measurements on two towers, one over bare soil and one over an olive tree, and their associated soil temperature/humidity/heat flux profiles at two corresponding pits, one under the bare soil mast and one under the tree (Fig. 2.8).

The soil and vegetation components of the energy budget and the top-meter water budget were monitored during three contrasting years (one dry, one wet, and one moderately dry, [34]). During the wet season, the water budget of the top first meter is closed—that is, the soil moisture storage variation corresponds roughly to evapotranspiration. But during the drier years, there is a deficit indicating deep root extraction and/or capillary rises of water from the deeper layer. We also evidenced the nonnegligible extraction of water by roots in the top 5 cm of soil. The ISBA (Interface Sol Biosphère Atmosphère, [35]) LSM was applied to this system. It was demonstrated that the partitioning of the evaporation and the transpiration according to a vegetation fraction cover computed using the turbid medium law (Beer Lambert), as is classically done for most LSMs, underestimates the transpiration. Accounting for the dense crown by multiplying the weighing factor by a factor of four (the ratio between the whole surface of the crown and its projection on the ground) allowed us to represent correctly the transpiration from isolated trees in ISBA (Fig. 2.9).

2.5.2 Regional water use in a mixed dry/irrigated agricultural landscape

Using spatially distributed models to infer ET space-time variability is becoming more widespread, but the evaluation of ET estimates, at both plot and regional scales, remains difficult. In the Kairouan area, we have used flux tower measurements on irrigated wheat/barley to test and calibrate the SAMIR model [36] at plot scale. SAMIR is an LSM based on the FAO-56 method that computes irrigation using a simple model of

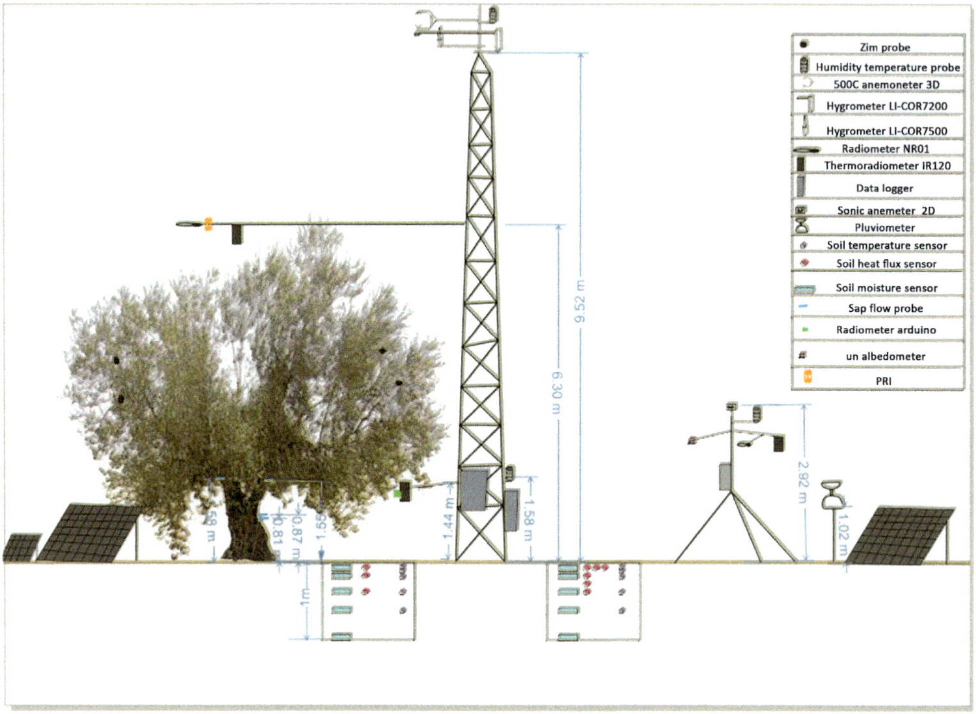

Fig. 2.8 Kairouan area (Nasrallah) olive tree experimental setup with two measurement masts. *(From W. Chebbi, G. Boulet, V. Le Dantec, Z. Lili Chabaane, P. Fanise, B. Mougenot, H. Ayari, Analysis of evapotranspiration components of a rainfed olive orchard during three contrasting years in a semi-arid climate, Agric. Forest Meteorol. 256–257 (2018) 159–178.)*

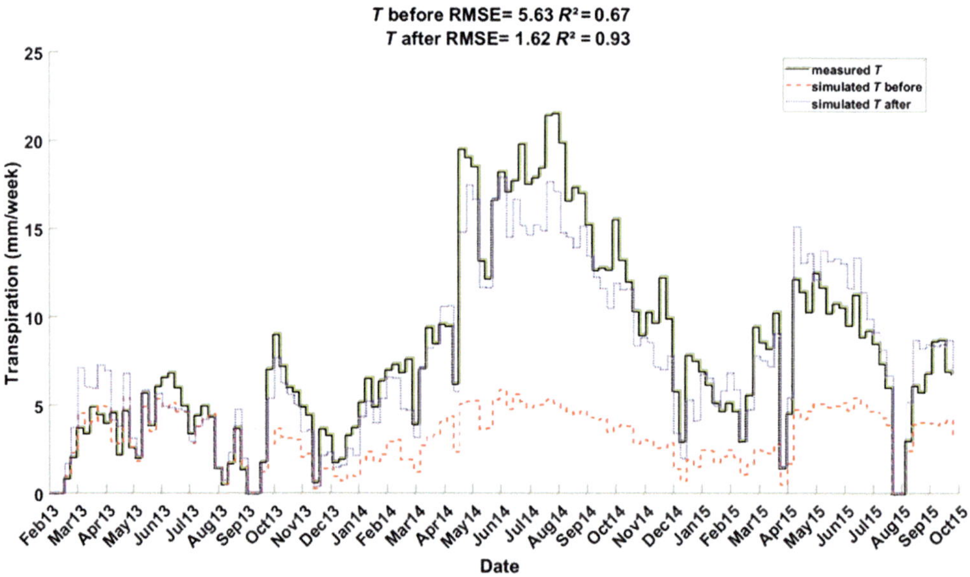

Fig. 2.9 Observed and simulated (ISBA) transpiration before (default) and after modification of the model inputs to account for effective transpiring area and lateral subsurface inflow during the wettest year (2014).

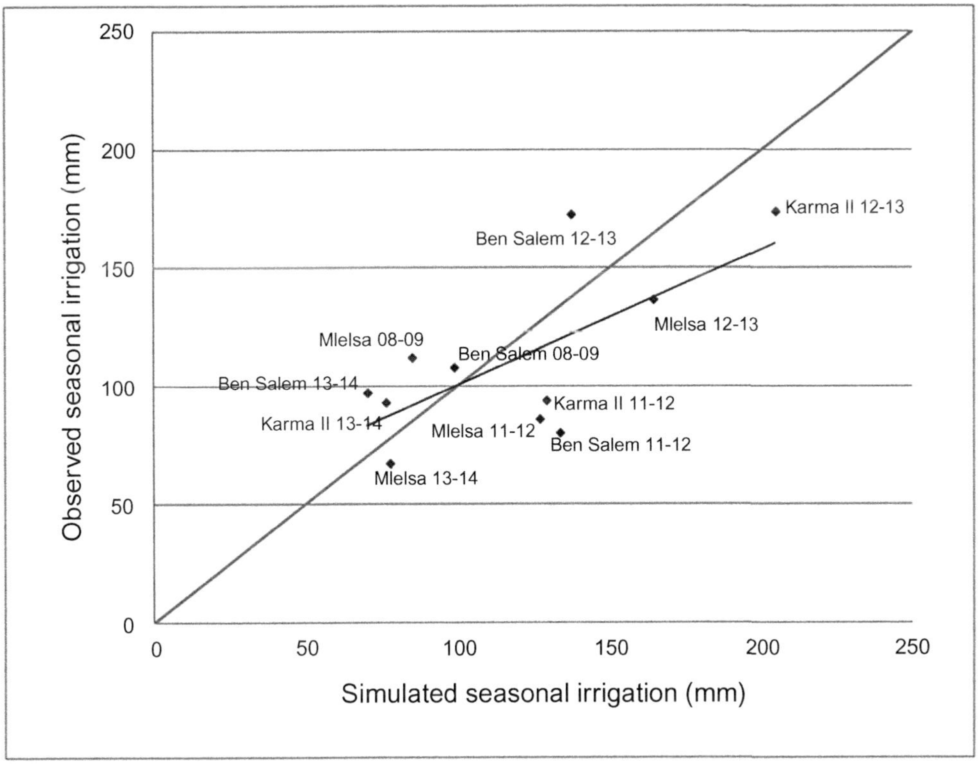

→ 25% mean absolute difference

→ Overall difference: 111 mm irrigation observed for 118 mm simulated

Fig. 2.10 Observed versus simulated (SAMIR) irrigation depths over the winter crop growing period.

irrigation practices. SAMIR was run with time series taken from SPOT5, Landsat, and SPOT4take5 high-resolution time series of NDVI. We have collected pumping volumes aggregated at monthly scales to check the relevance of the irrigation depths simulated by SAMIR [37]. Pumping extraction at seasonal scale was inferred with a small bias (111 mm vs. 118 mm, Fig. 2.10), but the timing especially at the beginning and in summer (lack of specific modeling of summer crops such as chili) was not realistic.

Another means to evaluate areally averaged ET is through the use of scintillometers. Scintillometry classically measures sensible heat flux integrated between the emitter and the receiver along a transect. For the Kairouan area, we have set up over 2.5 years an eXtra Large Aperture Scintillometer (XLAS) over a transect of 4 km, allowing us to validate not only the SAMIR model over the measurement footprint but also the SPARSE model [11] forced with the km resolution MODIS surface temperature product. This study showed that the evaporation component is not yet well constrained in the SAMIR

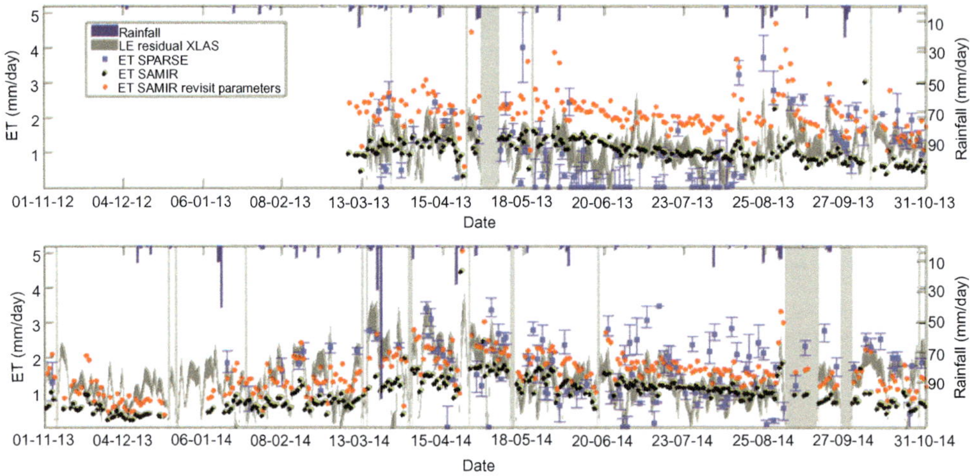

Fig. 2.11 ET time series showing estimates derived from the scintillometer XLAS (*gray area*), the energy balance model SPARSE (*blue dots*), and the water balance model SAMIR with parameters obtained from calibration on EC (*black dots*) or revisited (*red dots*).

model formulation and that the XLAS helped modify the evaporation reduction coefficient in the FAO-56 method used by SAMIR (Fig. 2.11). On the other hand, daily ET extrapolated from SPARSE estimates at the satellite overpass on days when a cloud-free MODIS image was available was found to be relatively reliable, but the presence of clouds prevented us from obtaining ET estimates over long periods, mainly in winter.

This long-term study has proven the complementarities of ET derived from energy and water budget approaches: the energy budget to constrain better the water status, the water budget to provide estimates when the energy budget methods fail (e.g., homogeneous conditions in the contextual approaches) or when there is not enough data (cloudy conditions, especially in winter, for instance).

2.5.3 Irrigation scheduling at the plot scale in the Tensift region (Morocco)

Improving irrigation scheduling has been identified as one of the main option for water saving in the Mediterranean area as long as the water demand in the future cannot be met by other technologies such as seawater desalinization. Within this context, as accurate estimations of root-zone soil moisture at the plot scale are not available over large areas, ET can be considered as a good proxy of water availability for crop development. Indeed, the gap between actual ET and potential ET (ET of the same surface under the same climate but in well-watered conditions) is directly related to the water availability in the soil.

ET models vary in their mathematical and conceptual representation, and, more specifically, in the way the water status of crops is assessed. As noted before, LSMs are appealing tools as they compute the water conditions of crops directly by solving the soil water balance. However, they require a large number of parameters to describe the soil and vegetation characteristics even for the most simpler models such as the FAO-56 dual crop approach. They also require an accurate description of water inputs including irrigation amounts and timing that are difficult to assess over large irrigated perimeters at the plot scale. By contrast, since surface temperature and water status are coupled when water stress occurs, the radiative surface temperature has been used as an indirect proxy of the water status of crops in combination with instantaneous energy balance modeling. As a first step toward irrigation scheduling, the abilities of the FAO-56 dual crop coefficient and the Two-Source Energy Balance (TSEB) model to predict water stress (i.e., the complement to 1 of the ratio between actual ET and potential ET) were compared on several instrumented wheat plots in the Haouz plain (Marrakech, Morocco) where irrigation amounts and timing were monitored (Fig. 2.12). The performances of both approaches to detect stresses appeared quite similar, at least for moderate water stresses.

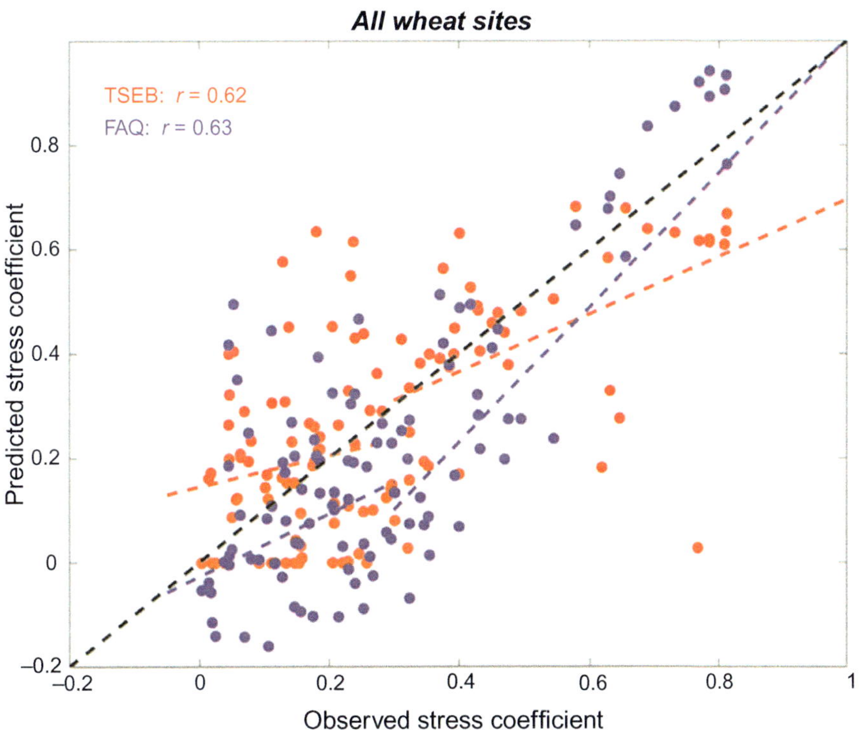

Fig. 2.12 Comparison of the ability of the FAO-56 dual crop coefficient (solving a soil water budget) and the Two-Source Energy Balance (TSEB) model (radiative surface temperature is taken as an indirect proxy of the crop hydric status) to predict a crop water stress index (the complement to 1 of the ratio between actual and potential ET) over wheat plots (Haouz plain, Morocco).

A real-time experiment of irrigation scheduling was implemented on two wheat plots located near Marrakech based on the FAO-56 dual crop coefficient approach in collaboration with the Cadi Ayyad University of Marrakech (UCAM, Morocco). Indeed, even if energy budget approaches are promising for an operational detection of crop water stress, surface temperature satellite products are still not available with the sensors in orbit at a spatial resolution and with a revisit time compatible with crop monitoring at the field scale. The agricultural practices were similar for the two plots apart from irrigation scheduling: one was irrigated by the farmer based on his own knowledge, while scheduling on the other was based on the FAO-56 water budget. The obtained results were promising, with 6% less of grain yields but an improvement of 9% on water use efficiency. The experiment was described by Le Page et al. [38] and in a short documentary that can be seen on YouTube (https://www.youtube.com/watch?v=p2ZlqBarOgk). Beyond the scientific issues, the collaboration with the local office in charge of the agricultural water management (Office Régional de Mise en Valeur Agricole du Haouz) was of prime importance to perceive how farmers can make this new information their own. Based on the results obtained on the field, a new tool for irrigation scheduling dedicated to farmers was developed, named the Satellite for Irrigation (SatIrr; http://osr-cesbio.ups-tlse.fr/Satirr); this was applied in various irrigated plots from Morocco to Lebanon.

Research is ongoing in attempts to merge simple water budget and energy budget approaches, taking advantage of remote sensing observations in the optical, thermal, and microwave domains. The main objective is to improve the prediction of evapotranspiration partition between evaporation, the water amount that is lost for the plant, and transpiration, which contributes to plant productivity. Indeed, future irrigation scheduling tools should be based on the transpiration part only.

2.5.4 Mediterranean forest

Flux measurements are performed in the frame of the ICOS European Research Architecture over various ecosystems (Integrated Carbon Observation System: http://www.icos-etc.eu/icos). In the South of France, two sites are covered by typical Mediterranean forests of mostly evergreen trees, *Quercus ilex* L. and *Pinus halepensis* Mill. The *Q. ilex* forest in Puechabon is located 35 km west of the city of Montpellier (43°75′N, 3°6′E, 250 m. a.s.l.). Mean annual values for precipitation and temperature calculated over the hydrological year from September 2000 to August 2016 were 807 mm, with a high annual variability (maximum 1400 mm and minimum 500 mm) and 14.5°C. The mean annual reference evapotranspiration was 1213 mm, with a lower variability than for precipitation (maximum 1350 mm and minimum 1080 mm). The difference between precipitation and reference evapotranspiration indicates a large climatic water deficit almost every year. The experimental site was described by Rambal et al. [39]. Trees are 5 m high with understory vegetation composed of sparse bushes of 2 m height. The soil developed over limestone and has a high proportion of stones and rocks (75% in the first 50 cm, more than 90% below). The plants' roots develop within the superficial soil layers, where a large fraction of the root mass is found, and along the cracks in the underlying limestone down to several meters (some roots were seen at a depth of 4.5 m). We derived evapotranspiration from eddy covariance measurements of latent heat flux available within the FLUXNET database ORNLDAAC2017 [40, 41].

Evapotranspiration follows a time dynamic strongly related to solar radiation, the quantity of vegetation, and the availability of soil water (Fig. 2.13). The available energy, in particular solar radiation, together with air temperature, is the main driver of climatic demand and reference evapotranspiration, so that the climate demand peaks in spring and summer. The development of vegetation and the water availability limit evapotranspiration with respect to ET_0. Actual evapotranspiration is low in winter. It increases in spring before decreasing at the beginning of summer. The peak of ET occurs in early summer, when there is no water limitation and vegetation activity and climatic demand is high. In summer, evapotranspiration is limited by the water reserve as the soil dries up, and ET remains low until fall and the return of rain. After that return, there is no more water stress and ET follows ET_0, which decreases until winter. Because of the relatively low level of vegetation cover, ET is generally lower than ET_0 in winter while water availability is high. Indeed vegetation fraction cover never exceeds 0.5, limiting transpiration. At the same time, the presence of coarse elements in the first soil layer limits evaporation. At the annual scale (Fig. 2.14), ET is significantly lower than ET_0 (between 25% and 50%). It is worth noting that annual ET_0 is increasing with time in relation to the temperature increase in the south-east of France, which is around 0.5 K every 10 years since 1980 [42]. At the same time, ET is not showing a clear trend, so that the ET/ET_0 ratio tends to decrease. The conservative behavior of annual ET indicates that the groundwater storage is a clear limiting factor of ET (as well as limiting vegetation growth). ET is also

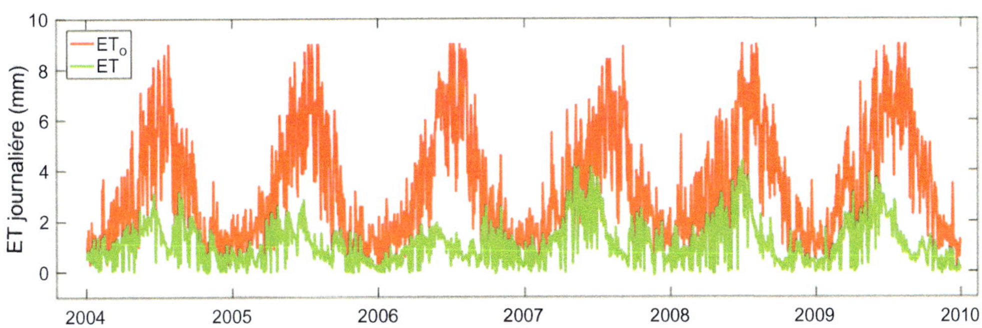

Fig. 2.13 Reference evapotranspiration (ET_0) and measured evapotranspiration (ET) from January 2004 to December 2009 in Puechabon (Fr). NB: ET_0 was obtained from the SAFRAN meteorological reanalysis that provide meteorological data at an 8 km spatial resolution over France (MeteoFrance product). Measurements were obtained from the Fluxnet database.

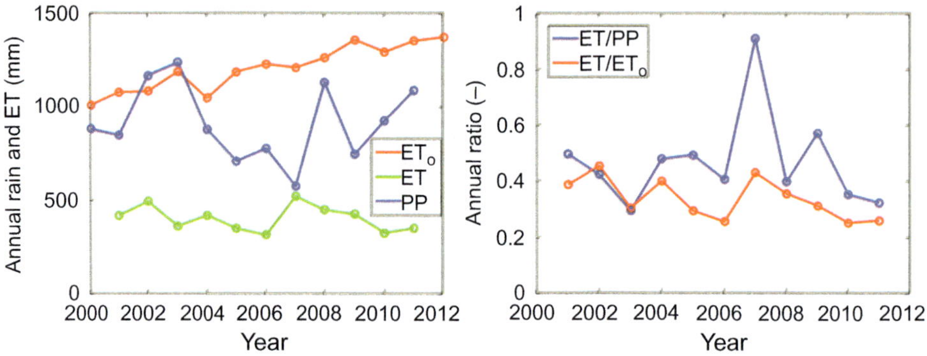

Fig. 2.14 Annual evolution of ET_0, ET, and precipitations (PP) and ET/ET_0 and ET/PP ratios between 2000 and 2012 in Puechabon.

significantly lower than rainfall (PP), with ET/PP ratios usually ranging around 0.4, indicating a very high karstic recharge level as in other karst systems in south-east France (e.g., [41]). However, in 2007, which was a very dry year with annual precipitation of 570 mm, the ET/PP ratio increased to above 0.9. That year ET was larger than average, which may be explained by a high level of ET_0, particularly in spring, and the large rainfall events that were occurring in fall 2016, implying very high ET during winter (Fig. 2.13).

Q. ilex experiences a strong water stress in summer, which results in low ET and very low plant water potential. In some situations, the plant hydraulic system is strongly affected and trees may suffer embolism, which can lead to death in some cases. This is particularly significant today, while the increase in air temperature leads to an increase of vapor pressure deficit and then of the climatic demand for transpiration. Tree death leads to a lower cover fraction and thus to a new ecological equilibrium in favor of regrowth when moisture is again available.

2.5.5 Evapotranspiration from various ecosystems in the Crau area (SE France)

Daily ET maps at kilometric spatial resolution for the Crau-Camargue pilot site in the lower Rhône Valley [17, 43] were produced from MODIS data using the EVASPA tool [44]. EVASPA has been designed to produce continuous mapping of daily ET by combining several sources of data (in particular MODIS TERRA and AQUA albedo, LAI, vegetation indices, and surface temperature products) and several surface energy balance models (S-SEBI [24]; triangle method [45]; see model outlines in previous sections in this chapter). The area is mainly flat and presents a wide variety of surfaces including dry and irrigated grasslands, wetlands, and various crops. The climate is Mediterranean, with irregular precipitation (the annual precipitation range is between 350 mm and 1100 mm, with an

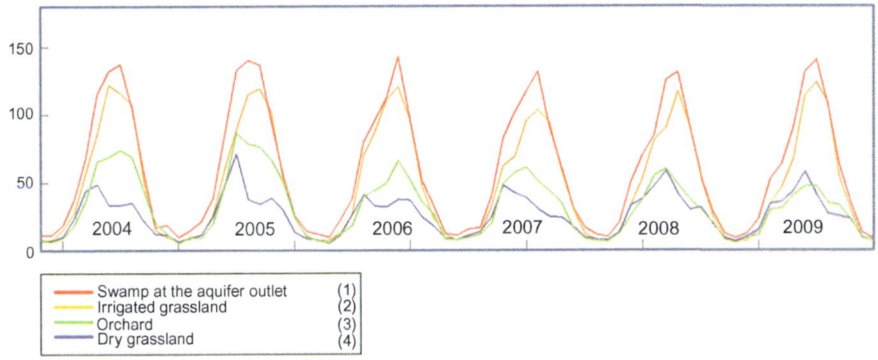

Fig. 2.15 ET over the Crau area for several land cover types. *(From A. Olioso, M. Bahir, B. Gallego Elvira, S. Garrigues, M. Mira, A. Chanzy, M. Weiss, O. Marloie, G. Boulet, Monitoring evapotranspiration over the Crau Aquifer from remote sensing and flux tower data, in: International conference MISTRALS "Statement and Prospects for society," 20–22 October 2015, Marseille, France, 2015. https://prodinra.inra.fr/record/409600.)*

average of 550 mm), long dry periods in spring and summer, and strong winds. Evapotranspiration is presented over several types of surfaces in Fig. 2.15:

1. wetlands at an outlet of the Crau underlying aquifer: they continuously receive water from the aquifer and from the drainage and irrigation channel networks;
2. irrigated grasslands: they are managed with a high level of irrigation amount (up to 2000 mm from spring to the end of summer) using simple border irrigation systems (flooding) [46]. Irrigation water is withdrawn from the Durance River through a very dense network of channels and a large part of the irrigation water (up to 80%) is drained to the aquifer [42];
3. orchards: they mostly consist of peach orchards; drip irrigation systems are used, the irrigation water being drawn from the underlying aquifer. One large orchard farm was bankrupted and the orchards were abandoned in 2005. A restoration program was set up to retrieve the original steppic landscape and the orchards were removed in 2009; and
4. steppic areas ("coussoul" areas): they consist of very short grass developing on soils with a very high level of stone (80%), so that the water availability is very low. This ecosystem is one of the driest ones in France.

The level of ET depends on the amount of available water (depending on precipitation pattern), on the development of vegetation, and on the amount of available solar energy. The year 2007, which was a very dry one, presented the lowest ET levels. Low levels were also obtained in 2008, but they resulted from a very cloudy year, in particular in spring, which limited the level of incoming radiation and then ET. In general, wetlands and irrigated areas, which receive large amounts of water, have high evapotranspiration throughout the year. Irrigated grasslands have a lower ET than wetlands, which may be explained by several factors: irrigation starts only in spring; grass mowing, three times a year, and sheep grazing in winter limit the amount of vegetation at some times, inducing a lower ET; and part of the 1 km MODIS pixels for grasslands are not fully covered by irrigated grassland. Dry grasslands have a very low evapotranspiration, as they almost never exceed 50 mm per month (300–350 mm per year). ET from orchards decreased over the period presented in Fig. 2.15 as the orchards were abandoned, dried down, or died as irrigation was stopped and the trees suffered a drastic lack of water. At the end of the period, orchard ET was similar to dry grass ET.

2.6 Conclusion

ET is strongly dependent on moisture availability in the Mediterranean region. Soil moisture is often a limiting factor and, in association, vegetation is usually sparsely distributed to maximize rainfall water availability, leaving large soil patches interacting directly with the atmosphere. ET is also a key element of sustainable water resources management, and an important proxy for stress monitoring and assessing the vulnerability of rainfed agrosystems or natural ecosystems to drought. Fortunately, methods to compute ET benefit from the increasing precision (less uncertainty, better space/time resolutions) in estimating driving factors such as reanalysis data or vegetation extent and typology through remote sensing. Data assimilation of high-resolution remote sensing products is certainly an important pathway to produce robust estimates of plant water health, plant water uptake, and subsequently to drought or fire hazard monitoring in this region.

References

[1] T.A. McMahon, M.C. Peel, L. Lowe, R. Srikanthan, T.R. McVicar, Estimating actual, potential, reference crop and pan evaporation using standard meteorological data: a pragmatic synthesis, Hydrol. Earth Syst. Sci. 17 (2013) 1331–1363.

[2] G. Allen, L. Pereira, D. Raes, M. Smith, Crop Evapotranspiration—Guidelines for Computing Crop Water Requirements, FAO Irrigation Papers, Rome, Italy, 1998 (300 pp.).

[3] C. Ollivier, N. Mazzilli, A. Olioso, K. Chalikakis, S. Carrière, C. Danquigny, C. Emblanch, Karst recharge-discharge semi distributed model to assess spatial variability of flows, Sci. Total Environ. 703 (2020) 134368, https://doi.org/10.1016/j.scitotenv.2019.134368.

[4] S. Carrière, N.K. Martin St-Paul, C. Belmys Cakpo, N. Patris, M. Gillon, K. Chalikakis, C. Doussan, A. Olioso, M. Babic, A. Jouineau, G. Simioni, H. Davi, The role of deep vadose zone water in tree transpiration during drought periods in karst settings—Insights from isotopic tracing and leaf water potential, Sci. Total Environ. 134332 (2020), https://doi.org/10.1016/j.scitotenv.2019.134332.

[5] P.D. Colaizzi, N. Agam, J.A. Tolk, S.R. Evett, T.A. Howell, P.H. Gowda, S.A. O'Shaughnessy, W.P. Kustas, M.C. Anderson, Two-source energy balance model to calculate E, T and ET: comparison of Priestley-Taylor and penman-Monteith formulations and two time scaling methods, Trans. ASABE 57 (2014) 479–498.

[6] S. Garrigues, A. Olioso, J.-C. Calvet, E. Martin, S. Lafont, S. Moulin, A. Chanzy, O. Marloie, S. Buis, V. Desfonds, N. Bertrand, D. Renard, Evaluation of land surface model simulations of evapotranspiration over a 12-year crop succession: impact of soil hydraulic and vegetation properties, Hydrol. Earth

Syst. Sci. 19 (2015) 3109–3131.www.hydrol-earth-syst-sci.net/19/3109/2015/. https://doi.org/10.5194/hess-19-3109-2015.

[7] B. Duchemin, R. Hadria, S. Erraki, G. Boulet, P. Maisongrande, A. Chehbouni, R. Escadafal, J. Ezzahar, J.C.B. Hoedjes, M.H. Kharrou, S. Khabba, B. Mougenot, A. Olioso, J.C. Rodriguez, V. Simonneaux, Monitoring wheat phenology and irrigation in Central Morocco: on the use of relationships between evapotranspiration, crops coefficients, leaf area index and remotely-sensed vegetation indices, Agric. Water Manag. 79 (2006) 1–27, https://doi.org/10.1016/j.agwat.2005.02.013.

[8] D.G. Miralles, T.R.H. Holmes, R.A.M. De Jeu, J.H. Gash, A.G.C.A. Meesters, A.J. Dolman, Global land-surface evaporation estimated from satellite-based observations, Hydrol, Earth Syst. Sci. 15 (2) (2011) 453–469, https://doi.org/10.5194/hess-15-453-2011.

[9] C. Albergel, S. Munier, A. Bocher, B. Bonan, Y. Zheng, C. Draper, D.J. Leroux, J.-C. Calvet, LDAS-monde sequential assimilation of satellite derived observations applied to the contiguous US: an ERA-5 driven reanalysis of the land surface variables, Remote Sens. 10 (2018) 1627.

[10] G. Boulet, J.-P. Lagouarde, 10. Bilan d'énergie des surfaces continentales et utilisation de la température de surface, Télédétection pour l'observation des surfaces continentales, ISTE, 2016.

[11] G. Boulet, B. Mougenot, J.-P. Lhomme, P. Fanise, Z. Lili-Chabaane, A. Olioso, M. Bahir, V. Rivalland, L. Jarlan, O. Merlin, B. Coudert, S. Er-Raki, J.P. Lagouarde, The SPARSE model for the prediction of water stress and evapotranspiration components from thermal infra-red data and its evaluation over irrigated and rainfed wheat, Hydrol. Earth Syst. Sci. 19 (2015) 4653–4672, https://doi.org/10.5194/hess-19-4653-2015.

[12] J.M. Norman, W.P. Kustas, K.S. Humes, Source approach for estimating soil and vegetation energy fluxes in observations of directional radiometric surface temperature, Agric. Forest Meteorol. 77 (1995) 263–293.

[13] F. Castelli, D. Entekhabi, E. Caporali, Estimation of surface heat flux and an index of soil moisture using adjoint-state energy balance, Water Resour. Res. 35 (1999) 3115–3125.

[14] M.J. Russell, M. Nunez, M.A. Chladil, J.A. Valiente, E. Lopez-Baeza, Conversion of nadir, narrowband reflectance in red and near-infrared channels to hemispherical surface albedo, Remote Sens. Environ. 61 (1997) 16–23.

[15] F. Jacob, A. Olioso, Derivation of diurnal courses of albedo and reflected solar irradiance from airborne POLDER data acquired near solar noon, J. Geophys. Res. Atmos. 110 (2005), https://doi.org/10.1029/2004jd004888.

[16] W. Lucht, C.B. Schaaf, A.H. Strahler, An algorithm for the retrieval of albedo from space using semi-empirical BRDF models, IEEE Trans. Geosci. Remote Sens. 38 (2000) 977–998.

[17] M. Mira, A. Olioso, B. Gallego-Elvira, D. Courault, S. Garrigues, O. Marloie, O. Hagolle, P. Guillevic, G. Boulet, Uncertainty assessment of surface net radiation derived from Landsat images, Remote Sens. Environ. 175 (2016) 251–270, https://doi.org/10.1016/j.rse.2015.12.054.

[18] J.R. Garratt, B.B. Hicks, Momentum, heat and water vapour transfer to and from natural and artificial surfaces, Q. J. R. Meteorol. Soc. 99 (1973) 680–687, https://doi.org/10.1002/qj.49709942209.

[19] A. Verhoef, H.A.R. De Bruin, B.J.J.M. Van Den Hurk, Some practical notes on the parameter kB-1 for sparse vegetation, J. Appl. Meteorol. 36 (1997) 560–572, https://doi.org/10.1175/1520-0450(1997)036<0560:spnotp>2.0.co;2.

[20] G. Boulet, A. Olioso, E. Ceschia, O. Marloie, B. Coudert, V. Rivalland, J. Chirouze, G. Chehbouni, An empirical expression to relate aerodynamic and surface temperatures for use within single-source energy balance models, Agric. Forest Meteorol. 161 (2012) 148–155, https://doi.org/10.1016/j.agrformet.2012.03.008.

[21] W.P. Kustas, B.J. Choudhury, K.E. Kunkel, L.W. Gay, Estimate of the aerodynamic roughness parameters over an incomplete canopy cover of cotton, Agric. Forest Meteorol. 46 (1989) 91–105.

[22] J.P. Lhomme, N. Katerji, J.M. Bertolini, Estimating sensible heat flux from radiometric temperature over crop canopy, Bound.-Layer Meteorol. 61 (1992) 287–300, https://doi.org/10.1007/bf02042936.

[23] Z. Su, The surface energy balance system (SEBS) for estimation of turbulent heat fluxes, Hydrol. Earth Syst. Sci. 6 (2002) 85–99.

[24] G.J. Roerink, Z. Su, M. Menenti, S-SEBI: a simple remote sensing algorithm to estimate the surface energy balance, Phys. Chem. Earth Part B-Hydrol. Oceans Atmos. 25 (2000) 147–157.

[25] Y.J. Xiong, S.H. Zhao, F. Tian, G.Y. Qiu, An evapotranspiration product for arid regions based on the three-temperature model and thermal remote sensing, J. Hydrol. 530 (2015) 392–404.

[26] O. Merlin, J. Chirouze, A. Olioso, L. Jarlan, A. Chehbouni, G. Boulet, An image-based four-source surface energy balance model to estimate crop evapotranspiration from solar reflectance/thermal emission data (SEB-4S), Agric. Forest Meteorol. 184 (2014) 188–203, https://doi.org/10.1016/j.agrformet.2013.10.002.

[27] W.G.M. Bastiaanssen, M. Menenti, R.A. Feddes, A.A.M. Holtslag, A remote sensing surface energy balance algorithm for land (SEBAL). 1. Formulation, J. Hydrol. 212–213 (1998) 198–212.

[28] R.G. Allen, M. Tasumi, R. Trezza, Satellite-based energy balance for mapping evapotranspiration with internalized calibration (METRIC)—Model, J. Irrig. Drain. Eng.-ASCE 133 (2007) 380–394, https://doi.org/10.1061/(asce)0733-9437(2007)133:4(380).

[29] G. Boulet, I. Braud, M. Vauclin, Study of the mechanisms of evaporation under arid conditions using a detailed model of the soil-atmosphere continuum. Application to the EFEDA I experiment, J. Hydrol. 193 (1997) 114–141.

[30] P. Lehmann, O. Merlin, P. Gentine, D. Or, Soil texture effects on surface resistance to bare-soil evaporation, Geophys. Res. Lett. 45 (10) (2018), https://doi.org/10.1029/2018gl078803. 398–310, 405.

[31] M. Shabou, B. Mougenot, Z. Chabaane, C. Walter, G. Boulet, N. Aissa, M. Zribi, Soil clay content mapping using a time series of Landsat TM data in semi-arid lands, Remote Sens. 7 (2015) 6059–6078.

[32] E. Delogu, G. Boulet, A. Olioso, B. Coudert, J. Chirouze, E. Ceschia, V. Le Dantec, O. Marloie, G. Chehbouni, J.-P. Lagouarde, Reconstruction of temporal variations of evapotranspiration using instantaneous estimates at the time of satellite overpass, Hydrol. Earth Syst. Sci. 16 (2012) 2995–3010.

[33] P.C. Guillevic, A. Olioso, S.J. Hook, J.B. Fisher, J.-P. Lagouarde, E.F. Vermote, Impact of the revisit of thermal infrared remote sensing observation on evapotranspiration uncertainty—A sensitivity study using AmeriFlux data, Remote Sens. 11 (2019) 573, https://doi.org/10.3390/rs11050573.

[34] W. Chebbi, G. Boulet, V. Le Dantec, Z. Lili Chabaane, P. Fanise, B. Mougenot, H. Ayari, Analysis of evapotranspiration components of a rainfed olive orchard during three contrasting years in a semi-arid climate, Agric. Forest Meteorol. 256–257 (2018) 159–178.

[35] J. Noilhan, J.F. Mahfouf, The ISBA land surface parameterisation scheme, Glob. Planet. Change 13 (1996) 145–159.

[36] V. Simonneaux, M. Lepage, D. Helson, J. Metral, S. Thomas, B. Duchemin, M. Cherkaoui, H. Kharrou, B. Berjami, A. Chehbouni, Estimation spatialisée de l'Evapotranspiration des cultures irriguées par télédétection. Application à la gestion de l'Irrigation dans la plaine du Haouz (Marrakech, Maroc), Sécheresse, numéro spécial eau et zone arides 20 (2009) 123–130.

[37] S. Saadi, V. Simonneaux, G. Boulet, B. Raimbault, B. Mougenot, P. Fanise, H. Ayari, Z. Lili-Chabaane, Monitoring irrigation consumption using high resolution NDVI image time series: calibration and validation in the Kairouan plain (Tunisia), Remote Sens. 7 (2015) 13005.

[38] M. Le Page, J. Toumi, S. Khabba, O. Hagolle, A. Tavernier, M. Kharrou, S. Er-Raki, M. Huc, M. Kasbani, A. Moutamanni, M. Yousfi, L. Jarlan, A life-size and near real-time test of irrigation scheduling with a Sentinel-2 like time series (SPOT4-Take5) in Morocco, Remote Sens. 6 (11) (2014) 11182–11203.

[39] S. Rambal, M. Lempereur, J.M. Limousin, N.K. Martin-StPaul, J.M. Ourcival, J. Rodriguez-Calcerrada, How drought severity constrains gross primary production (GPP) and its partitioning among carbon pools in a *Quercus ilex* coppice? Biogeosciences 11 (6855) (2014) 6869, https://doi.org/10.5194/bg-11-6855-2014.

[40] A. Olioso, C. Ollivier, N. Martin-St Paul, G. Simioni, M. Weiss, P. Guillevic, O. Marloie, S. Carrière, H. Davi, F. Huard, Monitoring vegetation fraction cover of French Mediterranean forests for evapotranspiration and water stress mapping, in: ESA—Living Planet Symposium, 13–17 May 2019, Milan, Italy, 2019, https://prodinra.inra.fr/record/484932.

[41] C. Ollivier, Caractérisation et la spatialisation de la recharge des aquifères karstiques, 2019 (Thèse de doctorat en Sciences de l'Univers. Soutenue le 27-05-2019 à l'Universté d'Avignon).

[42] A. Olioso, R. Lecerf, A. Chanzy, F. Ruget, F. Huard, A. Baillieux, P. Rossello, P. Lecharpentier, F. Trolard, F. Charron, S. Ruy, M. Alkassem Alosman, A.-L. Cognard-Plancq, B. Seguin, D. Courault, B. Gallego-Elvira, S. Garrigues, Bilan hydrique des agrosystèmes de Crau face aux changements globaux, in: L. Tatin, A. Wolff, J. Boutin, T. Dutoit (Eds.), Ecologie et conservation

d'une steppe méditerranéenne. La plaine de Crau, 2013. Coordination. Edition Quae, Paris, France. ISBN-13 9782759220823, pp. 266–288. 4 planches couleurs hors texte https://prodinra.inra.fr/record/490818.

[43] A. Olioso, M. Bahir, B. Gallego Elvira, S. Garrigues, M. Mira, A. Chanzy, M. Weiss, O. Marloie, G. Boulet, Monitoring evapotranspiration over the Crau aquifer from remote sensing and flux tower data, in: International conference MISTRALS "Statement and Prospects for society." 20–22 October 2015, Marseille, France, 2015, https://prodinra.inra.fr/record/409600.

[44] B. Gallego-Elvira, A. Olioso, M. Mira, S.R. Castillo, G. Boulet, O. Marloie, S. Garrigues, D. Courault, M. Weiss, P. Chauvelon, O. Boutron, EVASPA (EVapotranspiration assessment from SPAce) tool: an overview, Procedia Environ. Sci. 19 (2013) 303–310, https://doi.org/10.1016/j.proenv.2013.06.035.

[45] R. Tang, Z.-L. Li, B. Tang, An application of the Ts–VI triangle method with enhanced edges determination for evapotranspiration estimation from MODIS data in arid and semi arid regions: implementation and validation, Remote Sens. Environ. 114 (2010) 540–551, https://doi.org/10.1016/j.rse.2009.10.012.

[46] M. Alkassem Alosman, S. Ruy, S. Buis, P. Lecharpentier, J.C. Bader, F. Charron, A. Olioso, An improved method to estimate soil hydrodynamic and hydraulic roughness parameters by using easily measurable data during flood irrigation experiments and inverse modelling, Water 10 (2018) 1581.

Further reading

S.D. Carrière, C. Doussan, J. Ruffault, F. Pimont, G. Simioni, I. Scotti, F. Courdier, H. Davi, N.K. Martin-StPaul, K. Chalikakis, J.M. Limousin, C.B. Cakpo, Impact of local soil and subsoil conditions on inter-individual variations in tree responses to drought: insights from electrical resistivity tomography, Sci. Total Environ. 698 (2020) 134247, https://doi.org/10.1016/j.scitotenv.2019.134247.

CHAPTER 3

Mountain hydrology in the Mediterranean region

María J. Polo[a], Rafael Pimentel[a], Simon Gascoin[b], Claudia Notarnicola[c]
[a]Fluvial Dynamics and Hydrology Group, University of Cordoba-Andalusian Institute for Earth System Research, Córdoba, Spain
[b]CESBIO, Université de Toulouse, CNES/CNRS/INRA/IRD/UPS, Toulouse, France
[c]Institute for Earth Observation, EURAC, Bozen-Bolzano, Italy

3.1 Introduction

The definition of "mountain areas" is not easy to formulate with uniform quantitative criteria for all over the world, given the different mountain environments that can be found on different latitudes, continents, and climates. However, their shared features are clearly identified from a qualitative point of view, and a threshold altitude of 1000 m above sea level is commonly adopted to differentiate mountains from lowlands on a global scale. This comprises approximately a 27% of the emerged area in the world that provides, nevertheless, between one and two thirds of the global freshwater discharge, which can reach more than a 90% of the runoff on the catchment scale [1]. This fact coined the term "water towers" for mountains, with a relevant role for water supply to the population in different parts of the globe [2]. Particularly, Mediterranean mountains constitute the wettest areas of catchments ranging from mild to arid climates, and behave as natural reservoirs for water uses downstream, largely dependent on this water supply during the late spring and summer seasons. The importance of mountains on a global scale was first officially recognized by the United Nations in 1992 in the Agenda 21 [3].

The vulnerability of mountains to global warming in the context of water resources has been reported in different works [3–5]. Moreover, mountains constitute singular areas to monitor and detect climate-related impacts due to their orography and the strong dependence of climate on height even over not large areas; the resulting biodiversity can be extremely large in Mediterranean high mountains, where coastal and alpine ecosystems can be found in distances ranging from tenths to some hundreds of kilometers. The Mediterranean climate is also found in other regions than those surrounding the Mediterranean Sea, mainly in coastal areas of South America and West US, South Africa, and Western and South Australia; in these, mountain areas are found in California and part of the Chilean Andes. Fig. 3.1 shows the distribution of the major Mediterranean mountain areas over the world [6, 7].

Water Resources in the Mediterranean Region
https://doi.org/10.1016/B978-0-12-818086-0.00003-0

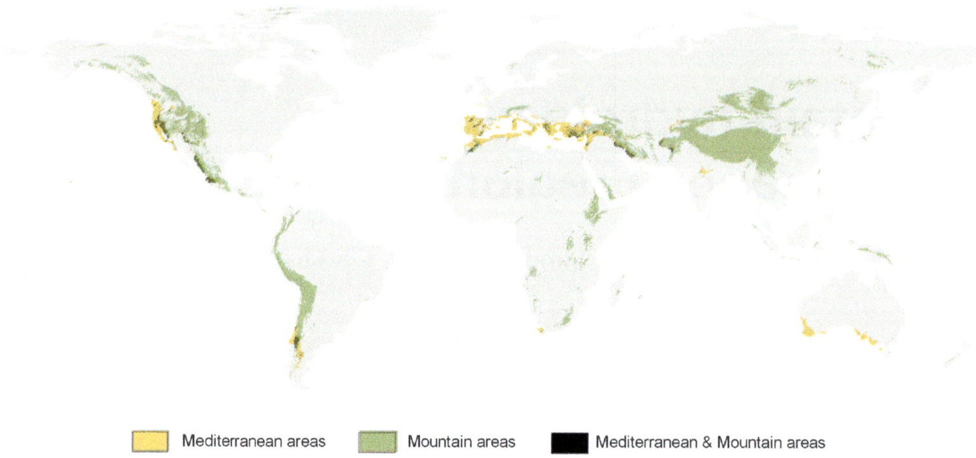

Fig. 3.1 Distribution of Mediterranean-climate (*yellow*) and mountain (*green*) areas in the world; mountain areas within the Mediterranean-climate zone are highlighted in black. *(Source: D. Karagulle, C. Frye, R. Sayre, S. Breyer, P. Aniello, R. Vaughan, D. Wright, Modeling global Hammond landform regions from 250-m elevation data, Transactions GIS 21 (2017) 1040–1060; F. Rubel, M. Kottek, Observed and projected climate shifts 1901–2100 depicted by world maps of the Köppen-Geiger climate classification, Meteorol. Z. 19 (2010) 135–141.)*

Mountain hydrology has been paid comparatively little attention in the scientific literature on Mediterranean-climate hydrology, with relevant exceptions [8–10]. As an example, Fig. 3.2 shows the evolution of scientific work focused on Mediterranean hydrology (MH), and Mediterranean mountain hydrology (MMH), from an automatic search by simple keywords in a digital scientific database [11]. With some examples in the late 60s of last century, it is after the deep drought period faced by Mediterranean

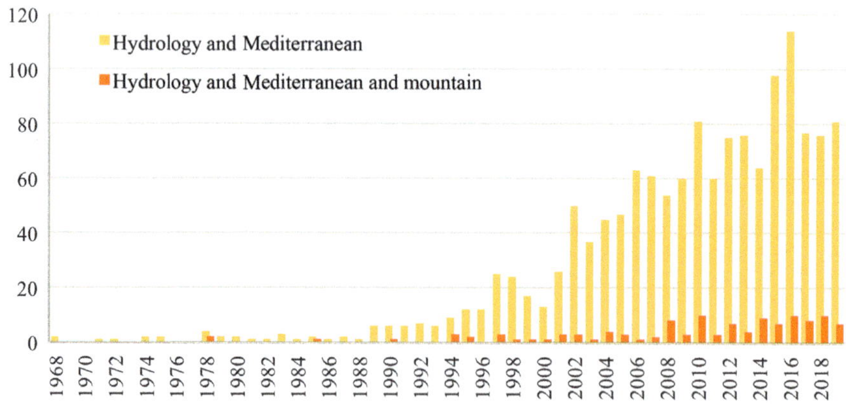

Fig. 3.2 Presence of Mediterranean mountain hydrology in a scientific publication framework. *(Source: www.scopus.com, 2019 (last accessed on 12 September).)*

countries in the early 90s that MH works launch a significant increasing evolution, being MMH just a marginal fraction of them. The development of MMH takes place at the beginning of the 21st century following some reference reviews and after international recognition of mountains as hot spots in a context of change [2]. Despite the presence of Mediterranean mountain sites in the world might be scarcely representative on a global basis, their singularities and value as biodiversity reservoirs and global change observatories make these areas key sites as field laboratories to better understand nonlinearity of hydrological processes under changing conditions in a warmer regime, and anticipate likely conditions in the future in colder regions as well as early proposing of adaptation plans in Mediterranean and other semiarid catchments in the world.

The conjunction of climate variability under Mediterranean conditions and the significant topographic gradients commonly found mountain areas results in a complex spatiotemporal pattern of hydrologic signatures in these sites. This requires special attention to be paid to both time and spatial scales to monitor and model hydrological processes in Mediterranean mountain areas, which makes it challenging to assess the impacts of potential future climate under different emission scenarios. This chapter is devoted to present the hydrological features of these areas and the major approaches to understand their evolution from the current experience facing an uncertain future of water-related ecosystem services' sustainability. For this, a short review of the current knowledge on long-term trends and the still open relevant questions is complemented with examples of key methods and tools to monitor selected processes in different areas, and the requirement of services to cope with climate variability and water resource availability from an operational perspective. Final reflections are also provided as key points at the end of the chapter.

3.2 State of the art, current research needs and challenges for the future

Different works have reviewed the hydrological role of mountains or studied mountain catchments all over the world, with a large fraction focusing on snow and glaciers properties and dynamics, and its impacts on the hydrological regime. More recently, Viviroli and Weingartner [12] provide a general insight of mountain hydrology, and compare different extensively monitored catchments in the Alps and the Rhine basin to highlight the key drivers of the hydrological regime, and to focus on transfer of scales to other regions in the world. Regarding Mediterranean-climate mountain areas, Latron et al. [8] provided an extensive review of their hydrological features from different works in small research catchments distributed over Europe and the United States. Their conclusion of needs for process-oriented research and modeling that could reproduce the significant seasonality in Mediterranean areas was also found in other works [9, 13]. The exhaustive review of snow hydrology in Mediterranean regions by Fayad et al.

[10] is a reference work that has put the light on the needs for better accuracy of the climate forcing representation to get deeper insight on the snowpack dynamics in these areas, and for bridging the gap between snow processes and their interactions with other surface and subsurface processes. Moreover, the impacts of global warming on these areas have received an increasing interest during the last decades, with not only similarities but also differences in the long-term trends found in different sites. Glaciers are present in a very small fraction of catchments within the Mediterranean region, which is another relevant difference from temperate and high latitude cold regions.

3.2.1 Observed trends of climatic drivers and hydrological descriptors

2002 was nominated by the United Nations as the International Year of the Mountains [14]. Just one year after, Beniston [2] published a reference review of the likely impacts of global warming in mountain regions, which also discussed its consequences for relevant sectors in these areas. The expected shift from snowfall to rainfall, added to enhanced torrentiality, and higher potential evaporation rates propagate throughout the cascade of processes in the water cycle, and affect the patterns of soil moisture, groundwater recharge and runoff production, flood frequency, dry spells occurrence and duration, among others, which ultimately influences the ecosystem dynamics and water resource availability and supply. Changes of land use during the last century in many Mediterranean mountain areas have also affected the evapotranspiration volumes and spatial patterns, and the hydrological partitioning, which makes the attribution of change a complex task [15]. During the last decades of the 20th century, different works report warming trends that are not always associated to significant decreases of precipitation, with difference between the northern and southern Mediterranean regions; however, the negative trends found for snowfall and earlier ablations in high altitudes, and the increasing rates of potential evapotranspiration, do impact the streamflow regime, with decreasing trends that may be due to different combination of processes depending on the local dominating conditions at each site [4, 16–20].

3.2.2 Projections of future climate scenarios on mountain areas

Global Circulation Models (GCMs) cannot generally assess climate in mountain areas due to their coarse spatial resolution that is not capable to capture the significant spatial scales related to and derived from topography. Regional Circulation Models (RCMs) bridge this scaling problem between the global and local scales, and have improved the performance in mountain areas in temperate and high latitudes [2]. However, in Mediterranean regions, the sparsely distributed occurrence of precipitation, and the extreme seasonality, still poses a constraint for regional models to adequately simulate snowfall occurrence; this uncertainty propagates through the hydrological processes cascade and impacts the accuracy of RCM and their projections on local scales. Moreover, climate and ocean teleconnections greatly affect Mediterranean-type regions, and their mountains [21], but these effects are not well reproduced by global and regional models. Nevertheless, on a regional basis, most of the models project a positive and negative trend of temperature and precipitation, respectively, in the Mediterranean areas by the end of the 21st century, accompanied by a significant decrease of snow occurrence and persistence, and of streamflow in mountain areas [22]. On a local basis, a wide spatial variability can be found, and some works point out to nonsignificant trends of annual precipitation volumes under some milder scenarios of emissions, but rather a significant trend toward torrentiality of snowfall and precipitation in general, with higher intensities and earlier snowmelt that greatly impact the seasonality of the streamflow regime [23]. Implications for water resource planning and management, and operational decision-making processes, are clear and require strong efforts to develop climate and hydrological services capable to deal with current and future variabilities, and adaptation plans for the likely future water scarcity scenario.

3.2.3 Major scientific questions to be solved

Despite the significant work done so far, there are still pending questions that require an additional effort to address what the future evolution of the hydrological regime will be in Mediterranean mountain areas. The following points aim to address the major issues identified as: (i) what climate trends are found in the highest areas of Mediterranean mountains and how uniform are their vertical gradients uphill?; (ii) what is the feedback of snowpack shifts from a hydrological point of view in these areas?; (iii) how will vegetation adapt to changing conditions and what impacts will this bring into the hydrological processes on the local and regional scales?; (iv) what are the significant scales of change and how can they be included in the generation of improved projections of future climate under different scenarios?

3.3 Major approaches to address key questions in mountain hydrology

Mountain hydrology in general, and particularly in Mediterranean regions, is not always feasible to assess from a single approach given the intrinsic variability that stems from abrupt topography. Especially, understanding the complex interactions of terrain, climate, and vegetation in these areas usually requires extensive modeling that, at the same time, requires high-resolution data sets. Global warming during the last decades and its expected impacts on mountain areas have spotted the need for increasing monitoring efforts and research on processes interactions; however, long-term studies face the challenges of lacking data in the highest (and usually wettest) areas in Mediterranean catchments, and the evolution of soil uses from the lowlands upward, with agriculture, tourism, and other managed developments displacing natural conditions. Usually a combination of approaches is required to understand and track the evolution of the hydrological regime on both the local and regional scales, whatever the time scale of assessment is. The following sections briefly address this issue.

3.3.1 Modeling approaches and significant scales: Requirements and constraints

In Mediterranean catchments, hydrological modeling can become a complex issue due to the large number of variability sources that coexist, some of them inter-related on different time and spatial scales: climate regime, topography, vegetation, soil types, and land uses. Most of the widely used models have been tested in temperate regions, and usually result in poor performance during the dry periods [13]. Additionally, mountain areas in Mediterranean regions are frequently found close to the coastline, which on one hand modifies climatic patterns, and on the other creates a significant topographic gradient that underlies many of those variability sources. Particularly, the presence of snow in areas surrounded by mild and semiarid climate conditions constitutes one of the most relevant features of the hydrological response of Mediterranean mountain catchments, which makes them different from snow-dominated areas in temperate and northern regions. This is the reason behind the appropriate performance of hydrological models that may be found, especially with a lumped approach, when streamflow is the target variable on a seasonal or annual basis, but failing to adequately reproduce the hydrological regime on an event-based scale; other variables different from flow are complex to satisfactorily simulate. Furthermore, the amount of data required for not only feeding (spatial variability of vegetation and soil characteristics and use) and driving (spatial and time variability of climatic) but also calibrating and validating models makes the simulation of hydrological processes on short- to medium-time scales a challenging task for researchers and technicians related to operational purposes [13].

Bearing this in mind, the topography adds complexity to mountain hydrology modeling; the spatial scales on which hydrological processes can be significant usually require a high spatial resolution level (not larger than 500 to 250-m pixel size, and preferably around tenths of meters in very steep areas) to characterize drivers, conditions, and results during transient periods. The needed information is rarely available and spatial interpolation algorithms require sound topographic corrections or parameter assessment [24, 25] that also rely on local data availability. Remote sensing constitutes one of the most important sources of information on the large scale (from 100 to 1000 km^2) in these areas, although limited by the length of the time series for long-term analyses, and by the spatial resolution of the oldest sensors with daily revisiting time. The recent launch of high spatial resolution satellite sensors and development of other remote sources of acquisition, like unmanned vehicles or terrestrial cameras, offer a promising tool for covering the gap of monitoring in these areas and provide better basis for hydrological modeling. However, topography still remains a challenge to face for many relevant applications of these techniques. Finally, when snow presence dominates the seasonal hydrological response, a physically based approach is usually required for an appropriate modeling of the accumulation-ablation cycles in Mediterranean regions. The following sections address further all the identified drawbacks.

3.3.2 Systematic on-site data collection of relevant variables: Pros and cons from the research and operational sides

Precipitation and air temperature, together with other weather variables such as wind, air humidity, and global radiation, and river flow are the most frequently monitored variables in the context of hydrology and water resource management. Soil moisture and groundwater head level can be found at some sites, but they are not usually included in standard networks nor do the monitoring points cover a large area in a systematic approach, but rather they are mostly found in the context of research or remediation actions. Evapotranspiration measurements can be found in agricultural areas, but systematic estimations are practically reduced to potential evapotranspiration calculations from measurements of weather variables. Other components of the water balance such as interception, subsurface flow, and evaporation from groundwater resources are barely monitored and mainly associated to research initiatives. The absence of dedicated networks is a crude reality in mountain areas, where the occurrence of precipitation as rainfall and snowfall, the local gradients of air temperature due to the effects of steep slopes, and the high exposure to wind additionally result in large uncertainties of the observations.

This lack of systematic recording of relevant information is explained not only by the costs but also due to the challenges associated to the scaling of point measurements of such variables to large areas. The spatial variability of the soil characteristics such as porosity, structure, texture, organic matter content, and etcetera, with great influence on the soil moisture and vegetation dynamics and regime, is more pronounced in mountains than in agricultural and other modified areas. This adds to the variability of the precipitation and atmospheric conditions regime and spatial distribution, that is, the major drivers of change on a local scale, in a feedback loop with interrelationships that contribute to a great dependence of variables and nonlinearity. In Mediterranean catchments, the influence of evapotranspiration, interception, and groundwater-runoff and river flow exchanges are critical to quantify the water cycle on different time scales, and to understand how global warming is impacting and will impact the local and regional hydrological regimes, and water resource availability. Moreover, in mountain areas, the lack of coverage of large-scale observation networks, especially in the southern Mediterranean regions, results in three relevant impacts, that are not independent: (i) the gradients with altitude of relevant drivers, such as temperature and to some extent precipitation, and key state variables, such as soil moisture, cannot be adequately addressed, which finally affects the spatial assessment of the hydrological budget on different scales; (ii) the estimation of the partition between snow- and rainfall [26] is subject to a significant uncertainty due to both undercatch of precipitation by pluviometers (especially under windy conditions during cold periods) and modeling, since in most of the cases this is achieved from adopting a threshold temperature approach; (iii) the under- or overestimation of precipitation and snowfall distortions further steps in hydrological studies, with poor models' performance in some cases, calibrated parameters out of their credible range of values, or closure of the energy and water balances being done on real evapotranspiration or groundwater on the catchment scale.

Research activity is usually behind most of the existent not large but well-covered areas in terms of hydrological monitoring; the main disadvantages are related to the time framework of research funding, with strong dependence on the individual willingness of the teams to maintain the research line and budget limitations. The benefits of these relevant sources of information are clear since they provide the scientific community with rigorous data sets to retrieve other relevant data and develop further analyses; however, they cannot substitute a large-scale monitoring network on the grounds of operational processes, which largely rely on standard precipitation, temperature, and flow-level measures, and snow depth records in some instances, and empirical or statistical algorithms to support the decision-making process.

3.3.3 Experimental research catchments: The advantages of an integrated approach

Beyond the provision of data to track the evolution of target variables, the design of monitoring networks requires a process-oriented approach in order to provide the most efficient information to rigorously answer the different questions that still need answer. Experimental research catchments, despite their usually small size, constitute "outdoors laboratories" [27] that have been providing the scientific hydrology community with integrated data sets to close the water budget on the basin scale and quantify the regime of the different components of the energy and water balance [28], and support other related studies (erosion, ecology, pollution, etc.). Many of them were established during the First International Hydrological Decade (1965–74) and some of the currently well-known interactions in the catchment system would not have been reached without this long-term integrated approach focused on processes, like the forest impacts on the hydrological regimen on different time scales [29]. Mountain areas pose an additional effort for the maintenance of these research sites due to complex topography and access, extreme climate conditions, lack of energy sources or communication facilities, which results in a lack of long-term experimental basins when compared to other topical areas, such as agricultural hydrology, and within these areas, Mediterranean regions are underrepresented.

Llorens et al. [20] summarized 30 years of continuous monitoring of hydrological processes in the Vallcebre Research Catchments, in the south-eastern Pyrenees, whose results are a good example of a multidisciplinary approach to better understand hydrological processes in Mediterranean mountain areas and a long-term data set to gain insight on the major drivers of variability, and change, and their significant scales, and provide experimental data to calibrate and validate hydrological models, derive trend analysis or test new instrumentation, among others. High-frequency (subhourly) records of meteorological variables and soil water content were complemented with monitoring of streamflow and groundwater level, with a special focus on the vegetation interaction with the water cycle in one of the most complete monitoring design to study rainfall interception and partitioning through the canopy, and transpiration. Furthermore, specific field manual and automatic sampling is carried out to assess specific processes (e.g., sediment transport during floods or tracers in water). However, this strong effort and commitment to the continuity of the recording is feasible on small-sized catchments (tens of square kilometers); mountain hydrology on a regional scale requires a combination of these small-scale field laboratories and monitoring systems efficient on the spatial and time scales, such as remote sensing, together with data-fusion algorithms capable of merging the long-term record of previous sensors with the fine spatial resolution of the recent constellations of satellites for a rigorous trend analysis.

Moreover, different initiatives aim at gathering similar approaches all over the world in order to facilitate intercomparison activities, data sharing, and other networking advantages. Within the context of mountain research, the International Network of Alpine Research on Catchment Hydrology (INARCH; http://www.usask.ca/inarch/ [30]) from the GEWEX Hydroclimatology Panel (https://www.gewex.org/) comprises mountain experimental catchments all over the world devoted to snow hydrology. The International Center of Integrated Mountain Development (ICIMOD; http://www.icimod.org/) supports work and joint efforts for a sustainable future for the people and ecosystems in the mountain areas in countries of the Himalayan region. More recently, the Mountain Research Initiative (http://www.mountainresearchinitiative.org/) aims to frame global change research and transfer activity on mountain areas, coleading actions like the GEO-GNOME Project (Group on Earth Observations Global Network for Observations and Information in Mountain Environments [31]), and their contribution to the Sixth Assessment Cycle of Intergovernmental Panel on Climate Change. Although these examples are not constricted to Mediterranean regions, they constitute a global framework to not only discuss but also share similarities and specificities of mountain areas, and offer scientific assessment for social and economic impacts of climate change; undoubtedly, Mediterranean mountain areas play a key role for an early assessment of potential changes, that may foster and enable proof-of-concept adaptation actions in other temperate and cold regions in the world.

3.3.4 Remote sensing: Consolidated role and promising opportunities of monitoring

Hydrological processes present a high variability across different scales. To quantify this dynamic observations from ground stations, remotely sensed imagery, and physical-base model simulations have been extensively addressed on several mountain areas in the last decades [32–35]. The capability to monitor larger areas makes remote sensing quite relevant for capturing the variability and evolution during the year of the snow cover, especially at high elevation, where the presence of ground data can be limited. Together with snow, other hydrological parameters such as soil moisture, evapotranspiration, surface water, and glaciers have been constantly retrieved and monitored from Earth Observation data in mountain areas [36–38]. These parameters have been addressed and analyzed by remote sensing tools operating in different domains of the electromagnetic spectrum going from the visible–near infrared to microwave. For the optical domain, thanks to the LANDSAT and NASA's Earth Observing System (EOS) programs, the interest moved from image analysis to the use and exploitation of added-value products. In fact, with the launch of the satellite Terra in late 1999 and Aqua in 2002 with the Moderate Resolution Imaging Spectroradiometer (MODIS) sensor, the era of available products for a broad range of parameters at global level started. These programs helped to exploit the big potential of remote sensing for hydrological and climate models [38]. In this domain, obscuration due to cloud presence and ground resolution remain as major challenges for mountain areas, especially in Mediterranean regions where the spatial variability of land use, covers, and topography pose additional challenges for data retrieval. For example, these areas can be characterized by a high variability such as short snow cover duration and fast melting processes, which need timely acquisitions and a higher spatial resolution in order to be able to capture such dynamics. Among the sensors working in this domain, it is also worthwhile to mention the Advanced Very High Resolution Radiometer (AVHRR). These three sensors provide different advantages and disadvantages when dealing with mountain areas. AVHRR and LANDSAT have the longest time series being available; whereas the main limitation for AVHRR is the ground resolution of 1000 m, for LANDSAT the disadvantage is the revisiting time of 16 days, which may limit some applications. As an example, these sensors are not able to match the requirements set by the Global Climate Observing System (GCOS) of the World Meteorological Organization (WMO) that considers a ground resolution of 100 m and daily acquisition as the optimal conditions to properly monitor snow cover in complex terrain (http://www.wmo.int/pages/prog/gcos/). Similar considerations for ground resolutions are valid for other parameters such as vegetation cover or evapotranspiration. Notwithstanding these limitations, several studies dedicated to Mediterranean mountain areas have been carried out, indicating the importance of long time series of data to detect changes and quantify water resources [39–41].

From 2014, with the launch of the Sentinel-1 satellite, which is part of the Copernicus program funded by the European Space Agency (ESA) and the European Commission (EU), a new era has opened for the monitoring of hydrological parameters [42], especially in mountain areas due to its higher spatial resolution. This program has put in orbit up to now two Synthetic Aperture Radar (SAR) sensors at high ground resolution (around 20 m), Sentinel-1A and 1B, two multispectral optical sensors at high ground resolution

(10–20 m), Sentinel-2A and Sentinel-2B, and two multispectral optical sensors at medium resolutions (300 m), Sentinel-3A and 3B. While the first two satellites have a revisiting time of around 5 days, the latter has a daily revisiting time. In this sense, it will represent at European level the continuity with MODIS acquisitions. In this direction, improved products based on Sentinel-3 have been recently developed for grain size and albedo [43].

Some recent publications have already showed the importance of the improved ground and time resolution of the Sentinel satellites to monitor snow changes and characteristics. Up to now, the use of SAR images was mainly addressed to monitor the presence of wet snow [44], since the signal in the microwave domain is very sensitive to small percentage of water in the snowpack. Despite the promising results to track the initial snowmelt stages, the detection of dry snow, useful for the quantification of the snow water equivalent (SWE), has produced controversial results. The signal from sensors operating with wavelengths from L-band (around 20 cm) to X-band (around 2 cm) interacts limitedly with the dry snowpack [45]. However, some studies showed the feasibility to link time series of backscattering coefficients to snow depth [46], to snow water equivalent [47, 48], and to detect the melting phases [49]. Further investigations are needed in this direction to fully understand the complex physical interaction of the radar signal with the different snowpack properties. The use of radar and its exploitation in mountain areas for hydrological parameters (e.g., snowmelt, soil moisture) is of high importance because of its high ground resolution (normally around few tenth of meters), and the independence from cloud cover. Due to the properties of the microwave signal, these images have been successfully applied for monitoring the inundated areas as well as movements related to rock glaciers and landslides. In this case, the main drawback is still the geometrical distortions that can be quite relevant in mountain areas, and this is valid for both active and passive microwave sources.

Sentinel-2 observations provide the opportunity to monitor the snow cover area at unprecedented spatial temporal resolution every 5th day at 20 m resolution [50], which opens new perspectives in terms of data assimilation in distributed snow models, especially in Mediterranean mountains where the snow cover is highly variable in space and time [51]. The stereo satellite method [52] is also a new option to retrieve snow depth above the tree line in specific mountain catchments with good accuracy at metric resolution. However, it relies on imagery from very high resolution satellite sensors, which are typically operated by private companies and therefore the image acquisition can be expensive. Synergies between Sentinel-1, Sentinel-2, and Sentinel-3 should be further explored to leverage the full potential of the Copernicus program for Mediterranean hydrology and, specifically, in these mountain areas.

The time series from the previous generation of satellites (e.g., MODIS) and the upcoming with the Copernicus will offer an unprecedented chance to support the monitoring of hydrological parameters from space. Considering the requirements posed to properly address the heterogeneity of mountain environment, the effort should be directed toward the development of techniques and methods, which can fuse high and medium resolution sensors to obtain time series at high temporal and ground resolution. Moreover, the further exploitation of the radar data can go in the direction of quantifying processes related to snow accumulation, snowmelt, and water availability in soils. Beyond the traditional Earth Observation tools, a plethora of other systems such as CubeSats, Unmanned Aerial Vehicles (UAVs), terrestrial photography, and smartphone technologies can provide high-resolution information that complements the Earth Observation acquisitions.

3.4 Applications of major approaches in representative locations in the Mediterranean region

The previous sections have addressed the singularity of mountain hydrology in the Mediterranean region, and its sources, and the major issues necessary for understanding and monitoring processes that are key for ecosystem and water resources sustainability. Representative examples of the major approaches to study these processes are now presented, focusing on two elements that are relevant in mountains: snow and vegetation. These experiences bring insight on the challenges for further assessing future changes that global warming may bring to these singular mountain areas.

3.4.1 Monitoring snowpacks in Mediterranean mountains

In north Mediterranean countries (France, Spain, Italy, etc.), the snowpack is routinely monitored by a number of public and private agencies using automatic stations (Fig. 3.3) and manual surveys. The main difficulty is the dispersion of the datasets between national and regional weather services, water agencies, hydropower companies, etc. As a result, even the most recent catalog of available in situ snow measurements remains largely incomplete over the Mediterranean region (European Snow Booklet, [57]). The EU-funded Pyrenean Climate Change Observatory allowed the collection and curation of a valuable transnational dataset of snow depth measurements across the Pyrenees. However, such effort remains isolated and limited to the period of the project realization so that the update of the database is not guaranteed in the future. Two sites of the WMO Solid Precipitation Intercomparison Experiment [53] are located near the Mediterranean region (Formigal, Spain; Forni Glacier, Italy); however, the influence of the Mediterranean climate is rather weak at these sites. Formigal is located in the western Pyrenees, where the climate is strongly influenced by the Atlantic Ocean, while Forni Glacier is characterized by a high alpine climate. In addition, both sites are located in areas of wet climate and therefore are not representative of the semiarid climate that is found in the southern Mediterranean region.

Fig. 3.3 Examples of snow measurement sites in Mediterranean mountains: (A) WMO-SPICE network (Formigal, Pyrenes, Spain) [53]; (B) TREMA network, picture copyright *LMI TREMA* (Atlas, Morocco) [54]; (C) MZA, picture copyright *Cesbio/O-LIFE* (Mzaar, Mount Lebanon, Lebanon) [55]; (D) Guadalfeo network (Sierra Nevada, Spain) [56].

Operational networks are virtually absent in south and east Mediterranean mountains, despite the acknowledged contribution of snowmelt to runoff and groundwater recharge in countries like Lebanon and Morocco. In situ snow measurements rely on international research program and not on operational agencies. In Morocco, snow monitoring stations were installed as part of a long-term collaboration between the Centre d'Etudes Spatiales de la Biosphère (CESBIO), the French National Research Institute for Sustainable Development (IRD), and the University Cadi Ayyad [54, 58]. In Lebanon, a snow monitoring plan was implemented by CESBIO/IRD, Lebanese National Council for Scientific Research and Saint Joseph University, including three automatic weather stations in the high elevation region of Mount Lebanon [55]. The lack of operational and permanent snow monitoring systems in the southern Mediterranean region contributes to create a "temperate bias" in the current knowledge of snow processes. Many snow models or snow routines in hydrological models were primarily developed and evaluated in temperate alpine regions or arctic regions; hence they may poorly represent important processes for Mediterranean mountain snowpacks like evaposublimation [59] or liquid water percolation [60, 61], among others.

There are many in situ observations in the western USA and the Central Andes of Chile and Argentina where the climate is also Mediterranean. These sites provide opportunities to extend our knowledge about the Mediterranean snow processes [10]. In addition, existing monitoring sites in the Mediterranean region should be more systematically included in international studies, like model intercomparison exercises, to make sure that the conclusions drawn from these studies are also applicable to the Mediterranean region. In particular, the Guadalfeo monitoring site in the Spanish Sierra Nevada [56] and the Rheraya pilot catchment in the High Atlas [62] stand out as valuable experimental sites to evaluate the validity of snowpack models and remote sensing techniques in semiarid Mediterranean regions.

3.4.2 The role of vegetation in mountain hydrology

Vegetation plays a relevant role in the water cycle that becomes dominant in many situations in Mediterranean regions, especially under nondisturbed conditions such as usually found in mountain areas, where the global and local climatic scales interact with the spatial patterns of covers distribution resulting in a coevolution of vegetation and hydrological mutual controls [63]; this makes it difficult to identify the individual contribution of vegetation and climate to the variability of the hydrological regime. First, interception redistributes precipitation over the soil surface, affecting both its volume and intensity, with impacts on the infiltration-runoff processes and the erosivity of rainfall; however, the final effects are largely dependent on the type of vegetative cover and the seasonal and annual precipitation regime. Llorens and Domingo [64] analyzed rainfall partitioning by vegetation in different Mediterranean sites in Europe and concluded that shrub species could reduce throughfall up to a 49% of the volumes found in temperate areas, whereas differences were not significant for tree species. Differences are also found between deciduous and perennial tree covers, with a higher variability of the former on different time scales [65]. Due to the limited storage capacity of vegetation, highest interception fractions are associated to low-intensity rainfall events, whereas high-intensity storms usually result in negligible interception volumes on the event scale; in all cases, interception's final impact is a water loss to the atmosphere due to evaporation from the canopy. The intrinsic variability of the Mediterranean climate on different time scales is the major source of variability for interception volumes [66]. However, in the intersection of the snow and vegetation domains in mountain areas, snowfall interception effects can significantly vary since the higher storage capacity of covers of water as snow produces its final fall off or melting to the soil, with reduced loss to the atmosphere; moreover, drifting by wind may easily transport the intercepted snow out of the watershed. Although this solid interception is not as relevant as in other cold regions since most of the snowpack accumulates above the tree line [10], sublimation may be relevant from the stored snow in Mediterranean mountains, especially during dry and cloudless interevent periods [59], which might suggest potential increases of water loss to the atmosphere from the intercepted snow, especially in a warmer and drier future. Nonetheless, the complex interaction of different variability sources makes it difficult to foresee the expected impacts since the observed trend toward torrentiality in many Mediterranean areas would result in lower interception fractions.

The occurrence of dry summers in Mediterranean mountain areas enhances the soil control of transpiration due to limited water content. The use of groundwater resources by tree species during the dry spell causes a significant reduction of water storage and summer-autumn streamflow, and impacts runoff generation. Partial or severe damage of trees may occur, without full recovery of transpiration after the cessation of long dry periods [67]. Thus, evapotranspiration strongly controls both soil water content and groundwater storage on both the seasonal and interevent scales, affected by this "memory" effect of antecedent dry years on the interannual scale, with a local source of variability associated to soil characteristics and vegetation patterns. This is absolutely relevant for land use planning and vegetation restoration/management in these areas, including adaptation actions to global warming. Projection analysis in different sites suggests a likely scenario of longer summer dry periods in the future [23, 67], which poses a dangerous threat in Mediterranean regions for both the ecosystems in mountain areas and the population downstream.

Finally, in these areas dew can constitute a significant water source for vegetation, especially under semiarid conditions, due to the large temperature interval during the day, with different experiments over the world to test further dew (and fog) harvesting technologies [68]. Some work estimated the decrease in water harvest from dew in the Mediterranean region up to 27% in the projections of future climate scenarios by 2080 [69], with a credible impact on water resources on a local community basis.

3.4.3 Energy balance components in Mediterranean high mountains and impact on streamflow

Fayad et al. [10], from their revision of snow hydrology in Mediterranean mountain areas, highlighted the major facts related to the partitioning of energy fluxes that are found in these sites when compared to temperate and northern latitudes. Radiation mostly drives the snowmelt regime; however, heat fluxes constitute a significant contribution during certain events, such as heat waves and rain-on-snow events, which are much more frequent in these regions, and during the final ablation at the end of the snow season, which may occur during the spring to the early summer. Moreover, sublimation can be a large fraction of the snow ablation, especially in years with different accumulation-ablation cycles, being this more representative in the highest altitudes. Herrero et al. [70] estimated sublimation up to a 40% of the annual ablation at 2500 m a.s.l. in Sierra Nevada (Spain) from an energy balance modeling approach. This preliminary result was further confirmed by field work carried out in this experimental site facing the Mediterranean Sea in southern Spain [56] during the following 7-year period. The conclusions [59] refined the estimation

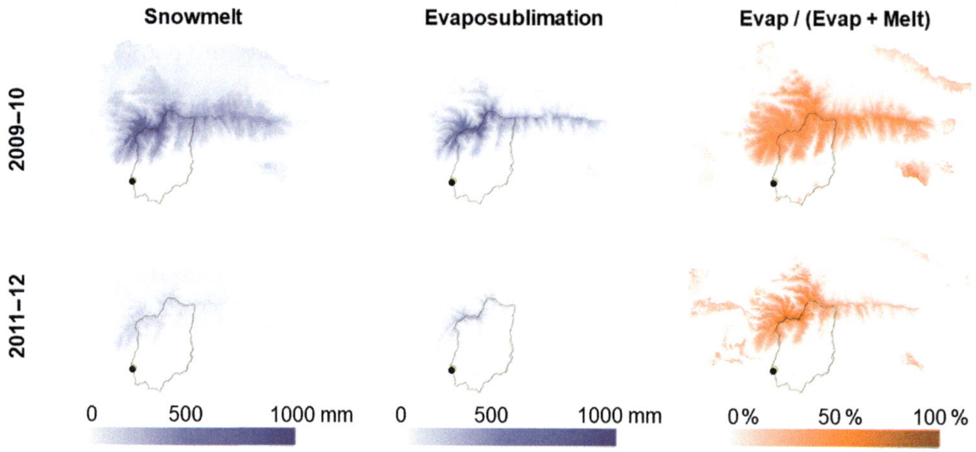

Fig. 3.4 Spatial distribution of the cumulative annual snowmelt (left) and evaposublimation (center) in Sierra Nevada (Spain) during wet (2009–10) and dry (2011–12) hydrological years, estimated from the results of Herrero and Polo [59], and the percentage of evaposublimation on the total annual ablation (right). The black line is the boundary of a gauged (black point) subbasin: 2009–10, annual river flow, snowmelt, and evaposublimation are 353, 81, and 48 hm^3, respectively (evaposublimation up to 37% ablation = 14% river flow); 2011–12, annual river flow, snowmelt, and evaposublimation are 31, 14, and 7 hm^3, respectively (evaposublimation up to 32% ablation = 22% river flow).

of the evaposublimation losses within the range of 24%–33% of the total annual ablation, with high annual variability depending on the seasonal occurrence of snowfall and mild weather events (Fig. 3.4). This work coined the term "evaposublimation" since, in fact, the atmospheric conditions during these states favor the coincidence of liquid water in the surface of the porous snowpack being evaporated and solid water sublimation, which is very complex to differentiate in practice. The results are in the range found in other Mediterranean areas, like the Atlas Mountains (44% [71]; 7%–20% [72]) or Israel (46–82% [73]), and other semiarid regions in the world [59]. Despite evaposublimation rates being one order of magnitude lower than snowmelt rates (0.5 against 5 mm/h, respectively), evaposublimation is a continuous flux from the snowpack, which is enhanced during cold and dry periods, whereas snowmelt exhibits an intermittent behavior. This is the main reason underlying the significant fraction of evaposublimation in semiarid regions, since snowmelt only takes place during warmer periods in winter; nonetheless, it is the dominating flux during spring and early summer. Conversely, on a seasonal basis, evaposublimation can rise up to a 50% of the monthly ablation during the coldest months, that is, December to February [59]. The most adverse scenario for the persistence of snowpacks during the cold season is an early accumulation of snow followed by a dry and sunny period, conditions that are not rarely found in Mediterranean mountains. Evaposublimation directly impacts the river flow volumes, and can raise to a significant fraction on an annual basis, especially in dry years, as highlighted from the example in Fig. 3.4.

Microscale processes, such as interaction with microtopography [74] and vegetation [75], rain-on-snow events [76], or the occurrence of Saharan dust transport to northern areas in the Mediterranean region [77], additionally play a key role for triggering snowmelt processes, especially during the warm season. The future perspective of warmer conditions together with a higher torrentiality of precipitation poses a credible threat for snowpacks in Mediterranean regions [78], since these expected shifts of climate involve longer dry periods during fall and winter, and significantly warmer springs with higher frequency of stormy events. Altogether, both sublimation and snowmelt are very likely to increase, which would result in larger fractions of water lost to the atmosphere from the snowpack, and shorter duration for the ablation process to be completed. The associated reductions of the snow season and the available water for the summer streamflow are expected to be accompanied by an increase of flood events in spring. However, it is difficult to forecast whether these impacts would be irreversible since the warmest regime could reduce the partition between snow- and rainfall, lowering the total amount of potential evaposublimation if snow reduces its occurrence and accumulation. In any case, the fluvial regime will suffer significant shifts on both the seasonal and annual scales, with large uncertainty on the sign of the expected change, very likely to be highly variable depending on the local conditions of each basin.

3.5 From research to decision-making: Needs and challenges for the cogeneration of operational services

Despite the enormous effort during the last decades to implement and develop monitoring networks in Mediterranean mountain areas, most of the currently available climate and hydrological services operating on a global or regional scale still fail to efficiently reach the different stakeholders related to water and land use planning and management, and society. Different underlying reasons (weaknesses) can be identified, beyond the already mentioned difficult access and climate conditions. The increasing availability of cost-efficient big-data exploitation tools can partially ameliorate this lack of service offer, but their applicability is restricted to stationary conditions in the area of interest, which leaves out the challenges posed by climate change and human modification of the territory. As the following points highlight, relevant scale issues require attention from the scientific community to bridge the gap between the current observational capacity and the practical generalized exploitation of the records. Finally, in this context, including the end users in the development of tailored reliable (credible) solutions and tools within a cogeneration framework is key in the pathway toward a successful and sustainable climate service market in Mediterranean mountain areas, which is one major challenge for science and transfer to society.

3.5.1 Monitoring systems

Beyond the increasing availability of remote sensing sources of data, the monitoring coverage of mountain areas still requires improvement in the highest altitudinal bands, which are key zones to capture the topographic gradients of precipitation and temperature and thus to precisely characterize rainfall and snowfall fields on at least a daily basis. In Mediterranean catchments, this is one of the major sources of uncertainty of any hydrological assessment [79], whatever the target application is; gauged streams are not usual in these high areas either. The variability of the local fluvial regime can be significant on the daily, weekly, seasonal, annual, and decadal scales, and a precise representation of the spatial patterns of the precipitation fields (i.e., good density of monitoring sites plus an appropriate spatial resolution of the digital elevation model of the study area) is required to minimize overestimation of the runoff from the propagation of errors associated to the precipitation volumes [79].

3.5.2 Forecasting systems

All the reported relevance and singularities of mountain areas in Mediterranean regions make it crucial to generate reliable forecasting systems for climate-related issues, and specifically, water-related sectors. The level of confidence generally reached by weather forecasting systems on the short-term has built in parallel trust on their skills in society, with decreasing accuracy from temperature forecast to precipitation occurrence, snow line altitude, and precipitation volume estimations. Flow volumes are still uncertain to forecast on the short term, and most of the decision-making on a local basis is made on empirical or statistical approaches from in-situ antecedent states and real-time rainfall measurements. However, in southern and eastern Mediterranean areas forecasting has a recent development and the lack of automated weather stations in the mountains still poses a constrain to adequately forecast extreme events like flash floods. Seasonal forecasts of climate and streamflow have been receiving attention during the last decade (https://climate.copernicus.eu/seasonal-forecasts), being their skill still far from the standard ranges associated to short-term forecast on a regional or local basis in Mediterranean areas. Scale issues already addressed in this chapter are the major underlying reason, but research effort is being currently done (www.claraproject.eu; www.prosnow.org) to improve their accuracy by means of user-tailored climate services.

For mountain areas, some examples of forecasting systems from research actions that have reached a proof-of-concept stage are worth mentioning, like the global framework for forecasting streamflow in snowy regions from MODIS data, SnowCloudHydro [80], that included two Mediterranean-type watersheds as test cases, or the local system for quasi-real time assessment of the water volumes in the snowpacks in the Spanish Sierra Nevada from automated weather stations and physically-based modeling, GMS-SNOWMED [71] (www.uco.es/dfh/snowmed), which also provides an estimation of the daily snowmelt and evaposublimation volumes that can be used to forecast streamflow changes in the short-term and anticipate dry spells on a weekly basis by local stakeholders. Other forecasting examples deal with climate-related impacts in Mediterranean areas, such as the seasonal predictions of Fire Weather Index [81] in the Mediterranean Europe which, despite showing significant skills for prediction only at 1 month lead time just in some areas of south-eastern Europe, is a first step toward the development of generalized services that can feed advanced local services in the future.

3.5.3 Adaptation to future climate scenarios

The assessment of global projections of future climate scenarios requires a detailed downscaling in the Mediterranean region, and particularly in their mountain areas, to provide some reliable information to stakeholders from a quantitative point of view. Nevertheless, the reported trends in different sites point out that torrentiality and aridness will increase their domain on both the spatial and time domains in these areas, which requires a thorough and early inclusion in environmental, economic, and social agendas in the political, national, regional, and local frameworks. Impacts have been already outlined in previous sections. Development of monitoring and forecasting systems can bridge the existing gaps in most of the current operational frameworks dealing with snow, water availability, extreme events, and other derived issues. However, a warmer future will also bring easier access to mountains and open expectations for new developments [82]. Beyond the impacts of a future potentially shorter (and more irregular) snow season on winter sport activities and mountain tourism, higher pressures on the agricultural systems [83] in the lowlands of mountain areas are very likely to result in migration of farming uses uphill occupying domains less dominated by snowpacks but still wetter than lower areas. Moreover, the expected negative impact of global warming on summer coastal tourism could bring a lengthening of the tourism season in southern Mediterranean countries [84], with higher demands of water and a potential shift toward mountain activities. In a context of less global availability of water, caution on this occupation should be taken, and participation actions must be enhanced for an integrated adaptation planning. In the context of the recent High Mountain Summit event (Geneve, 2019; https://highmountainsummit.wmo.int), chaired by the World Meteorological Organization, a call for action "to foster high-level dialogue and engage decision-makers and local actors to develop a roadmap to science-based, user-driven knowledge and information systems supporting sustainable development and risk reduction in mountain and downstream regions" has been launched. Mediterranean mountain areas are sentinels of the climate and human-driven changes, and undoubtedly require urgent plans toward early detection of critical changes and successful adaptation results.

3.6 Relevant mountain areas in other Mediterranean-like regions in the world

Relevant mountain areas in other regions in the world with Mediterranean climate conditions are mostly found in western America, close to the sea. Not only do they share climatic patterns, but also dominant land uses and population pressures, and water scarcity of seasonal and annual scales. Similarities of climate and hydrological trends are found between both regions in the world with varying impacts [10], although influenced by different large-scale climate forcing. For example, the relevance of evaposublimation in the ablation of snowpacks has been also reported [59] in the Andes (30%–90%), California (20%–32%), or Colorado (17%–43%). Besides El Niño-Southern Oscillation, in the Andes, the Pacific Ocean is the most influencing teleconnection both on the decadal and interdecadal scales, whereas in the Mediterranean region the Atlantic Ocean oscillations affect the seasonal and interannual regimes on such scales. The Mediterranean Andes region exhibits a gradient from their two latitudinal neighboring regions: the northern subtropical desert Andes and the southern wet temperate Andes. The Andes is considered a regional water tower for many regions to the south and different works have identified impacts of global warming in consonance with those reported in the Mediterranean region mountains. González-Reyes et al. [19] sum up the observed changes during the 20th century as a significant positive shift of the mean air temperature, with an increase of the 0°C isoline altitudinal band, a negative but not significant trend of the precipitation, and a significant decrease of the glacier area. The northern subregion is identified as the most vulnerable to climate impacts on water availability, with a transition toward the south, and an apparently increasing impact of the Pacific oscillations during the last decades of the 20th century.

In northern America Mediterranean areas, similar climate shifts are found [2] and even in apparently water-rich sites water scarcity is a growing issue downstream mountain areas [5]. The decrease in runoff reported in California [85] by impacts of global warming on snowmelt, clearly reflects the supply role of mountains in catchments where water-related conflicts are recurrent enhanced by the increasing occurrence of droughts. The relevance of understanding the complex interactions in Mediterranean mountain areas is a paramount issue to propose and manage adaptation actions, in which lessons learnt from previous research and application expertise in areas with more monitoring and modeling efforts must be shared and built over in those with more recent recording and development.

3.7 Key points and main conclusions

The intrinsic variability of hydrological processes in Mediterranean regions is enhanced by topography in mountain areas, which constitute "water towers" for population and ecosystem services. Their vulnerability to global warming is mainly driven by the expected shift of snowfall to rainfall, less persistent snowpacks, increased evapotranspiration rates and torrentiality, and loss of biodiversity, among others. Critical issues to cope with the challenge of sustainability of these highly valuable environments beyond the 21st century include efforts to improve the monitoring resources, both ground and remotely sensed, and enhance the sharing of knowledge from different sites of the globe, and foster the cogeneration of advanced climate services capable to deal with the spatial scale effects and the uncertainty of forecasts. For this, end users and citizen participation, together with land use policies that face the likely demands to come as access to mountains gains feasibility, are essential requirements in sustainable and efficient adaptation plans. International initiatives like the High Mountain Summit allow a place for Mediterranean mountains to be specially considered, as early warning sites for potential future to come across highlands in temperate and high latitudes.

References

[1] H.P. Liniger, R. Weingartner, M. Grosjean, Mountains of the World: Water Towers for the 21st Century. A contribution to global freshwater management, Mountain Agenda, University of Berne, Berne, 1998.

[2] M. Beniston, Climatic change in mountain regions: a review of possible impacts, Clim. Chang. 59 (2003) 5–31.

[3] United Nations, Earth Summit: Agenda 21, The United Nations Programme of Action From Rio. The final text of agreements negotiated by governments at the UNCED (3-14 June 1992), UN, Rio de Janeiro, 1992.

[4] J.I. López-Moreno, M. Beniston, J.M. García-Ruiz, Environmental change and water management in the Pyrenees: facts and future perspectives for Mediterranean mountains, Glob. Planet. Chang. 61 (2008) 300–312.

[5] A.W. Nolin, Perspectives on climate change, mountain hydrology, and water resources in the Oregon Cascades, USA, Mt. Res. Dev. 32 (2012) S35–S46.

[6] D. Karagulle, C. Frye, R. Sayre, S. Breyer, P. Aniello, R. Vaughan, D. Wright, Modeling global Hammond landform regions from 250-m elevation data, Trans. GIS 21 (2017) 1040–1060.

[7] F. Rubel, M. Kottek, Observed and projected climate shifts 1901–2100 depicted by world maps of the Köppen-Geiger climate classification, Meteorol. Z. 19 (2010) 135–141.

[8] J. Latron, P. Llorens, F. Gallart, The hydrology of Mediterranean mountain areas, Geogr. Compass 3 (2009) 2045–2064.

[9] C. de Jong, I. Gürer, A. Rimmer, A. Shaban, M. Williams, Climate and hydrology, in: I.N. Vogiatzakis (Ed.), Mediterranean Mountain Environments, Wiley, 2012, pp. 87–113.

[10] A. Fayad, S. Gascoin, G. Faour, J.I. López-Moreno, L. Drapeau, M. Le Page, R. Escadafal, Snow hydrology in Mediterranean mountain regions: a review, J. Hydrol. 551 (2017) 374–396.

[11] www.scopus.com, 2019 (last accessed on 12 September).

[12] D. Viviroli, R. Weingartner, The hydrologica significance of mountains: from regional to global scale, Hydrol. Earth Syst. Sci. 8 (2004) 1016–1029.

[13] M. Merheb, R. Moussa, C. Abdallah, F. Colin, C. Perrin, N. Baghdadi, Hydrological response characteristics of Mediterranean catchments: a review, Hydrol. Sci. J. 61 (2016) 2520–2539.

[14] United Nations, Proclamation of the International Year of the Mountains, Report on the 1998 UN General Assembly Meeting (New York), UN, New York, 1998.

[15] J.I. López-Moreno, S. Beguería, J.M. García-Ruiz, Trends in high flows in the Central Spanish Pyrenees: response to climatic factors or to land-use change? Hydrol. Sci. J. 51 (2006) 1039–1050.

[16] S. Beguería, I.J. López-Moreno, A. Lorente, M. Seeger, J.M. García-Ruiz, Assessing the effects of climate oscillations and land use changes on streamflow in the Central Spanish Pyrenees, Ambio 32 (2003) 283–286.

[17] J. Chenoweth, P. Hadjinicolaou, A. Bruggeman, J. Lelieveld, Z. Levin, M.A. Lange, E. Xoplaki, M. Hadjikakou, Impact of climate change on the water resources of the eastern Mediterranean and Middle East region: modeled 21st century changes and implications, Water Resour. Res. 47 (2011)W06506.

[18] N. Peleg, E. Morin, H. Gvirtzman, Y. Enzel, Rainfall, spring discharge and past human occupancy in the Eastern Mediterranean, Clim. Chang. 112 (2012) 769–789.

[19] A. González-Reyes, J. McPhee, D.A. Christie, C.L. Quesne, P. Szeijner, M. H. Masiokas, R. Villalba, A.A. Muñoz, S. Crespo, Spatiotemporal variations in hydroclimate across the Mediterranean Andes (30°–37°S) since the early twentieth century, J. Hydrometeorol. 18 (2017) 1929–1942.

[20] P. Llorens, F. Gallart, C. Cayuela, M. Roig-Planasdemunt, E. Casellas, A.J. Molina, M.M. de las Heras, G. Bertrán, E. Sánchez-Costa, J. Latron, What have we learnt about Mediterranean catchment hydrology? 30 years observing hydrological processes in the Vallcebre Research Catchments, Geophys. Res. Lett. 44 (2018) 475–502.

[21] A. Marchane, L. Jarlan, A. Boudhar, Y. Tramblay, L. Hanich, Linkages between snow cover, temperature and rainfall and the North Atlantic Oscillation over Morocco, Clim. Res. 69 (2016) 229–238.

[22] J.M. García-Ruiz, J.I. López-Moreno, S. Vicente-Serrano, T. Lasanta-Martínez, S. Beguería, Mediterranean water resources in a global change scenario, Earth-Sci. Rev. 105 (2011) 121–139.

[23] M.J. Pérez-Palazón, R. Pimentel, M.J. Polo, Climate trends impact on the snowfall regime in Mediterranean mountain areas: future scenario assessment in Sierra Nevada (Spain), Water 10 (2018) 720.

[24] C. Aguilar, J. Herrero, M.J. Polo, Topographic effects on solar radiation distribution in mountainous watersheds and their influence on reference evapotranspiration estimates at watershed scale, Hydrol. Earth Syst. Sci. 14 (2010) 2479–2494.

[25] C. Aguilar, M.J. Polo, Generating reference evapotranspiration surfaces from the Hargreaves equation at watershed scale, Hydrol. Earth Syst. Sci. 15 (2011) 2495–2508.

[26] Y. Host, P. Chevallier, A. Coudrain, Y. Lejeune, P. Etchevers, Relationship between precipitation and air temperature: comparison between the Bolivian Andes and the Swiss Alps, Hydrol. Sci. J. 50 (2005) 989–997.

[27] D. Tetzlaff, S.K. Carey, J.P. McNamara, H. Laudon, C. Soulsby, The essential value of long-term experimental data for hydrology and water management, Water Resour. Res. 53 (2017) 2598–2604.

[28] T. Burt, J. McDonnell, Whither field hydrology? The need for discovery science and outrageous hydrological hypotheses, Water Resour. Res. 51 (2015) 5919–5928.

[29] I.R. Calder, Blue Revolution II, Integrated Land & Water Resources Management, Earthscan, London, 2005.

[30] J.W. Pomeroy, M. Bernhardt, D. Marks, Water resources: research network to track alpine water, Nature (correspondence) 521 (2015) 32.

[31] C. Adler, E. Palazzi, A. Kulonen, J. Balsiger, G. Colangeli, D. Cripe, N. Forsythe, G. Goss-Durant, Y. Guigoz, J. Krauer, D. Payne, N. Pepin, M. Peralvo, J. Romero, R. Sayre, M. Shahgedanova, R. Weingartner, M. Zebisch, Monitoring mountains in a changing world: new horizons for the global network for observations and information on mountain environments (GEO-GNOME), Mt. Res. Dev. 38 (2018) 265–269.

[32] F. Hüsler, T. Jonas, M. Riffler, J.P. Musial, S. Wunderle, A satellite-based snow cover climatology (1985–2011) for the European Alps derived from AVHRR data, Cryosphere 8 (2014) 73–90.

[33] M. Huss, B. Bookhagen, C. Huggel, D. Jacobsen, R.S. Bradley, J.J. Clague, M. Vuille, W. Buytaert, D. R. Cayan, G. Greenwood, B.G. Mark, A.M. Milner, R. Weingartner, M. Winder, Toward mountains without permanent snow and ice, Earth's Future 5 (2017) 418–435.

[34] M. Beniston, D. Farinotti, M. Stoffel, L.M. Andreassen, E. Coppola, N. Eckert, A. Fantini, F. Giacona, C. Hauck, M. Huss, H. Huwald, M. Lehning, J.I. López-Moreno, J. Magnusson, C. Marty, E. Morán-Tejéda, S. Morin, M. Naaim, A. Provenzale, A. Rabatel, D. Six, J. Stötter, U. Strasser, S. Terzago,

C. Vincent, The European mountain cryosphere: a review of its current state, trends, and future challenges, Cryosphere 12 (2018) 759–794.

[35] Y. Fan, M. Clark, D.M. Lawrence, S. Swenson, L.E. Band, S.L. Brantley, P.D. Brooks, W.E. Dietrich, A. Flores, G. Grant, J.W. Kirchner, D.S. Mackay, J.J. McDonnell, P.C.D. Milly, P.L. Sullivan, C. Tague, H. Ajami, N. Chaney, A. Hartmann, P. Hazenberg, J. McNamara, Hillslope hydrology in global change research and Earth system modeling, Water Resour. Res. 55 (2019) 1737–1772.

[36] D.P. Lettenmaier, D. Alsdorf, J. Dozier, G.J. Huffman, M. Pan, E.F. Wood, Inroads of remote sensing into hydrologic science during the WRR era, Water Resour. Res. 51 (2015) 7309–7342.

[37] L. Pasolli, C. Notarnicola, G. Bertoldi, L. Bruzzone, R. Remelgado, F. Greifeneder, G. Niedrist, S. Della Chiesa, U. Tappeiner, M. Zebisch, Estimation of soil moisture in mountain areas using SVR technique applied to multiscale active radar images at C band, IEEE J. Sel. Top. Appl. Earth Obs. Remote Sen. 8 (2015) 262–283.

[38] M.F. McCabe, M. Rodell, D.E. Alsdorf, D.G. Miralles, R. Uijlenhoet, W. Wagner, A. Lucieer, R. Houborg, N.E.C. Verhoest, T.E. Franz, J. Shi, H. Gao, E.F. Wood, The future of Earth observation in hydrology, Hydrol. Earth Syst. Sci. 21 (2017) 3879–3914.

[39] T. Raziei, I. Bordi, L.S. Pereira, A snow variability analysis in Iran in relation to global warming and climate change, Eur. Water 59 (2017) 45–51.

[40] F.A. Saavedra, S.K. Kampf, S.R. Fassnacht, J.S. Sibold, Changes in Andes Mountains snow cover from MODIS data 2000–2016, Cryosphere 12 (2018) 1027–1046.

[41] M.W. Baba, S. Gascoin, L. Jarlan, V. Simonneaux, L. Hanich, Variations of the snow water equivalent in the Ourika Catchment (Morocco) over 2000–2018 using downscaled MERRA-2 data, Water 10 (2018) 1120.

[42] Z. Malenovský, H. Rott, J. Cihlar, M.E. Schaepman, G. García-Santos, R. Fernandes, M. Berger, Sentinels for science: Potential of Sentinel-1, -2, and -3 missions for scientific observations of ocean, cryosphere and land, Remote Sens. Environ. 120 (2012) 91–101.

[43] A. Kokhanovsky, M. Lamare, O. Danne, C. Brockmann, M. Dumont, G. Picard, L. Arnaud, V. Favier, B. Jourdain, E. Le Meur, B. Di Mauro, T. Aoki, M. Niwano, V. Rozanov, S. Korkin, S. Kipfstuhl, J. Freitag, M. Hoerhold, A. Zuhr, D. Vladimirova, A.K. Faber, H.C. Steen-Larsen, S. Wahl, J.K. Andersen, B. Vandecrux, D. van As, K.D. Mankoff, M. Kern, E. Zege, J.E. Box, Retrieval of snow properties from the Sentinel-3 Ocean and Land Colour Instrument, Remote Sens. 11 (2019) 2280.

[44] T. Nagler, H. Rott, E. Ripper, G. Bippus, M. Hetzenecker, Advancements for snowmelt monitoring by means of Sentinel-1 SAR, Remote Sens. 8 (2016) 348.

[45] F.T. Ulaby, D. Long, Microwave Radar and Radiometric Remote Sensing, University of Michigan Press, Ann Arbor, 2014.

[46] H. Lievens, M. Demuzere, H. Marshall, R.H. Reichle, L. Brucker, I. Brangers, P. de Rosnay, M. Dumont, M. Girotto, W.W. Immerzeel, T. Jonas, E.J. Kim, I. Koch, C. Marty, T. Saloranta, J. Schöber, G.J.M. De Lannoy, Snow depth variability in the Northern Hemisphere mountains observed from space, Nat. Commun. 10 (2019) 4629.

[47] S. Pettinato, E. Santi, M. Brogioni, S. Paloscia, E. Palchetti, C. Xiong, The potential of COSMO-SkyMed SAR images in monitoring snow cover characteristics, IEEE Geosci. Remote Sens. Lett. 10 (2013) 9–13.

[48] S. Paloscia, S. Pettinato, E. Santi, M. Valt, COSMO-SkyMed image investigation of snow features in alpine environment, Sensors 17 (2017) 84.

[49] C. Marin, G. Bertoldi, V. Premier, M. Callegari, C. Brida, K. Hürkamp, J. Tschiersch, M. Zebisch, C. Notarnicola, Use of Sentinel-1 radar observations to evaluate snowmelt dynamics in alpine regions, Cryosphere Discuss. 175 (2019) (in review).

[50] S. Gascoin, M. Grizonnet, M. Bouchet, G. Salgues, O. Hagolle, Theia snow collection: high-resolution operational snow cover maps from Sentinel-2 and Landsat-8 data, Earth Syst. Sci. Data 11 (2019) 493–514.

[51] M.W. Baba, S. Gascoin, L. Hanich, Assimilation of Sentinel-2 data into a snowpack model in the High Atlas of Morocco, Remote Sens. 10 (2018) 1982.

[52] R. Marti, S. Gascoin, E. Berthier, M. de Pinel, T. Houet, D. Laffly, Mapping snow depth in open alpine terrain from stereo satellite imagery, Cryosphere 10 (2016) 1361–1380.
[53] World Meteorological Organization, WMO Solid Precipitation Intercomparison Experiment (SPICE) (2012-2015), WMO, Geneve, 2018.
[54] L. Jarlan, S. Khabba, S. Er-Raki, M. Le Page, L. Hanich, Y. Fakir, O. Merlin, S. Mangiarotti, S. Gascoin, J. Ezzahar, M.H. Kharrou, B. Berjamy, A. Saaïdi, A. Boudhar, A. Benkaddour, N. Laftouhi, J. Abaoui, A. Tavernier, G. Boulet, V. Simonneaux, F. Driouech, M. El Adnani, Remote sensing of water resources in semi-arid Mediterranean areas: the joint international laboratory TREMA, Int. J. Remote Sens. 36 (2015) 4879–4917.
[55] A. Fayad, S. Gascoin, G. Faour, P. Fanise, L. Drapeau, J. Somma, A. Fadel, A. Al Bitar, R. Escadafal, Snow observations in Mount Lebanon (2011–2016), Earth Syst. Sci. Data 9 (2017) 573–587.
[56] M.J. Polo, J. Herrero, R. Pimentel, M.J. Pérez-Palazón, The Guadalfeo monitoring network (Sierra Nevada, Spain): 14 years of measurements to understand the complexity of snow dynamics in semiarid regions, Earth Syst. Sci. Data 11 (2019) 393–407.
[57] A. Haberkorn, European snow booklet—an inventory of snow measurements in Europe. EnviDat (2019), https://doi.org/10.16904/envidat.59.
[58] A. Chehbouni, R. Escadafal, B. Duchemin, G. Boulet, V. Simonneaux, G. Dedieu, B. Mougenot, S. Khabba, H. Kharrou, P. Maisongrande, O. Merlin, A. Chaponnière, J. Ezzahar, S. Er-Raki, J. Hoedjes, R. Hadria, A. Abourida, A. Cheggour, F. Raibi, A. Boudhar, I. Benhadj, L. Hanich, An integrated modelling and remote sensing approach for hydrological study in arid and semi-arid regions: the SUDMED Programme, Int. J. Remote Sens. 29 (2008) 5161–5181.
[59] J. Herrero, M.J. Polo, Evaposublimation from the snow in the Mediterranean mountains of Sierra Nevada (Spain), Cryosphere 10 (2016) 2981–2998.
[60] J.M. Pflug, G.E. Liston, B. Nijssen, J.D. Lundquist, Testing model representations of snowpack liquid water percolation across multiple climates, Water Resour. Res. 55 (2019) 4820–4838.
[61] A. Fayad, S. Gascoin, The role of liquid water percolation representation to estimate snow water equivalent in a Mediterranean mountain region (Mount Lebanon), Hydrol. Earth Syst. Sci. Discuss. 514 (2019) (in review).
[62] Y. Hajhouji, V. Simonneaux, S. Gascoin, Y. Fakir, B. Richard, A. Chehbouni, A. Boudhar, Modélisation pluie-débit et analyse du régime d'un bassin versant semi-aride sous influence nivale. Cas du bassin versant du Rheraya (Haut Atlas, Maroc), La Houille Blanche 3 (2018) 49–62.
[63] P. Gómez-Giráldez, C. Aguilar, M.J. Polo, Natural vegetation covers as indicators of the soil water content in a semiarid mountainous watershed, Ecol. Indic. 46 (2014) 524–535.
[64] P. Llorens, F. Domingo, Rainfall partitioning by vegetation under Mediterranean conditions. A review of studies in Europe, J. Hydrol. 335 (2007) 37–54.
[65] A. Muzylo, P. Llorens, F. Domingo, Rainfall partitioning in a deciduous forest plot in leafed and leafless periods, Ecohydrology 5 (2012) 759–767.
[66] P. Llorens, J. Latron, M. Álvarez-Cobelas, J. Martínez-Vilalta, G. Moreno, Hydrology and biogeochemistry of Mediterranean forests, in: D.F. Levia, D. Carlyle-Moses, T. Tanaka (Eds.), Forest Hydrology and Biogeochemistry: Synthesis of Past Research and Future Directions, Ecological Studies, vol. 216, Springer, Dordrecht, 2011, pp. 301–320.
[67] P. Llorens, R. Poyatos, J. Delgado, I. Oliveras, F. Gallart, A multi-year study of rainfall and soil water control son Scots pine transpiration under Mediterranean mountain conditions, Hydrol. Process. 24 (2010) 3053–3064.
[68] K.F. Kaseke, L. Wang, Fog and dew as potable water resources: maximizing harvesting potential and water quality concerns, GeoHealth 2 (2018) 327–332.
[69] M. Tomaszkiewicz, M.A. Najm, D. Beysens, I. Alameddine, E.B. Zeid, M. El-Fadel, Projected climate change impacts upon dew yield in the Mediterranean basin, Sci. Total Environ. 566 (2016) 1339–1348.
[70] J. Herrero, M.J. Polo, A. Moñino, M.A. Losada, An energy balance snowmelt model in a Mediterranean site, J. Hydrol. 371 (2009) 98–107.
[71] O. Schulz, C. de Jong, Snowmelt and sublimation: field experiments and modelling in the High Atlas Mountains of Morocco, Hydrol. Earth Syst. Sci. 8 (2004) 1076–1089.

[72] A. Boudhar, G. Boulet, L. Hanich, J.E. Sicart, A. Chehbouni, Energy fluxes and melt rate of a seasonal snow cover in the Moroccan High Atlas, Hydrol. Sci. J. 61 (2016) 931–943.
[73] R. Sade, A. Rimmer, M.I. Litaor, E. Shamir, A. Furman, Snow surface energy and mass balance in a warm temperate climate mountain, J. Hydrol. 519 (2014) 848–862.
[74] R. Pimentel, J. Herrero, M.J. Polo, Subgrid parameterization of snow distribution at a Mediterranean site using terrestrial photography, Hydrol. Earth Syst. Sci. 21 (2017) 805–820.
[75] J. Revuelto, V. Vionnet, J.I. López-Moreno, M. Lafaysse, S. Morin, Combining snowpack modeling and terrestrial laser scanner observations improves the simulation of small scale snow dynamics, J. Hydrol. 533 (2017) 291–307.
[76] K.N. Musselman, F. Lehner, K. Ikeda, M.P. Clark, A.F. Prein, C. Liu, M. Barlage, R. Rasmussen, Projected increases and shifts in rain-on-snow flood risk over western North America, Nat. Clim. Chang. 8 (2018) 808–812.
[77] B. Di Mauro, R. Garzonio, M. Rossini, G. Filippa, P. Pogliotti, M. Galvagno, U. Morra di Cella, M. Migliavacca, G. Baccolo, M. Clemenza, B. Delmonte, V. Maggi, M. Dumont, F. Tuzet, M. Lafaysse, S. Morin, E. Cremonese, R. Colombo, Saharan dust events in the European Alps: role in snowmelt and geochemical characterization, Cryosphere 13 (2019) 1147–1165.
[78] J.I. López-Moreno, S. Gascoin, J. Herrero, E.A. Sproles, M. Pons, E. Alonso-González, L. Hanich, A. Boudhar, K.N. Musselman, N.P. Molotch, J. Sickman, J. Pomeroy, Different sensitivities of snowpacks to warming in Mediterranean climate mountain areas, Environ. Res. Lett. 12 (2017)074006.
[79] M. Egüen, C. Aguilar, J. Herrero, A. Millares, M.J. Polo, On the influence of cell size in physically-based distributed hydrological modelling to assess extreme values in water resource planning, Nat. Hazards Earth Syst. Sci. 12 (2012) 1573–1582.
[80] E.A. Sproles, R.L. Crumley, A.W. Nolin, E. Mar, J.I. López-Moreno, SnowCloudHydro—a new framework for forecasting streamflow in snowy, data-scarce regions, Remote Sens. 10 (2018) 1276.
[81] J. Bedia, N. Golding, A. Casanueva, M. Iturbide, C. Buontempo, J.M. Gutiérrez, Seasonal predictions of fire weather index: paving the way for their operational applicability in Mediterranean Europe, Clim. Serv. 9 (2018) 101–110.
[82] G. Dono, R. Cortignani, D. Dell'Unto, P. Deligios, L. Doro, N. Lacetera, L. Mula, M. Pasqui, S. Quaresima, A. Vitali, P.P. Roggero, Winners and losers from climate change in agriculture: insights from a case study in the Mediterranean basin, Agric. Syst. 146 (2016) 65–75.
[83] T.P.L. Nguyen, L. Mula, R. Cortignani, G. Seddaiu, G. Dono, S.G. Virdis, M. Pasqui, P.P. Roggero, Perceptions of present and future climate change impacts on water availability for agricultural systems in the western Mediterranean region, Water 8 (2016) 523.
[84] F. Giorgi, P. Lionello, Climate change projections for the Mediterranean region, Glob. Planet. Chang. 63 (2008) 90–104.
[85] S.M. Jepsen, T.C. Harmon, M.W. Meadows, C.T. Hunsaker, Hydrogeologic influence on changes in snowmelt runoff with climate warming: numerical experiments on a mid-elevation catchment in the Sierra Nevada, USA, J. Hydrol. 533 (2016) 332–342.

CHAPTER 4

Irrigation and water governance

Carles Sanchis-Ibor[a], François Molle[b], Marcel Kuper[c,d]
[a]Centro Valenciano de Estudios del Riego (CVER), Universitat Politècnica de València, València, Spain
[b]Institut de Recherche pour le Développement (IRD), UMR G-Eau, Univ Montpellier, Montpellier, France
[c]Cirad, UMR G-Eau, Univ Montpellier, Montpellier, France
[d]IAV Hassan II, Rabat, Morocco

4.1 Introduction

The Mediterranean climate is characterized by very low rainfall during the warm season, which means water is lacking when it is most needed. This singularity hinders the productive capacity of regional agro-systems and has historically stimulated the use of irrigation by Mediterranean societies in order to increase land profitability, to ensure food security, or simply to survive (in many areas of North Africa, for instance, a productive agriculture is hardly possible without irrigation). Irrigation has become a consubstantial part of Mediterranean cultures, societies, and natures, along with pastoralism and rain-fed agriculture (cereals, pulses, olives, vineyards), through a historical construction dating back many thousands of years. Insufficient local resources have also necessitated water works, such as the many aqueducts constructed by the Romans and, later, large-scale storage facilities.

Mediterranean irrigated landscapes have developed through a long process of "technological aggradation." Irrigation is a palimpsest where it is possible to read the influence of the Arab agronomic revolution, feudal territorial organization, or the footprint of the foundations of monasteries. These successive layers are particularly visible in those territories affected by the historical fluctuations of the shifting Christian-Muslim boundary, which facilitated the mixture of different water culture traditions that are still present in infrastructure, practices, and lexical borrowings.

Numerous local communities developed a myriad of small- and large-scale irrigation systems, but centralized powers also guided the construction of some emblematic waterworks, such as the Nile Valley/Delta canals. During the twentieth century, as part of nation-building projects, the modern state became increasingly involved in irrigation development, mainly through the construction of large dams and large-scale irrigation schemes, or the modernization of existing ones. This development model was subsequently challenged, particularly from the 1970s to 1980s onward, by the private exploitation of groundwater resources by hundreds of thousands of wells and tube wells. All in all, irrigation development in the Mediterranean gradually mobilized an enormous proportion of both surface and groundwater resources, leading to basin closure due to the over-allocation of these resources. In addition, increased competition over water

Water Resources in the Mediterranean Region
https://doi.org/10.1016/B978-0-12-818086-0.00004-2

resources led to the marginalization of certain irrigation communities and farmers. This is the case, for instance, in public surface-water irrigation schemes where the annual water allocation has become insufficient, and for groundwater exploitation where the drop in water tables has excluded certain categories of farmer. The response of the public authorities in most countries has so far mainly focused on seeking alternative water resources rather than restricting water use for irrigation. Once surface water was fully mobilized, the attention was focused on groundwater, while, more recently, treated wastewater and desalinated water have received more attention.

A looming water crisis has been diagnosed for the Mediterranean over the past three decades [1, 2], and the need to drastically reduce agricultural water demand is routinely underlined. The importance of water governance, state reforms, legal development, and stakeholder participation are also emphasized [3]. However, there are few success stories on water governance in the Mediterranean, in particular related to irrigation. In this chapter, we explore the reasons for the gap between the general consensus on the need for better water governance and stark field realities showing the continued degradation of water resources.

4.2 The making of Mediterranean irrigation

4.2.1 Traditional irrigation

Historically, irrigation has been both a mechanism of societies' adaptation to their territory and an instrument of the transformation of nature. Since antiquity, Mediterranean cultures have taken advantage of hydrological processes to generate agricultural benefits. The channeling of recurrent Nile River floods to water riverine land is probably one of the most well-known and productive strategies of adaptation to natural dynamics. Since the Protodynastic period (5200–5000 BP) and up to the construction of Aswan High Dam, Egyptian agriculture was closely attuned to the cycle of the river flow to use its waters and sediments. This is the most ancient form of irrigation in the Mediterranean. The diversion technique was also applied, on a minor scale, in numerous North African *wadis* and Iberian *ramblas*, where flash floods were diverted by rudimentary weirs (*ougoug, sadd, boqueres*, *derramadores*, etc.) to wet arid/semiarid lands with turbid waters through spate irrigation [4].

At the same time, irrigation and drainage channels were used to transform ecosystems and reclaim wetlands. The Etruscan culture developed wetland drainage techniques that were disseminated beyond the Italic peninsula by the Romans to create new agricultural lands and settlements [5]. Since classical antiquity, Mediterranean plains and deltas (Nile, Evros, Po, Rhone, Ebro) have been artificialized through successive drainage and irrigation development works. Far from these coastal environments, in the higher valleys or desert margins, adapted irrigation systems were developed to overcome harsh topographic or climate conditions. Mediterranean mountains host numerous small irrigation systems, which supply agricultural land in narrow valleys and steep slopes through ingenious terrace systems, documented since the medieval period on both northern and southern shores [6], and also on the islands [7]. Undoubtedly, oases are one of the most singular manifestations of this traditional water culture. The *foggaras* (or *khettaras, qanats*, etc.) are quasi-horizontal galleries dug to capture the aquifer and bring the water to oases, such as Kharga (Egypt), the Gourara (Algeria), or the Tafilalet (Morocco) [4]. This technique was also widely disseminated through Mediterranean Spain, where 6230 *minas* have been recently inventoried [8].

The sustainability of these community-managed irrigation systems is closely linked to age-tested institutions [9, 10]. The Arab *jmâa* is the collective institution that ruled water management in North African *seguias* [11]. This institutional formula is similar to the *alhamas*, which ruled collective water use in the rural settlements of Al-Andalus. In essence, this Arabic model has evident parallels with the organizations developed in the medieval kingdoms of Spain and France at both municipal (*consells or universitats* in the Aragón Crown and Occitània) and channel levels (communes of the València séquies). This latter organization, separated from the municipal institutions, was enshrined in the first Spanish Water Law (1866) by Rodríguez de Cepeda to create the irrigation communities (named initially *sindicatos de riego*, and *comunidades de regantes* after 1879) and it was also brought to the French legislation (Water Law of 1865) by Jaubert de Passa to create the *Associations Syndicales* (ASA). By contrast, the *jmâa* was hardly ever recognized in North Africa, and formal water user associations, generally not functional, were imposed instead by the state from the 1990s onward (see Section 6.3.2). This showed the state's determination to obtain a formal entry point into community-managed irrigation systems, where it had hitherto played only a minor role [12]. More generally, the institutions of communal irrigation systems have shown a capacity to solve internal conflict but have become increasingly dependent on external actors, in particular the state (see, for instance, Riaux [13] for a comparative analysis of France and Morocco). They have also failed to upscale the principles of collective management to catchment or aquifer level so as to create nested organizations to share and govern the entire resource system.

Traditional Mediterranean irrigation schemes are robust socio-ecological systems which, in many cases, have been in existence for several centuries. However, many are currently threatened by a combination of various processes. First, urban sprawl may degrade or destroy traditional systems, replacing agricultural land with residential, tourist, or industrial uses. Transportation infrastructure fragments the irrigated area, dividing and isolating farms and leaving the remaining smaller landholdings more vulnerable to new dangers [14, 15]. Second, the ownership structure combined with heritage rules often led to small and fragmented plots, hindering the competitiveness of farmers. Smallholding is predominant in the majority of these traditional systems, and farming is also affected by declining product prices and the weakening of local markets linked to globalization. Third, numerous traditional systems have been affected by state-led irrigation development, aquifer overexploitation through private tube wells, and river basin closure processes, inducing subtle water grabbing processes that favor recently developed irrigated areas or other economic sectors (see Pascon's work [16] for an early analysis of the impact of state-led irrigation development on community-managed irrigation systems in the Haouz Plain, Morocco). The demise of a large number of *khettara* (Morocco) or *foggara* systems (Algeria) further illustrates this issue [17]. Fourth, rural out-migration and farmer aging particularly impact small, traditional mountain irrigation systems, causing problems of intergenerational continuity, land and infrastructure abandonment, or even the breakdown of the transmission of traditional knowledge.

This crisis has triggered some reactions from experts and institutions who defend the heritage and cultural values, and the social and ecological functions of these systems. They claim a new role for these systems in the context of the redesigning of urban and peri-urban territories, rural development strategies, and food security discourses. Some of these local initiatives have achieved international recognition, such as UNESCO World Heritage Sites (the Elx Palm Grove, the Valencia Water Court, and the Palestinian Battir hill landscape), and FAO Globally Important Agricultural Heritage Systems (e.g., the Siwa Oases in Egypt, the Ghout System in Algeria, and the Gafsa Oases in Tunisia). At the same time, some communities in the Po Valley have developed plans and projects to preserve the cultural heritage and natural value of their canals, involving both users and public administrations [18]. In Spain, the defenders of medieval irrigation gardens (*huertas*), in Granada, Zaragoza, Murcia, and València, among others, have joined in an umbrella organization (Intervegas) to demand legal measures for land protection and traditional irrigation conservation and improvement.

4.2.2 The hydraulic mission of the 20th century

Rivers would be worth less if they dragged golden sands: infinite treasures roll day and night along their beds, and we, foolish, let them get lost in the abysses of the Ocean Joaquín Costa [19]

The twentieth century witnessed an unprecedented expansion of the Mediterranean's irrigated land. According to current estimations [20], irrigated areas multiplied almost fivefold between 1900 and 2000, increasing from 5.8 to 25.3 Mha (Table 4.1). Most of this expansion took place after World War II, when irrigated land increased at a rate of 0.32 Mha per year in the region (0.07 Mha per year between 1900 and 1950).

This remarkable development was mainly the result of the involvement of the modern state in the mobilization of water resources and the promotion of irrigation—a sociopolitical process known as the *hydraulic mission* [21] that has also taken place in other regions of the world. The roots of this phenomenon can be traced back to the Enlightenment period, when the enthusiasm for science and technical progress created a collective awareness of the human capacity to subdue the forces of nature. The development of engineering as a reputed profession, and its incorporation into the administration of the nascent modern state, provided moral and technical means to national governments to promote the use of water resources as a valuable economic asset.

In the nineteenth century, irrigation expanded in some Mediterranean countries as part of nation-building projects. Indeed it was an essential part of the modern Egypt envisioned by Mehmet Ali, under whose government (1805–48) numerous dykes, deep channels, and waterwheels were built to divert Nile waters during low water periods to 1.4 Mha of land.

Table 4.1 Area equipped for irrigation (thousands of hectares) in the Mediterranean between 1900 and 2016.

	1900	1910	1920	1930	1940	1950	1960	1970	1980	1990	2000	2005	2016
Albania	25	26	27	28	31	39	145	284	371	423	340	362	354
Algeria	23	64	105	146	187	229	231	238	253	455	569	980	1360
Egypt	2310	2327	2345	2362	2379	2397	2698	2952	3000	3189	3422	3500	3714
France	252	293	335	376	417	459	500	767	1370	2100	2634	2678	2830
Greece	100	131	162	199	275	230	425	730	961	1195	1451	1594	1530
Israel	15	18	21	24	28	47	136	172	200	206	194	225	225
Italy	1300	1407	1625	1875	2125	2376	2844	3172	3500	3858	3892	3973	4004
Jordan	10	11	12	14	15	20	30	34	37	63	77	80	107
Lebanon	11	14	16	19	22	25	40	68	86	86	104	104	104
Macedonia	4	5	5	5	5	13	89	104	116	136	128	128	128
Morocco	27	27	27	27	301	574	848	920	1217	1258	1443	1490	1530
Portugal	19	34	49	64	131	214	446	652	734	878	792	617	548
Spain	1000	1204	1375	1418	1461	1530	1875	2379	3029	3402	3735	3858	3923
Syria	65	76	86	97	175	314	520	451	539	693	1211	1428	1310
Tunisia	103	103	102	102	101	101	100	200	243	300	394	418	476
Turkey	603	667	730	793	857	918	1133	1800	2700	4071	4745	5215	5215
Others	20	23	25	27	28	56	195	211	237	250	252	148	168
	5889	**6429**	**7048**	**7576**	**8539**	**9541**	**12,236**	**15,117**	**18,571**	**22,544**	**25,364**	**26,798**	**27,526**

Source: Modified from K. Freydank, S. Siebert, Towards Mapping the Extent of Irrigation in the Last Century: A Time Series of Irrigated Area Per Country, University of Frankfurt, 2008. 2005 and 2016 data obtained from FAOSTAT. Others include the Occupied Palestinian Territory, Andorra, Bosnia and Herzegovina, Serbia, Montenegro, Kosovo, Croatia, Slovenia, Malta, and Cyprus. Some countries (Spain, France, Portugal, Jordan, Syria, etc.) include areas not strictly Mediterranean. Also, the early irrigation development by irrigation communities seems to remain under the radar of the figures presented here. For example, Pascon [16], in his historical analysis of the Haouz plain in Morocco, shows that the extent of traditional irrigation systems (*khettara*, seguia) developed from the eleventh century onward is already more than the 27,000 ha presented in the table for the irrigated area of Morocco prior to 1940.

Beyond their social and economic impact, some of the projects executed in this period also had a symbolic value, as a representation of the strength, progress, and modernity of the new nations. This is also the case of the Canale Cavour, an 85 km channel built in only 3 years (1863–66) to irrigate 300,000 ha of the Piedmont with water from the Po River.

Perhaps the most explicit illustration of the hydraulic mission as an intellectual project is the Spanish *Regeneracionismo*, a national revivalist movement led by Joaquín Costa, which placed irrigation as the catalyst of the social and economic regeneration of the country after the loss of its colonies in Cuba and the Philippines. The materialization of Costa's hydro-social dream was not immediate but it provided a roadmap for the engineering projects proposed by Manuel Lorenzo Pardo (1933) during the Second Republic and executed mainly during the long Francoist dictatorship (1939–75) [22, 23].

Throughout the first half of the twentieth century, in Spain and other Mediterranean countries, the state considered irrigation as the most effective instrument to sustain rural population, supply internal food markets, and stimulate exports. Moreover, some political forces saw irrigation as an alternative to coercive land reforms and a strategy to control the territory. For example, these ideas were behind the development of the notion of *bonifica integrale* posited by Arrigo Serpieri during the Italian fascist period, which advocated for state backing of drainage and irrigation projects, and for the creation of new settlements to colonize land reclaimed from waterlogged areas [24]. Although these projects launched by Mussolini fared far below expectation [25, 26], they were taken as a model for the internal colonization promoted by the Francoist regime in Spain, based on dam building and groundwater development projects. Geopolitical concerns also boosted the Israeli hydraulic mission. From the late 1930s, the Zionist enterprise firmly promoted irrigation and large-scale water projects to consolidate its borders, provide food security, and absorb immigrants [27]. The construction of the National Water Carrier, completed in 1964, allowed for the expansion of irrigation, bringing water from Lake Tiberias as far south as the Negev Desert [28].

The capacity of irrigation to fix populations and facilitate territorial control was firmly embraced during the European colonization of northern Africa. Paraphrasing Wittfogel [29], the hydraulic mission was exported by the colonialist states as a form of "occidental despotism" [30]. In Morocco, France promoted irrigation under several forms. While in the 1930s the priority was establishing private investors in large-scale public schemes (e.g., the Beht irrigation system in the Gharb plain), in other areas (e.g., Haouz), French engineers improved traditional irrigation systems while limiting the rights to their water and bestowing the "surplus" water resources and the most fertile properties to settlers or members of the local elites [16]. Only in the Tadla scheme did the administration attempt to expand irrigation to improve the living conditions of the agro-pastoral society [16, 31]. In Tunisia, the colonial hydraulic mission included dams and large-scale public schemes (e.g., in the Medjerda valley), public boreholes, and attendant public irrigation schemes, while using nonrenewable groundwater in the oases in the south [32].

After independence, nationalist elites prolonged this hydraulic mission throughout northern Africa in order to alleviate rural poverty and fulfill the political goal of food self-sufficiency, as well as demonstrating the strength of the new states. Following the 1952 revolution, Nasser rekindled the idea of domesticating the Nile River with the High Aswan Dam (132 Bm^3), constructed between 1960 and 1970. The new dam completely eliminated the flood regime, allowed the wide-scale development of year-round cultivation, expanded the irrigated area by 0.5 Mha, and consolidated rice cultivation in 0.3 Mha. Dam building was also the basis for the promotion of irrigation in independent Morocco. In 1967, it introduced a vast hydraulic infrastructure program (Programme National de l'Irrigation, PNI) to bring one million hectares under irrigation before the year 2000, while setting an objective of "one dam per year" in 1986.

The management and execution of these hydro-social dreams were carried out by powerful hydraulic bureaucracies, moved by a civil engineering ethos and strong political support [33]. In Turkey, the General Directorate of State Hydraulic Works (Turkish acronym DSI) was created in 1954 on the model of the US Bureau of Reclamation. Similar administrations were created in other Mediterranean countries, such as Mekorot in Israel (1937), the Ministry of Agriculture in Tunisia (1958), the Ministry of Water Resources and Irrigation in Egypt, and the *Dirección General de Obras Hidráulicas* in Spain (1931). These administrative bodies shared their involvement in the development of water resources with four other categories of powerful actor, namely politicians, construction companies, landed elites, and development banks [34].

Since the 1980s, such agencies have been weakened by financial restrictions, neoliberal reforms, and environmentalist pressure [30]. Moreover, the hydraulic mission faced growing criticism among academics due to the impacts of dam policies (resettlement, sedimentary, and environmental issues), water quality degradation (pollution in rivers and wetlands), groundwater overexploitation, and a lack of participation. Despite such factors, and the challenge posed by climate change to water resource management, in recent decades, some countries, pressured by agricultural or construction lobbies and trapped in political inertia, have pressed ahead with (or retain on their agenda) large infrastructure projects. This is illustrated by Turkey's plan to bring 1.7 Mha under irrigation and to produce 27 billion kWh in its GAP project (Turkish acronym for Southeastern Anatolia Project) [35], while similar projects have been recently launched in other regions [36]. Irrigation expansion is also being pursued by the Morocco Green Plan (PMV) and Egypt's plan to reclaim 630,000 ha of desert land.

4.2.3 The emergence of a groundwater economy

Since the 1980s, irrigation development in the Mediterranean has increasingly been driven by the private exploitation of groundwater resources (Fig. 4.1), powered by hundreds of thousands of private wells and tube wells. This *groundwater economy* has often been encouraged by the state, as surface water resources had generally been fully mobilized already, through subsidies or a pragmatic/lenient policy of authorizing the drilling of wells. The wells and tube wells were installed inside existing public irrigation schemes, thereby compensating for the shortage of surface water. In Morocco, for example, it was

Fig. 4.1 From left to right and top to bottom: (A) Spillway of Ataturk Dam, Turkey (F. Molle); (B) Santillana Dam, Spain (Deensel); (C) Duckbill weir in a large-scale irrigation scheme in the Haouz, Morocco (F. Molle); (D) traditional flow partitioner of a *foggara*, Algeria (M. Herlem); (E) battery of pumps, Nile Delta (F. Molle); (F) well and on-farm reservoir for drip-irrigated extension, Zagora, Morocco (J.Y. Jamin); (G) plastic agriculture/drip irrigation, France (F. Molle); (H) Center Pivot irrigation, Aragon, Spain (F. Molle)

estimated that in 2012 over 180,000 ha within surface irrigation schemes were irrigated with groundwater [37]. In river basins like the Oum Er-Rbia, Tensift, or Souss, as well as in all public schemes of the Maghreb, Spain, southern Italy, or Syria (Ghab Valley), wells pepper the landscape, sometimes providing more water than is distributed by public canals (see Kuper et al. [38] for the graphic example of the Tadla scheme in Morocco). In other places, this phenomenon is more recent as, for example in the Nile Delta, where intensive well drilling has been observed in the past 10–15 years [39].

Wells have also proliferated in rainfed areas, where the supply of groundwater has allowed farmers to increase, and more importantly secure, their yields in times of drought. This has been observed in Morocco, Tunisia, and Yemen, but most prominently in the Beqaa Valley of Lebanon [40], the Orontes River basin of Syria [41], and most tellingly in Algeria, where the irrigated area increased from 228,000 ha in 1985 to slightly over a million ha in 2011, of which 88% depends on groundwater [37]. This trend developed at a time when state-led irrigation development was increasingly viewed as inefficient and costly and when surface water resources had been fully exploited. Structural adjustment programs and a neoliberal *zeitgeist* favored private initiatives and a shift toward intensive commercial agriculture. Groundwater has also been used to expand agriculture into arid or desert areas based on fossil water. Large agribusinesses in, for example, Algeria, Egypt, Jordan, and Morocco, as well as smaller local investors, have invested in tube wells to tap often fossil groundwater at great depths [42, 43].

In other words, most Mediterranean countries have witnessed a sharp increase in intensive groundwater use over the past 40 years. This has led to environmental degradation, including the consistent drop of aquifer levels, increased levels of salinity, and, more generally, the degradation of water quality in these aquifers, the drying of springs and rivers, and a subsequent loss of biodiversity [44]. Intensive groundwater use has also led to social inequity, as small-scale farmers were often unable to "follow" the decline of groundwater tables and were "pumped out" of the competitive agricultural markets of the groundwater economy [45].

The high cost of energy associated with groundwater irrigation is stimulating the expansion of solar-powered systems throughout the Mediterranean basin, which is perfectly aligned with recommendations for low-carbon irrigated agriculture. Egypt and Morocco encourage and even subsidize the expansion of photovoltaic technologies, and some Spanish regions have recently included such subsidies in their irrigation modernization programs. In some Spanish groundwater "irrigator communities," energy costs have dropped from €0.33 to €0.02 after installing the solar panels, a not-insignificant change in a context of decreasing fruit market prices. However, this has removed the economic restraint on groundwater depletion and enhanced "a strong risk of overexploitation, and even depletion of water resources" in some territories [46]. The difficulties that Mediterranean states have been seen to face in controlling groundwater expansion in recent decades do not augur well for water resource conservation if these solar-powered pumping systems proliferate.

4.3 Irrigation and the water cycle

4.3.1 Open and closed basins in the Mediterranean

The Mediterranean basin offers a striking contrast between its major rivers, fed by snowpack and/or rainfall on large mountain ranges, and its smaller coastal basins that are in general critically overexploited. The former category includes not only the Ebro (fed by the Pyrenees Mountains), the Rhône, and the Po (fed by the Alps) but also rivers like the Sebou and Moulouya in Morocco (fed by the Atlas and Rif Mountains) or the Ceyhan in Turkey. Smaller basins with extensive irrigation areas and other uses, such as the Jucar or Segura in Spain, the Chélif in Algeria, the Medjerda in Tunisia, the Guediz in Turkey, the Prinos in Greece, or the Orontes (Lebanon/Syria/Turkey), undergo permanent or periodic severe stress. The Nile River is a separate case that combines a massive average annual flow (~70 billion m^3 entering Egypt) with a situation of closure: although it still exhibits an outflow to the sea of around 12 billion m^3 per year, Molle et al. [47] have shown that most of this flow is needed to maintain the salt balance of both surface and groundwater and that potential "savings" are very limited.

What causes the closure of river basins (understood as a situation where unchecked net withdrawals exceeding demand, including that of ecosystems) for most of the year? The conventional script is that the main reasons for scarcity are population growth and climate change—a handy way to naturalize and depoliticize crises and to shifting the blame away from managers or decision-makers. This Malthusian view is conveniently echoed by the media and official reports and establishes itself as common knowledge. Dramatic dry spells periodically seem to reinforce this understanding.

While there is no question that such factors compound water shortages, the Malthusian dominant narrative is a convenient fig leaf for a process of basin "overbuilding" that has clear sociopolitical drivers that illustrate those identified in the literature [34, 48]. Large-scale public irrigation development has long been a favored policy of governments embracing the "hydraulic mission" for a series of ideological, economic, and political reasons, as recalled above. In many cases, that mission has been, and in many places is still being, continued far beyond economic rationality and what the availability of water would dictate. Moreover, states' encouragement of private groundwater exploitation has further compounded basin closure. Hydrologic realities are evaded or ignored, if not tampered with, to accommodate more water resource development. Climate change projections are also glossed over so as not to affect the scope for planning (e.g., in Morocco, Spain, and Lebanon).

The region has no shortage of examples of incremental but excessive water resource development, "optimistic" planning, and oversized-cum-idle new infrastructure. In Lebanon, for example, the government is pursuing the age-old project of developing the "Canal 800" irrigated area in the Litani Basin, despite its already critical state [40]. In Algeria, diversions to support failing irrigated areas are made based on water that is already allocated (see for example the Maqtaa Basin). Egypt is prey to the anarchical expansion of irrigation by both the state and individual farmers [42, 49].

Closure has also been driven by the chaotic spread of individual wells/tube wells described above. Fossil groundwater cases aside, groundwater usually flows to springs, wetlands, river beds, or the sea, and these fluxes are therefore intercepted and "diverted" by the wells before they naturally go back to the surface, thus contributing to overall basin closure.

Pressure over water resources in closed basins is often passed on to neighboring basins with more abundant resources, thus eventually contributing to the gradual closure of the latter. Algeria is probably the most striking example of a country commonly resorting to interbasin transfers (probably a reflection of its access to cash from the oil industry). Tunisia is also transferring an increasing share of the resources of its northern river basins [50], while France, Italy, Spain, Morocco, and Lebanon have implemented or are planning new water transfers [33]. Another way to augment supply is to reuse treated wastewater (see Section 6.3.3. and Chapter 5).

Basin closure also makes the interconnectedness of users across the hydrologic cycle more intense and visible. Since there is no "slack" in the system, and withdrawals or pollution at one point are not "dampened" by a larger water flow, externalities travel swiftly and water allocation increasingly resembles a zero-sum game. This question has been discussed at length, yet there is a ubiquitous misconception about how water can be "saved" in such circumstances. Improvements in water-use efficiency, for example through the use of drip irrigation—see next section—or the lining of canals, alter the pathways of water and generally impact return flow appropriators, unless these flows go to sinks and are unusable.

Potential water savings in the Mediterranean were estimated by UNDP Plan Bleu in the late 1990s at almost a quarter of the total water demand by 2025, with two thirds to be realized in the agricultural sector, and was taken as a regional target objective [51]. It was supposed to be achieved by raising average plot-level irrigation efficiency to 80%, reducing conveyance losses down to 10%, raising distribution efficiency in the domestic sector to 75%, and the rate of recycling in the industrial sector to 50%. More recent analyses have raised doubts about this flawed approach, found to be an "optimistic perspective" [52, 53].

4.3.2 Water-saving technology in irrigation. A second hydraulic mission?

The hydraulic mission was based on blind confidence in the capacity of technology to mobilize water resources, creating a perception of water abundance among managers and users. A similar sense of abundance prevails with farmers installing tube wells pumping groundwater that can run 24 h a day, 7 days a week. Basin closure processes uncovered the real limits of the resource's availability and paved the way for the renewal of water and irrigation policies. Hydraulic bureaucracies have resorted to technology to move these frontiers and to reopen the closed basins, by incorporating nonconventional water resources (desalinated, treated wastewater), transferring water between basins, or through the introduction of costly water-saving technology (pressurized systems) in the irrigation sector.

Due to its magnitude and impact, the introduction of pressurized irrigation might be the most significant technical change in the Mediterranean region since the Medieval Arab Agricultural Revolution [54]. Today, half of the irrigated land in the Mediterranean has introduced sprinklers or microirrigation (mainly drip systems) at farm level. Israel has pioneered this technological change (99.6% of the irrigated lands), whereas Spain has the largest pressurized area (2.6 Mha). Other European countries, such as France, Italy, and Greece, have a significant area with sprinkler irrigation (87%, 42%, and 53%, respectively), whereas drip irrigation is rapidly progressing in Morocco and Tunisia (20% and 45%). Although this process of technological change is probably reaching its ceiling in some EU countries, the largest irrigated areas of the Mediterranean (Egypt and Turkey) still have a significant potential (and plans) for the expansion of this technology (Fig. 4.2).

Mediterranean countries have bet on this technological shift. A second hydraulic mission has been launched over the last three decades aimed at renewing support for the irrigation sector through generous subsidies for the modernization of infrastructure [55]. A powerful discourse of modernity and efficiency, founded on scientific research and disseminated by international agencies, has permeated the different national irrigation plans in the region. Italy, Morocco, and Tunisia have subsidized up to 100% of the cost of the implementation of pressurized technology at farm level, replacing open canal systems, or even encouraging the opening of new irrigated areas using microirrigation. Spain has a double level of subsidies, depending on the administration involved (national or regional), ranging between 40% and 100%, and various financing plans. Both Morocco and Spain plan to keep this support in the future, increasing the pressurized area by 0.55

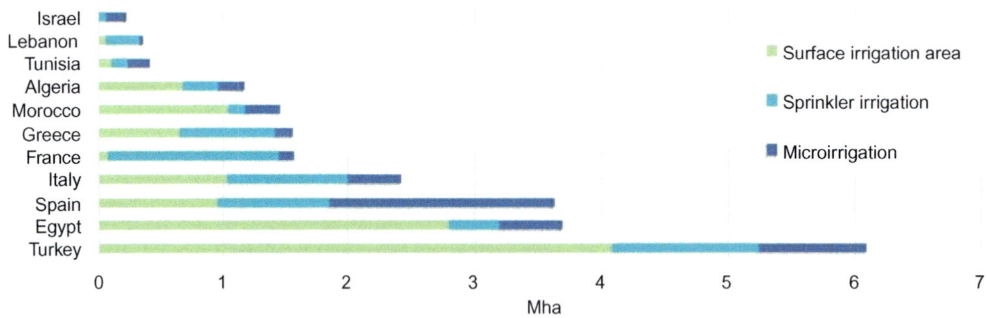

Fig. 4.2 Irrigation techniques in selected Mediterranean countries. *Sources: FAO, ICID, DSI, EASAC, and EUROSTAT.*

and 0.8 Mha, respectively, according to the *Programme National d'Économie d'Eau en Irrigation* and the *Estrategia Nacional de Regadíos 2018–2025*. In the Irrigation Plan of 2014 and the NWRP, Turkey and Egypt combine subsidies for pressurized networks with massive investment in infrastructure rehabilitation, canalette renewal, pump stations, and concrete lining of earthen canals [55].

Although this policy is generally welcomed by farmers, WUAs, and the irrigation industry, their respective objectives are not necessarily the same. In many regions, the adoption of sprinkler and microirrigation owes little to water-saving goals. These techniques provide other advantages well valued by farmers, such as a significant reduction in labor costs or increased crop production. Tunisia boasts impressive increases in yields and water productivity (70%) associated with drip adoption [32]. Such factors have also stimulated the adoption of pressurized irrigation in numerous areas of the EU, where increased yield and uniformity of fruit and vegetables (an important standard of supermarket demand) prompt the introduction of these technologies. The dominance of these other incentives creates a double discourse, whereby the administration and the boards of the WUAs justify the promotion of pressurized irrigation with reference to water-saving purposes, while farmers frequently pursue the installation of the systems for labor-saving and productivity [56, 57].

The discourse deployed by governmental agencies has been contested by recent research on the conceptualization of water-use efficiency [58, 59]. The promised efficiency of drip irrigation has been questioned on various grounds [56, 60]. First, far from experimental plots, on-farm studies have observed that irrigators do not always reduce the amount of water used in their fields. For example, Benouniche et al. [61] show for a case in Morocco that only about one third of the sample farms achieved the promised efficiencies of 80%–90%. All other farmers overirrigated, either to ensure good yields or due to poor-quality equipment. More than one third of the sample farms showed efficiencies comparable to gravity irrigation (50%–60%). Second, a decrease in water application does not necessarily translate into a decrease in water consumption, particularly where drip irrigation largely increases evapotranspiration [59]. Third, the decrease in water use stimulates the reallocation of the water theoretically saved to: (i) other purposes in the same basin (urban, industrial, environmental, or agricultural); (ii) water transfers to neighboring regions; (iii) an intensification of the crop pattern; (iv) a shift to more productive and water-intensive crops or varieties; (v) an extension of the irrigated area [62]. These features, sometimes combined, have been observed in several countries, such as the United States, Tunisia, Iran, China, Spain, and Morocco [63, 64]. In these cases, the adoption of drip irrigation can result in the well-known rebound effect at basin scale [65] that further increases water demand and consumption. The frequent consequence of modernization, that is, irrigation expansion, is not unavoidable, as can be seen in some specific geographical or socioeconomic contexts [66, 67]. The subsidized adoption of drip should come with reductions in water rights and the prohibition of expansion

[65, 68], but Mediterranean states have so far largely turned a blind eye to this question, favoring productive growth over sustainability.

In areas with no irrigation infrastructure, drip is frequently the vector of irrigation expansion, despite often being considered a demand-side measure [54]. In the olive groves of Andalusia (Spain), irrigation has significantly increased as a strategy to eliminate alternate bearing to increase yield. In the vineyards of southern France, irrigation has tripled since 2000, stimulated by subsidies, market requirements, and increasingly irregular climate, despite a decline in the total vineyard area [69]. On occasion, the state (e.g., Spain, Egypt) has stimulated the process by making the construction of new wells for the expansion of irrigation into rain-fed areas conditional upon the installation of pressurized distribution networks [70].

Pressurized irrigation can cause significant increases in energy costs, as reported in several regions of Spain and Morocco, particularly those areas using surface water [71–73]. In groundwater-fed areas, the cost can sometimes be compensated by the reduction in water abstraction and the fact that pumping already generates pressurized water. Yet in many cases, farmers have an intermediary storage facility first filled with groundwater from which water is pumped again to the drip system. Further, in areas using both groundwater and surface water for irrigation, groundwater withdrawals may decrease or stop when drip irrigation is introduced, because, in the absence of irrigation extension, the use of this technology reduces the use of water in the irrigated area, and farmers can meet the water demand (totally or partially) with the (cheaper) surface water [66, 74].

The environmental impact of these technological changes on Mediterranean ecosystems is difficult to assess because of the large number of factors involved, including the promotion of intensive agriculture and the use of chemical inputs associated with pressurized irrigation. However, an improvement in water quality outputs has been reported in Spain. After the implementation of pressurized irrigation, the mass of salts and N exported from the irrigated land to receiving water bodies downstream significantly decreased, without an increase in NO_3 or salt concentrations observed so far [75, 76]. However, the reduction in return flows can reduce the inflow to rivers or associated wetlands, altering their hydrological balance and natural habitats.

Microirrigation may also incur a loss of cultural heritage in some ancient hydraulic systems. Historical artifacts are replaced by pressurized devices, which render water flows "invisible" and rapidly deteriorate after being abandoned. Norms and practices are also altered, with traditional farming knowledge being replaced by engineering principles and centralized irrigation, water management, and decision-making. However, the processes of "bricolage" observed in different contexts show that farmers adapt the technology to their needs, and have, in some cases, encouraged young rural people to engage with farming [57, 77].

4.3.3 Nonconventional water resources for irrigation

The use of nonconventional water resources, such as desalinated water or treated wastewater, has paved the way for continued supply-side policies in some Mediterranean regions. These options provide an opportunity to reopen closed basins, expanding the offer by resorting to technologies that are easily embedded in greening discourses. Tunisia has long been a pioneer in this field and started reusing Tunis wastewater in the 1960s, but Israel is now at the vanguard of this technological change, integrating desalination and wastewater reuse (the latter currently supplying 45% of the water used in agriculture) [78]. Turkey had 200,000 ha irrigated with wastewater in 2004, and Morocco, Spain, and Algeria are developing plans to further exploit these resources to supply their irrigation systems (see Chapter 5 for a full treatment of this issue). In North Africa, however, wastewater is often used regardless of whether it has been treated and the necessary treatment does not always increase the available resource.

Both desalination and treated wastewater have problems that hinder their on-farm application. Treated wastewater requires strict quality control, particularly for the cultivation of vegetables, and increases soil salinity. Desalinated water has a high concentration of Boron and lacks certain ions that are important for plant growth. Its significant energy costs have negative impacts on GHG emissions and water prices. In numerous cases, both desalination and wastewater reuse have had an additional cost, since it is necessary to pump the water from desalination or treatment plants to the irrigation areas.

As is illustrated by the Israeli case [27], expensive desalinated water is mainly and massively used for urban supply, while the wastewater from these areas is the main source for the agricultural sector. Urban dwellers have the financial capacity to absorb the cost of desalination, allowing farmers to receive wastewater/freshwater at a reasonable price. Spain has developed a significant desalination capacity (31 plants with the capacity for an annual supply of 713 Mm^3), but it is still underused. The majority of the desalination plants were built as a political response to the cancellation of the Ebro Water Transfer, with no prior commitment from irrigation communities to purchase or use the water produced. Given its high cost, desalinated water is only regularly used in the intensive horticulture of Murcia and Eastern Almeria [79] (notwithstanding the plant under construction in southern Morocco), and is partially subsidized by the state during drought periods [80]. In these areas, farmers are planning two new, solar-powered desalination plants in order to offset the energy cost barrier.

4.3.4 Adaptation to climate change: anything but degrowth

The Mediterranean basin has been identified as one of the world's most vulnerable regions to climate change [81]. According to various IPCC scenarios, in the coming decades, mean temperatures will increase, particularly in summer; heat waves will be more frequent; mean rainfall will decrease; and drought periods will increase in frequency and intensity [82–84]. As a result, both evapotranspiration and water resource availability are likely to be significantly affected, worsening the conditions in which irrigation is managed and developed throughout the region. If the irrigated area remains unchanged, demand is expected to increase between 4% and 18% by the end of the 21st century, depending on the temperature rise—between 2°C and 5°C [85], although assessments based on canopy temperature temper these values [86]. This has critical implications for agricultural water management.

Most expected trends have already been observed in some Mediterranean countries. Since the 1960–70 period, the western Mediterranean has undergone a 1–1.5°C increase in mean temperature [87, 88], and the eastern Mediterranean has an average summer temperature 1.5° higher. Heat waves have broken extreme temperature records in numerous stations since the beginning of the 21st century, and droughts have occurred more frequently and intensively since 1950 [89–91]. Rainfall is expected to decrease 4% for each 1° of temperature rise in the region [92], and observed trends are consistent with these values (albeit with some regional specificities). Since 1970, rainfall markedly decreased in central Spain but remained stable on the Mediterranean coast, while, in Italy, it slightly increased in the Adriatic but moderately decreased in the Tyrrhenian regions. Rainfall decrease is still moderate, but there is a more significant reduction in the number of rainy days and an increase in torrentiality [93].

Therefore, climate change is expected to compound the structural water scarcity of numerous Mediterranean basins. These trends have been analyzed and a panoply of measures proposed and assessed. Frequently mentioned for the farm level are increasing irrigation efficiency with new technologies and improved management, altering crop patterns and rotations, and increasing storage capacity, alongside shifting to organic farming, introducing agricultural insurance, and diversifying household income [94–96]. At the same time, water augmentation policy measures, such as the use of nonconventional water resources (treated wastewater, desalinated water), interbasin transfers, or the conjunctive use of surface and groundwater, are repeatedly recommended as a means of meeting (irrigation) water demand at the national or regional scale. Such adaptation measures, both at farm and resource system levels, are already being applied or projected in numerous regions to cope with recurrent droughts or structural water deficits. Climate change is stimulating a rebranding and greening of the hydraulic mission, relaunching old, capital-intensive infrastructure, and paving the way for the dissemination of new "water-saving" technologies.

Indeed, in most Mediterranean countries, the promotion of microirrigation is the main policy put forward: the message that managers should look at actual water consumption (i.e., evapotranspiration) rather than withdrawals is understandably unpalatable to decision-makers seeking technological fixes. France has a policy of enforcing "ecological continuity" and environmental flows, but pressure from the agricultural sector has recently reopened the possibility of small- and medium-scale storage infrastructure, despite opposition from the Ministry of Environment and green NGOs. In some other countries, like Lebanon and Turkey, the expansion of irrigation is still planned or under construction [33]. Despite scientific evidence and wide media coverage that climate change is one of the most critical challenges facing the irrigation sector in the Mediterranean region, for many irrigators it ranks lowest among their concerns (see Richard-Ferroudji et al. [97] for France, and Ortega-Reig et al. [98] for Spain). They have already had to cope with water scarcity over recent decades and feel primarily vulnerable to market and sociopolitical uncertainties. Climate change, drought, and scarcity are concepts that farmers find hard to distinguish, especially when they are already struggling with insufficient water for irrigation.

4.4 Modes of water governance

4.4.1 Water, irrigation, and the state

Many definitions of water governance that have gained currency [3] and have often been taken on wholesale by countries emphasize issues of participation, transparency, trust, and engagement. Nonetheless, water governance in Mediterranean countries remains very state-centered. European countries arguably differ in that respect because of the gradual (yet still limited) progress made in terms of stakeholder involvement, largely due to the implementation of European regulations (such as the Water Framework Directive).

"Good governance" is generally construed as ensuring the conditions that enable the financial sustainability of the sector and effective Integrated Water Resource Management. State-centered governance is therefore characterized by the distribution of roles across the water-related government agencies, the (formal) laws and regulations, and by the kit of policy tools that is adopted. Most countries have passed water laws, which reflect many of the standard principles and "good practices" promoted internationally. Yet these principles, for example "user-pay" and "polluter-pay," often remain dead letter and are not, or not fully, implemented. This is particularly true of formal groundwater regulation in the Middle East and North Africa (MENA), which was "found to have minimal influence over water use policy and decision-making" [99]. Laws can sometimes be passed under a degree of pressure from international partners, who make loans or other agreements conditional upon policy changes; for example, with the Code de l'eau finally adopted by Lebanon's parliament in 2018 after almost 15 years of negotiation [40]. The "Europeanization" of EU member states, with influence reaching toward Turkey, has also driven many policy and legal changes. Noticeable progress has been made in European countries in recent decades, particularly since the implementation of the Water Framework Directive.

The distribution of roles within government agencies, and possibly elsewhere, is a key point of IWRM-inspired and state-centered governance structures. Ideally, a high-level committee would ensure the coordination of policies between the sectors having a bearing on water resources. Morocco, for example, has an interministerial committee and a Conseil Supérieur de l'Eau et du Climat (CSEC), but they have been deemed largely ineffective by the Conseil Économique, Social et Environnemental (CESE). The CESE [100] also pointed to many institutional dysfunctions within and between ministries, failures to properly implement the 1995 Water Law (e.g., with regard to the polluter-pay principle, safeguard and prohibition zones on overexploited aquifers, sanctioning, and so on), and insufficient coordination between departments concerned with water.

Policy contradictions between sectors, or a "lack of alignment," are probably a common feature of water governance in all countries [101]. According to the GWP [52], there is, in the Arab world, a "lack of integration between sectoral water-related policies which leads to fragmented programs and inefficient utilization of technical capacities and financial resources," and limited awareness of environmental issues. This speaks to the weakness or absence of a regulatory body to ensure the compatibility of multiple water uses with both the available resource and the health of aquatic ecosystems. The crux of water governance is the competition, and the consistency required, between agricultural uses, domestic and industrial uses, hydropower, recreational uses in certain cases, and environmental needs. In many countries, the promotion of commercial agriculture by lobbies combines with the political strategic importance of the sector for poverty alleviation, food security, and social stability to support irrigation development beyond hydrological/ecological limits. The regulatory body, where it exists, is often entrusted to the administration responsible for the environment, itself either confined to a weak ministry (Egypt, Lebanon, Turkey, etc.) or placed under a strong sectorial ministry (e.g., Public Works or Water and Energy). How the administrations in charge of water, energy/hydropower, irrigation, agriculture, and the environment feature in the government's flowchart says little about their respective overall influence. What matters is how actual bureaucratic and political power permeates and flows though this chart [54].

In all countries, the domestic (and often industrial) use of water is given priority. The most ubiquitous policy contradiction is where the water or environmental administration is pitted against that of agriculture/irrigation, both vying for the remainder of the resource. In Morocco, the Plan Maroc Vert (PMV) under the Ministry of Agriculture promotes the expansion and intensification of irrigated agriculture, even in areas where water tables are already dropping by 1 m a year due to overexploitation. In Egypt, the Ministry of Land Reform and Agriculture promises not only to expand desert agriculture by 600,000 ha but also to "modernize" agriculture in the "old lands" and to save 12 Bm^3, despite the Ministry of Water Resources and Irrigation demonstrating the impossibility of such a target based on a faulty up-scaling of theoretical improvements in plot-level efficiencies [34]. EU policies have also at times extended subsidies promoting water-intensive crops.

Excessive irrigation development comes with a lack of integration of environmental dimensions. Environmental impacts include that of stream and aquifer overexploitation on river base flows (e.g., Spain and Greece), or springs and wetlands (Azraq Oasis, Jordan;

Marrakech Palm Grove, Morocco; Ichkeul Lake, Tunisia; Konya Lake, Turkey; Tablas de Daimiel, Spain; Lake Karla or Argolis, Greece, etc.), and coastal areas (saline water intrusion in all countries). Misalignment between sectors does not have a rational bureaucratic solution, since it "reflects turf battles between ministers and ministries in the attempt to gain a bigger slice of the state budget and maximize bureaucratic or political power" [54].

While emphasis is often placed on the state and its policies, a situation of legal pluralism (regarding both land and water resources) frequently prevails in the region, notably its southern/eastern half. Thus it is important to focus not on formal regulations or informal practices alone, but rather on the multiple interactions between them [102, 103]. Legal pluralism means that access to land is often mediated by rules that reflect various, often overlapping, forms of tenure and ownership, while access to water resources, especially those managed by private or community initiatives, is split between social rules and perceptions and variable state legitimacy.

4.4.2 Water user associations and participatory water management

Participatory language now permeates all laws and policy documents. However, the participation of users in water management often remains minimal (although in Europe the Water Framework Directive has initiated a gradual change in this respect).

4.4.2.1 Irrigation schemes

As mentioned earlier, the Mediterranean has a long history of community-based irrigation systems, largely developed endogenously, on which "modern" associations with varying degrees of autonomy (higher in Spain than in France, for example) have been built. These include around 500 Italian Reclamation and Irrigation Consortia (*Consorzi di Bonifica e Irrigazione*), associations of landowners involved in collective water and land management, approximately 2000 irrigation user associations (ASAs) in France, managing over 300,000 ha of irrigated networks of all sizes, some dating back 500 years [104], and numerous irrigation communities (*comunidades de regantes*) in Spain. Particularly in Spain, these represent a powerful lobby in terms of water policy at the national level, actively participating in discussions around basin-level water planning, drought crisis, and interbasin water transfers.

In 2000, French and Spanish organizations launched the Euro-Mediterranean Irrigators Community (EIC)—a nonprofit association created to facilitate exchange of experiences and to act as a stakeholder at the European Commission in the Strategic Coordination Group of the Water Framework Directive. A total of 39 associations and national federations from nine Mediterranean countries (plus Germany) are currently members of this organization.

Notwithstanding their equally rich history of communal irrigation, water user associations (WUAs) established by the state in southern and eastern Mediterranean countries, deliberately ignoring existing irrigation communities (*Jmâa*), have been beset by a number of difficulties and constraints [105]. In Morocco, the lack of political will to empower associations is apparent in the law on WUAs, which includes a long list of duties ascribed to farmers, while the benefits are minimal to nil. In Tunisia, 1253 GDAs are responsible to a varying extent for a total of 213,000 ha of public irrigated area. While GDAs have more autonomy than Moroccan WUAs, their technical and financial management of irrigation networks, and their performance in terms of service quality, varies widely. Problems such as water shortages, insufficient funds for major repairs, and elite capture or political meddling have triggered successive reforms but without eliminating the problems. In Egypt, various experiments with WUAs from the tertiary to the district level, generally part of donor-funded projects, have failed to institutionalize a degree of participatory management [106]. In Turkey, in the early 1990s, the DSI started to transfer O&M responsibilities to a variety of collectives according to a model of irrigation management transfer (IMT) inspired by the Mexican example and largely motivated by an intent to externalize costs onto users.

In the southern countries, the overall record of such experiences is unconvincing. States (or their water bureaucracies) have dragged their feet and shown reluctance to devolve or share power with users [105]. The financial autonomy of WUAs has been a major problem, notably with regard to the long-term maintenance of key infrastructure.

4.4.2.2 Collective management of groundwater

Some farmer groups have been managing collective wells. In Turkey, groundwater irrigation cooperatives are responsible for 74% of the groundwater used for irrigation, over approximately 684,000 ha [36] and so are a portion of GDAs in Tunisia. Egypt and Morocco also have associations and cooperatives managing collective wells [107]. Spontaneous private initiatives by collectives to invest in shared wells can also be found in most countries [107]. In Europe, Spain stands out as having the highest occurrence of groundwater collectives [108].

Managing a collective well is one thing, but collectively managing an aquifer is a far more daunting challenge. A few, largely state-led, initiatives aimed at establishing a degree of comanagement are of particular note, although their success has so far been limited. In Bsissi, Gabès Governorate, Tunisia, a negotiation was started between a group of around 100 groundwater users and the local administration (CRDA) after tension erupted when the CRDA decided to close illegal wells in the area. The CRDA agreed to legalize illegal wells and connect them to the electricity grid (lowering pumping costs), as well as to subsidize the shift to microirrigation. In exchange, farmers would help to enforce the drilling ban and assist the CRDA in closing disused or abandoned wells [109, 110].

Morocco has been experimenting with "aquifer contracts" since 2004 in the Souss-Massa River Basin. There, too, the administration proceeded to close illegal wells, triggering social unrest and the launching of a consultation process [111]. An agreement was signed in 2007, whereby the state promised to build new reservoirs, study the possibility of desalination and wastewater treatment, subsidize a shift to drip irrigation, and regularize "'illegal" wells, against a freezing of the expansion of irrigated areas for citrus and vegetables, an increase in groundwater user fees, and the reinforcement of the water police. This contract was not properly enforced as the tension was diffused after a few years of good rainfall. *Contrats de nappe* were also considered in the Tensift Basin and Saïss Plain, but, in the latter, it resulted in supply-augmentation projects rather than a comanagement initiative [112].

In the highlands of Jordan, the Ministry of Water and Irrigation, supported by the German Cooperation Agency (GIZ), convened a "Highland Water Forum"—a multi-organizational dialogue mechanism with 60 stakeholders from the area (agricultural water users, government institutions, NGOs, and research institutions) [113]. But after 3 years and the production of an action plan, the forum's secretariat was shifted under the ministry, showing the ministry's resolve to keep control of the process rather than delegate power to users.

In France, *contrats de nappe* between the administration and local users organized in OUGC (*Organisme unique de gestion collective*), generally under the aegis of the chamber of agriculture, have been established in areas with overexploited aquifers with a view to restoring a balance between abstraction and availability of the resource. Experiences in the Clain and Beauce aquifers, for example, show a slow and gradual institutionalization of comanagement and the implementation of regulatory measures [114]. Spanish law has created *Juntas Centrales de Usuarios* or *Comunidades Generales* in those aquifers declared "overexploited". These institutions replicate the structure of the irrigator communities but incorporate all the individual or collective users of one aquifer (urban or agricultural, supply companies or irrigator communities). Users are thus forced to cooperate to handle the common resource and agree on an exploitation plan. While not a panacea, some positive experiences have been observed in terms of water resource conservation and water governance [108].

4.4.2.3 River basin management

The organization of water management at the river basin level has been disseminated as a global "good practice," reflecting IWRM principles. In EU countries, it has been enshrined in the Water Framework Directive (WFD). In Spain, *Confederaciones Hidrográficas* have been active for a century but with significant differences in their governance. The *Confederaciones Sindicales Hidrográficas* created in 1926 had a radical participatory structure, under the control of an Assembly of Users, but were suppressed and subdued under the rigid control of the engineers of the Ministry of Development during Franco's regime. Shortly after the return to democracy, the *Confederaciones* partially recovered some participatory functions, which subsequently fulfilled WFD requirements.

In France, the WFD's translation into national law entrusted the *Agences financières de bassin* (a type of river basin organization set up in 1964) with new missions aimed at the wider "good status" of water bodies. *Autorità di Bacino* (River Basin Authorities) and River Basin Districts in Greece have also developed river basin plans to rationalize allocation and ensure environmental flows; there is limited information available on their effectiveness.

Under a typical process of policy transfer, the injunction to form River Basin Organizations has spread globally and to countries like Morocco, Algeria, and Turkey in particular [115]. A study of the Tensift Basin, Morocco, has found that the basin agency had insufficient means to pursue its mandate and limited autonomy, the ministry having retained the decision-making power on most crucial issues [100, 116]. In Lebanon, the Litani River Authority struggles to maintain its power and has few features of a river basin management authority [40]. In Algeria, the ABHs (Agence de Bassin Hydraulique) are project agencies rather than regulators.

4.4.2.4 A trend toward corporatization

European Mediterranean countries have a decreasing number of aging farmers and an increasing mean landholding size, showing some land concentration. Where land is still available, agribusiness companies, capital ventures, and sometimes small entrepreneurs have expanded dramatically. In Egypt, 75% of all desert reclamation (for agriculture and urban projects) has relied on private investors and corporate modes of farming [117]. These capitalist ventures have in general benefited from generous government subsidies and financial incentives, and/or been encouraged by permissive state regulations and a lack of control of groundwater abstraction.

Egypt's desert development projects since the 1950s have included the development of agriculture outside of the existing oases as part of the "New Valley" project in the western desert. The government reclaimed thousands of hectares for irrigation near oases such as Kharga and Dakhla, and since the 1990s, further south near the border with Sudan (East and West Oweinat). Sometimes, public wells have been handed over to groups of farmers who manage them; sometimes the state is the direct beneficiary through parastatal companies or, more directly, the army [42, 117]. Although state-sponsored or private agribusiness dominates these kinds of enterprises across the region, there are also cases where small-scale investors are able to develop small- to medium-scale projects in the desert or arid lands, based on one or a few wells. This is the case in the highlands of Jordan, in Algeria, and in Morocco, and on the fringes of the Nile Delta. Amichi et al. [43] show in the case of intensive agricultural development in Algeria's Sahara that enabling conditions generally include road accessibility and relative proximity to urban centers or export facilities, modes of land tenure that make claims/access to land possible, and the possibility of obtaining well licenses (or authorities turning a blind eye to groundwater development).

These formulas also thrive in the coastal *plasticulture* landscapes (Albenga and Vittoria in Italy, Demre and Kumluca in Turkey, Hamla in Lebanon, Tahir in Algeria, and Tartus in Syria), where vegetables and fruits are intensively cultivated in greenhouses with drip lines and numerous workers, frequently in harsh and precarious conditions. Some of these areas have been developed by local investors, such as Biskra's *plasticulture* in Algeria's Sahara or Almeria's *Sea of Plastic* (Spain), where the initial investment of medium and small holders in the 1980s has morphed into an agro-technological cluster, where companies of very different sizes coexist and with a trickle-down effect on the regional economy. In other plasticulture clusters, such as the Souss-Massa (Morocco), foreign corporate capital has had a critical role in the expansion of intensive agricultural models, also with important impacts on regional incomes, but, as in Almeria, some associated social and environmental negative externalities.

Private companies are progressively penetrating the irrigation sector through other strategies, such as public-private partnerships (PPPs) or outsourcing contracts. In Guerdane, PPP procedures facilitated and subsidized access to land and water resources to foreign investors. While some farmers benefited from the project, presented as a success story by international development banks, most small-scale farmers have experienced water, fertile land, and development dispossession [118]. In Spain, private companies frequently penetrate the irrigation sector through Build-Operate-Transfer contracts. They either acquire irrigated areas under major water infrastructure projects or enter the sector when water user organizations involve a company to implement drip irrigation systems as part of modernization projects. The role of private companies in the market ranges from the mere outsourcing of specific tasks formerly carried out by farmers or hired staff, to the complete control of water management and billing procedures. This has created various hybrids between private companies and irrigator communities, in which the social control of water and collective decision-making processes can be severely weakened [119, 120].

4.5 Conclusions

Over the last millennium, water has been abstracted, stored, conveyed, and distributed through the fields of the Mediterranean basin by means of simple but ingenious technical devices, creating an array of varied waterscapes, valuable hybrids of nature, and culture. These historical socio-ecological systems, whose longstanding existence makes them a paradigmatic example of sustainability, currently face critical challenges due to the pressures of urbanization, fragmented smallholdings, and market volatility. These factors are compounded by the competition for water and other resources between modern irrigation systems and other sectors.

Mediterranean irrigated landscapes changed radically in the course of the twentieth century. The irrigation area doubled between 1900 and 1960, from 5.8 to 12.236 Mha, and doubled again just before 2000 (25.3 Mha), led by Spain between 1960 and 1980, and Turkey between 1980 and 2000 [20]. Growth in the irrigated area has been paralleled by an increase in dam storage capacity and groundwater abstraction. Hydraulic bureaucracies have placed irrigation development at the center of nation-building projects, guided by political support and top-down governance frameworks. While the hydraulic mission has succeeded in many Mediterranean regions in terms of sustaining rural development, food production, and exports, it has ultimately resulted in a process of basin closure, worse impacts of recurrent drought, and severely damaged aquatic ecosystems. This evidence, and the emergence of the sustainability paradigm, has caused a shift toward demand management. However, although widely promoted in national and international water policy documents, this approach has come up against hydrologic complexities and is often confined to contested "water-saving" measures.

Pressurized irrigation has been the most significant materialization of this policy shift. Microirrigation and sprinkler networks have rapidly expanded since the 1980s, as a supposed win-win alliance between farmers searching for improved production and reduced labor costs and the administration aiming to reduce water withdrawals, with the logical support of industry and agribusiness. With around 9 Mha installed in the Mediterranean over the last 40 years, the generalization of pressurized systems is the most significant recent change in irrigation technology. However, recent research has revealed that real water-savings at basin scale were rarely achieved [60], and, above all, that savings from reduced withdrawals/application were used to meet other demands, expanding irrigation or otherwise.

The shift to demand management has most often been partial or rhetoric. Large development or interbasin transfer projects are still being planned or in progress; the groundwater rush continues with deeper drilling and the use of fossil aquifers and is likely to be stimulated in the immediate future by solar energy. At the same time, desalination and wastewater recycling are surfacing as a technological fix to overcome—or further push—the limits of closed basins. Continued artificialization of the water cycle is pursued by hydraulic administrations that follow neoMalthusian principles and/or respond to political pressures from agricultural or other lobbies. Environmental factors continue to receive little attention, despite the WFD compelling EU countries to give increased consideration to ecological status and enforce environmental flows. The lack of coordination and integration of the different administrations involved in water management in each of the Mediterranean countries generates policy contradictions, hinders the adoption of environmental management principles, and sustains path-dependent development policies. As climate change looms on the horizon, current irrigation policies leave Mediterranean countries ill-prepared for the challenges ahead.

References

[1] A. Hamdy, M. Abu-Zeid, C. Lacirignola, Water crisis in the mediterranean: agricultural water demand management, Water Int. 20 (4) (1995) 176–187.

[2] G. Holst-Warhaft, T. Steenhuis, Losing paradise: The Water Crisis in the Mediterranean, Ashgate Publishing Limited, Farnham, 2012.

[3] P. Rogers, A.W. Hall, Effective Water Governance, TEC Background Papers 7, Global Water Partnership Technical Committee, Stockholm, 2003.

[4] J. Bisson, Mythes et réalités d'un désert convoité: le Sahara, L'Harmattan, Paris, 2003.

[5] C. Fassò, l'Irrigazione in Italia: ieri oggi e domani Conference proceedings I campi hanno sete, Associazione Nazionale Bonifiche ed Irrigazione and Food and Agriculture Organization, Roma, 2003.

[6] S. Asins, Linking Historical Mediterranean terraces with water catchment harvesting and distribution structures, in: J.P. Morel, et al. (Ed.), The Archaeology of Crop Fields and Gardens. Proceedings of the 1st Conference on Crop Fields and Gardens Archaeology, Bari, Edipuglia, 2006.

[7] A. Graziadei, Traditional irrigation in the shaping of cultural landscapes. The case study of Tricarico Southern Italy, in: Sanchis-Ibor, et al. (Ed.), Irrigation Society and Landscape, Tribute to Tom F. Glick, UPV, València, 2014, pp. 869–893.

[8] M. Antequera, Las galerías drenantes en el sector oriental y suroriental de la Península Ibérica. Identificación análisis y gestión patrimonial, PhD thesis, Universitat de València, 2015.

[9] E.W. Coward, Irrigation and Agricultural Development in Asia: Perspectives from the Social Sciences, Cornell University Press, Ithaca, NY, 1980.

[10] E. Ostrom, Institutions and common-pool resources, J. Theor. Polit. 4 (3) (1992) 243–245.

[11] N. Bouderbala, J. Chiche, A. Herzenni, P. Pascon, La question hydraulique v. 1: Petite et moyenne hydraulique au Maroc, Rabat, (1984).

[12] L. Bekkari, Yépez del Castillo I. L'appropriation du modèle d'association d'usagers de l'eau par une communauté villageoise du Moyen Atlas au Maroc, Cah. Agri. 20 (1–2) (2011) 73–77.

[13] J. Riaux, State Rules—Community Rules: The Local Governance of Water. Compared Anthropology of Two Ancient Irrigating Systems in Context of Public Intervention: The Ait Bou Guemez valley High Atlas Morocco and the plain of Vinça French Pyrenees Social Anthropology and ethnology, Ecole des Hautes Etudes en Sciences Sociales, EHESS, https://tel,archives-ouvertes, fr/tel-00420606, 2006.

[14] C. Sanchis-Ibor, Les terres de l'Horta de València. Crònica de la recent reducció superficial del regadiu històric, Afers Fulls de Recerca i Pensament 47 (2004) 11–128.

[15] R. Courtot, La huerta de Valencia, ¿Qué porvenir? in: J. Romero, M. Francès (Eds.), La Huerta de Valencia. Un paisaje cultural de futuro incierto, PUV, València, 2012, pp. 23–32.

[16] P. Pascon, Le Haouz de Marrakech, Éditions marocaines et internationals, Haouz (Morocco), 1977.

[17] S. Idda, B. Bonte, H. Mansour, S.A. Bellal, M. Kuper, Monument historique ou système bien vivant? Les foggaras des oasis du Touat Algérie et leur réalimentation en eau par pompage, Cah. Agri. 26 (5) (2017) 55007.

[18] C. Leibundgut, I. Kohn, European traditional irrigation in transition part II: traditional irrigation in our time—decline rediscovery and restoration perspectives, Irrig. Drain. 63 (2014) 273–293.

[19] J. Costa, La fórmula de la agricultura española, Imprenta de Fortanet, Madrid, 1912.

[20] K. Freydank, S. Siebert, Towards Mapping the Extent of Irrigation in the Last Century: A Time Series of Irrigated Area per Country, University of Frankfurt, 2008.

[21] T. Allan, The Middle East Water Question: Hydropolitics and the Global Economy, Tauris Academic Publications, London, 2001.

[22] E. Lopez-Gunn, 'Agua Para todos': the new regionalist hydraulic paradigm in Spain, Water Altern. 2 (3) (2009) 370–394.

[23] E. Swyngedouw, Modernity and hybridity: nature, regeneracionismo and the production of the Spanish waterscape 1890–1930, Ann. Assoc. Am. Geogr. 89 (3) (1999) 443–465.

[24] S. Misiani, C. Gómez, Construyendo la nación: Reforma agraria y modernización rural en la Italia del siglo XX, Prensas universitarias, 2017.

[25] M. Bandini, La "bonifica integrale" en Italia. La reforma del agro en Italia, Publicaciones del Inc., Madrid, 1952.

[26] E. Novello, La bonifica in Italia: Legislazione credito e lotta alla malaria dall'Unità al fascismo, Franco Angeli, Milano, 2003.

[27] E. Feitelson, The four eras of Israeli water policies, in: N. Becker (Ed.), Water Policy in Israel: Context Issues and Options, Springer, New York, 2013, pp. 15–32.

[28] E. Feitelson, A. Seltzer, R. Almog, Water history facets of landscape change in Israel/Palestine 1920–1970: a question of scale and periodization, Water Hist. 6 (2014) 265–288.

[29] K. Wittfogel, Oriental Despotism: A Comparative Study of Total Power, New Haven, 1957.

[30] F. Molle, P.P. Mollinga, P. Wester, Hydraulic bureaucracies and the hydraulic mission: flows of water flows of power, Water Altern. 2 (3) (2009) 328–349.

[31] P. Préfol, Prodige de l'irrigation au Maroc: le développement exemplaire du Tadla 1936–1985, Nouvelles Editions latines, Paris, 1986.

[32] A. Hamdane, Tunisia, in: F. Molle, C. Sanchis-Ibor, L. Avella (Eds.), Irrigation in the Mediterranean: Technologies Institutions and Policies, Springer, New York, 2019.

[33] F. Molle, C. Sanchis-Ibor, L. Avella Reus (Eds.), Irrigation in the Mediterranean: Technologies Institutions and Policies, Springer, New York, 2019.

[34] F. Molle, Nirvana concepts narratives and policy models: Insight from the water sector, Water Altern. 11 (2008) 131–156.

[35] L. Harris, Water and conflict geographies of the Southeastern Anatolia Project, Soc. Nat. Resour. 15 (8) (2002) 743–759.

[36] S. Topçu, A. Kibaroglu, Z. Kadirbeyoglu, Turkey, in: F. Molle, C. Sanchis-Ibor, L. Avella (Eds.), Irrigation in the Mediterranean: Technologies Institutions and Policies, Springer, New York, 2019.

[37] M. Kuper, N. Faysse, A. Hammani, T. Hartani, S. Marlet, M.F. Hamamouche, F. Ameur, Liberation or anarchy? The Janus nature of groundwater use on North Africa's new irrigation frontiers, in: Jakeman et al., (Ed.), Integrated Groundwater Management, Springer, New York, 2016, pp. 583–618.

[38] M. Kuper, A. Hammani, A. Chohin, P. Garin, M. Saaf, When groundwater takes over: linking 40 years of agricultural and groundwater dynamics in a large-scale irrigation scheme in Morocco, Irrig. Drain. 61 (2012) 45–53.

[39] D.E. El-Agha, A. Closas, F. Molle, Under the radar: dynamics of groundwater use in the Nile delta, Hydrogeol. J. 25 (2017) 1621–1631.

[40] M.H. Nassif, Analyse multiscalaire des politiques et de la gouvernance de l'eau dans le bassin du Litani Liban, PhD thesis, Université Paul Valery Montpellier, 2019.

[41] M. Saadé-Sbeih, A.H. Asaad, O. Shamali, F. Zwahlen, R. Jaubert, Groundwater balance politics: aquifer overexploitation in the Orontes River Basin, Water Altern. 11 (3) (2018) 663–683.

[42] D. Acloque, Conquérir le désert: recomposition des acteurs et des territoires agricoles en Egypte, PhD thesis, Université de Paris-Nanterre, 2019.

[43] F. Amichi, S. Bouarfa, M. Kuper, P. Caron, From oasis archipelago to pioneering eldorado in Algeria's Sahara. Irrig. Drain. (2019) https://doi.org/10.1002/ird.2308.

[44] M.R. Llamas, P. Martínez-Santos, Intensive groundwater use: Silent revolution and potential source of social conflicts, J. Water Resour. Plan. Manag. 131 (5) (2005) 337–341.

[45] F. Ameur, H. Amichi, M. Kuper, A. Hammani, Specifying the differentiated contribution of farmers to groundwater depletion, Hydrogeol. J. 25 (6) (2017) 1579–1591.

[46] FAO, The Benefits and Risks of Solar Powered Irrigation—A Global Overview, FAO, Rome, 2018.

[47] F. Molle, I. Gafaar, D.E. Al-Agha, E. Rap, The Nile delta's water and salt balances and implications for management, Agric. Water Manag. 197 (2018) 110–121.

[48] M. Reisner, Cadillac Desert: The American West and Its Disappearing Water, Viking, New York, 1986.

[49] F. Molle, Egypt, in: F. Molle, C. Sanchis-Ibor, L. Avella (Eds.), Irrigation in the Mediterranean: Technologies Institutions and Policies, Springer, New York, 2019.

[50] W. Ben Fraj, M. Elloumi, F. Molle, The politics of interbasin transfers: socio-environmental impacts and actors strategies in Tunisia, Nat. Res. Forum 431 (2019) 17–30.

[51] Plan Bleu, Sustainable solutions for water in the Mediterranean: managing scarcity and improving quality. Priority 1: "Improving water demand management", Report relating to target 1 MED 1-1, Mediterranean Process for the 6th WWF, 2012.
[52] GWP, La gestion de la demande en eau: l'expérience Méditerranéenne, Global Water Partnership, 2012.
[53] A. Chohin-Kuper, P.A. Garzón Delvaux, P. Strosser, Approche économique de la gestion de la demande en eau en Méditerranée: instruments économiques, Plan Bleu Valbonne. Les Cahiers du Plan Bleu 15, 2014.
[54] F. Molle, C. Sanchis-Ibor, Irrigation policies in the Mediterranean: trends and challenges, in: F. Molle, C. Sanchis-Ibor, L. Avella (Eds.), Irrigation in the Mediterranean: Technologies Institutions and Policies, Springer, New York, 2019.
[55] R. Khadra, J.A. Sagardoy, Irrigation Governance Challenges in the Mediterranean Region: Learning from Experiences and Promoting Sustainable Performance, CIHEAM/Springer, 2019.
[56] S. Van der Kooij, M. Zwarteveen, H. Boesveld, M. Kuper, The efficiency of drip irrigation unpacked, Agric. Water Manag. 123 (2013) 103–110.
[57] M. Ortega-Reig, C. Sanchis-Ibor, G. Palau-Salvador, M. García Mollá, L. Avellà-Reus, Institutional and management implications of drip irrigation introduction in collective irrigation systems in Spain, Agric. Water Manag. 187 (2017) 164–172.
[58] F.A. Ward, M. Pulido-Velazquez, Water conservation in irrigation can increase water use, Proc. Natl. Acad. Sci. U. S. A. 105 (2008) 18215–18220.
[59] C. Perry, P. Steduto, Does Improved Irrigation Technology Save Water? A Review of the Evidence, Food and Agriculture Organization of the United Nations, Cairo, 2017.
[60] R.Q. Grafton, J. Williams, C.J. Perry, F. Molle, C. Ringler, P. Steduto, et al., The paradox of irrigation efficiency, Science 361 (2018) 748–750.
[61] M. Benouniche, M. Kuper, A. Hammani, H. Boesveld, Making the user visible: analysing irrigation practices and farmers' logic to explain actual drip irrigation performance, Irrig. Sci. 326 (2014) 405–420.
[62] J.P. Venot, M. Kuper, M. Zwarteveen, Panda or Hydra? The Untold Stories of Drip Irrigation, Drip Irrigation for Agriculture: Untold Stories of Efficiency, Innovation and Development, Routledge, Abingdon, UK, 2017.
[63] M. Kuper, F. Ameur, A. Hammani, Unraveling the enduring paradox of increased pressure on groundwater through efficient drip irrigation, in: M. Kuper, J.P. Venot, M. Zwarteven (Eds.), Drip Irrigation for Agriculture, Routledge, 2017, pp. 85–104.
[64] F. Molle, O. Tanouti, Squaring the circle: impacts of irrigation intensification on water resources in Morocco, Agric. Water Manag. 192 (2017) 170–179.
[65] J. Berbel, C. Gutiérrez-Martín, J.A. Rodríguez-Díaz, E. Camacho, P. Montesinos, Literature review on rebound effect of water saving measures and analysis of a Spanish case study, Water Resour. Manag. 293 (2015) 663–678.
[66] C. Sanchis-Ibor, M. García-Mollá, L. Avellà, Effects of drip irrigation promotion policies on water use and irrigation costs in Valencia, Spain, Water Policy 19 (1) (2017) 165–180.
[67] M. Soto-García, V. Martínez-Álvarez, P.A. García-Bastida, F. Alcón, B. Martín-Górriz, Effect of water scarcity and modernisation on the performance of irrigation districts in South-Eastern Spain, Agric. Water Manag. 124 (2013) 11–19.
[68] A. Dumont, B. Mayor, E. López-Gunn, Is the rebound effect or Jevons paradox a useful concept for better management of water resources? Insights from the irrigation modernisation process in Spain, Aquat. Proc. 1 (2014) 64–76.
[69] L. Ponchant, C. Gary, J. Molenat, J. Prévot, D. Rollin, Ressources et Usages de l'eau sur le Bassin Versant de l'Hérault: Enjeux de gestion des ressources et d'adaptation des systèmes viticoles aux changements globaux, G-Eau Working Paper 2, Montpellier, 2017.
[70] S. Sese-Minguez, H. Boesveld, S. Asins-Velis, S. van der Kooij, J. Maroulis, Transformations accompanying a shift from surface to drip irrigation in the Cànyoles Watershed Valencia, Spain, Water Altern. 10 (1) (2017) 81–99.

[71] J.A. Rodríguez-Díaz, L. Pérez-Urrestarazu, E. Camacho-Poyato, P. Montesinos, The paradox of irrigation scheme modernization: more efficient water use linked to higher energy demand, Span. J. Agric. Res. 9 (4) (2011) 1000–1008.

[72] I. Fernández-García, J.A. Rodríguez-Díaz, E. Camacho-Poyato, P. Montesinos, J. Berbel, Effects of modernization and medium term perspectives on water and energy use in irrigation districts, Agric. Syst. 131 (2014) 56–63.

[73] M.R. Doukkali, C. Lejars, Energy cost of irrigation policy in Morocco: a social accounting matrix assessment, Int. J. Water Resour. Dev. 313 (2015) 422–435.

[74] M. García-Mollá, M.V. Ortega, C. Sanchis-Ibor, L. Avellà, The effects of irrigation modernization on the cost recovery of water in the Valencia Region Spain, Water Sci. Technol. Water Supply 143 (2014) 414–420.

[75] M.T. Jiménez-Aguirre, D. Isidoro, R. Barros, Effect of Irrigation Modernization on Water and Nitrogen Use Efficiency, EGU General Assembly 2014 Geophysical Research Abstracts 16, 2014, 775.

[76] R. Barros, D. Isidoro, R. Aragüés, Irrigation management nitrogen fertilization and nitrogen losses in the irrigation return flows of La Violada irrigation district Spain, Agric. Ecosyst. Environ. 155 (2012) 161–171.

[77] M. Benouniche, M. Zwarteveen, M. Kuper, Bricolage as innovation: opening the black box of drip irrigation systems, Irrig. Drain. 63 (5) (2014) 651–658.

[78] D. Lavee, E. Feitelson, J.E. Hadas, Israel, in: F. Molle, C. Sanchis-Ibor, L. Avella (Eds.), Irrigation in the Mediterranean: Technologies Institutions and Policies, Springer, New York, 2019.

[79] M.B. Bernabé-Crespo, E. Gil-Meseguer, J.M. Gómez-Espín, Desalination and water security in southeastern Spain, J. Polit. Ecol. 26 (2019) 486–499.

[80] C. Sanchis-Ibor, M. García-Mollá, T. Torregrosa, M. Ortega-Reig, M. Sevilla-Jiménez, Water transfers between agricultural and urban users in the region of Valencia Spain. A case of weak governance? Water Security **7** (2019) 100030.

[81] F. Giorgi, Climate change hot-spots. Geophys. Res. Lett. **33** (2006) L08707, https://doi.org/10.1029/2006GL025734.

[82] E. Hertig, Y. Tramblay, Regional downscaling of Mediterranean droughts under past and future climatic conditions, Glob. Planet. Chang. 151 (2016) 36–48.

[83] W. Cramer, J. Guiot, M. Fader, J. Garrabou, J.P. Gatusso, A. Iglesias, et al., Climate change and interconnected risks to sustainable development in the Mediterranean, Nat. Clim. Chang. 8 (2018) 972–980.

[84] D. Jacob, J. Petersen, B. Eggert, A. Alias, O.B. Christensen, L.M. Bouwer, et al., EURO-CORDEX: new high-resolution climate change projections for European impact research, Reg. Environ. Chang. 14 (2014) 563–578.

[85] M. Fader, S. Shi, W. Von Bloh, A. Bondeau, W. Cramer, Mediterranean irrigation under climate change: more efficient irrigation needed to compensate increases in irrigation water requirements, Hydrol. Earth Syst. Sci. 20 (2016) 953–973.

[86] S. Siebert, H. Webber, G. Zhao, F. Ewert, Heat stress is overestimated in climate impact studies for irrigated agriculture, Environ. Res. Lett. 12 (2017) 054023.

[87] J.C. Gonzalez-Hidalgo, D. Peña-Angulo, M. Brunetti, N. Cortesi, Recent trend in temperature evolution in Spanish mainland 1951–2010: from warming to hiatus, Int. J. Climatol. 36 (6) (2016) 2405–2416.

[88] A. Ribes, L. Corre, A. Gibelin, B. Dubuisson, Issues in estimating observed change at the local scale—a case study: the recent warming over France, Int. J. Climatol. 36 (11) (2016) 3794–3806.

[89] D.L. Hartmann, A.M.G. Klein, M. Tank, L.V. Rusticucci, S. Alexander, Y. Brönnimann, et al., Observations: atmosphere and surface, in: T.F. Stocker et al., (Ed.), Climate Change 2013: The Physical Science Basis, Cambridge University Press, 2013.

[90] M. Turco, N. Levin, N. Tessler, H. Saaroni, Recent changes and relations among drought vegetation and wildfires in the Eastern Mediterranean: the case of Israel, Glob. Planet. Chang. 151 (2016) 28–35.

[91] S.M. Vicente-Serrano, J.I. López-Moreno, S. Beguería, J. Lorenzo-Lacruz, A. Sánchez-Lorenzo, J.M. García-Ruiz, et al., Evidence of increasing drought severity caused by temperature rise in southern Europe, Environ. Res. Lett. 9 (2014) 044001.

[92] P. Lionello, L. Scarascia, The relation between climate change in the Mediterranean region and global warming, Reg. Environ. Chang. 18 (2018) 1481–1493.
[93] P. Alpert, S.O. Krichak, H. Shafir, D. Haim, I. Osetinsky, Climatic trends to extremes employing regional modeling and statistical interpretation over the E. Mediterranean, Glob. Planet. Chang. 63 (2008) 163–170.
[94] A. Iglesias, L. Garrote, Adaptation strategies for agricultural water management under climate change in Europe, Agric. Water Manag. 155 (2015) 113–124.
[95] A. Iglesias, R. Mougou, M. Moneo, S. Quiroga, Towards adaptation of agriculture to climate change in the Mediterranean. Reg. Environ. Chang. 111 (2011) 159–166, https://doi.org/10.1007/s10113-010-0187-4.
[96] K.S. Harmanny, Z. Malek, Adaptations in irrigated agriculture in the Mediterranean region: an overview and spatial analysis of implemented strategies, Reg. Environ. Chang. 19 (5) (2018) 1401–1416.
[97] A. Richard-Ferroudji, P. Garin, M. Matignon, L. Maton, M. Montginoul, J.D. Rinaudo, et al., Engager des agriculteurs à répondre à l'injonction d'adapter la gestion de l'eau au changement climatique. Une démarche de prospective participative avec des agriculteurs usagers des nappes du Roussillon (France), in: V. Ancey, B. Dedieu, M. Antona et al., Agir en situation d'incertitude en agriculture, Regards pluridisciplinaires au Nord et au Sud, Peter Lang Editions scientifiques internationals, 2013.
[98] M. Ortega-Reig, C. Sanchis-Ibor, M. García-Mollá, M. Pulido-Velázquez, C. Girard, P. Marcos, et al., Adaptación de la agricultura a escenarios de cambio global, Aplicación de métodos participativos en la cuenca del río Júcar, Economia Agraria y Recursos Naturales 18 (2) (2018) 29–51.
[99] M.T. Zeitoun, N. Allan, A. Al Aulaqi, A. Jabarin, H. Laamrani, Water demand management in Yemen and Jordan: addressing power and interests, Geogr. J. 178 (1) (2012) 54–66.
[100] CESE, *La gouvernance par la gestion intégrée des ressources en eau au Maroc: Levier fondamental de développement durable*. Version definitive. Conseil Economique Social et Environnemental, Auto-Saisine 15 (2014).
[101] G. Özerol, H. Bressers, Scalar alignment and sustainable water governance: the case of irrigated agriculture in Turkey, Environ. Sci. Pol. 45 (2015) 1–10.
[102] F. Cleaver, Development through bricolage: Rethinking institutions for natural resource management, Routledge, Abingdon, UK, 2012.
[103] M. Kuper, H. Amichi, P.L. Mayaux, Groundwater use in North Africa as a cautionary tale for climate change adaptation, Water Int. 42 (6) (2017) 725–740.
[104] S. Loubier, T. Ruf, P. Garin, France, in: F. Molle, C. Sanchis-Ibor, L. Avella (Eds.), Irrigation in the Mediterranean: Technologies Institutions and Policies, Springer, New York, 2019.
[105] W. Ghazouani, F. Molle, E. Rap, Water Users Associations in the Near-East Northern Africa Region, IFAD interventions and overall dynamics, IWMI, 2012.
[106] F. Molle, E. Rap, Brief Retrospective on Water User Organizations in Egypt. Water and Salt Management in the Nile delta, Project Report 3, IWMI, Cairo, 2013.
[107] A. Closas, F. Molle, Groundwater governance in the Middle-East and Northern-Africa Region, Report submitted to USAID, IWMI, 2016. http://gw-mena,iwmi,org/outputs/.
[108] M. García-Mollá, C. Sanchis-Ibor, L. Avellà, J. Albiac, D. Isidoro, S. Lecina, Spain, in: F. Molle, C. Sanchis-Ibor, L. Avella (Eds.), Irrigation in the Mediterranean: Technologies Institutions and Policies, Springer, New York, 2019.
[109] A. Frija, A. Chebil, S. Speelman, N. Faysse, A critical assessment of groundwater governance in Tunisia, Water Policy 16 (2014) 358–373.
[110] A. Hamdane, Le contrôle de l'utilisation des eaux souterraines et la gestion participative des nappes, Report to FAO, 2015.
[111] BRLi, Agroconcept, Gestion de la demande en eau dans le bassin méditerranéen—Exemple du Maroc—Cas d'étude du Souss Massa, AFD et Plan Bleu, 2013.
[112] K. Del Vecchio, S. Barone, Has Morocco's groundwater policy changed? Lessons from the institutional approach, Water Altern. 113 (2018) 638–662.
[113] A. Mesnil, N. Habjoka, The Azraq Dilema: Past Present and Future Groundwater Management, GIZ, Amman, 2012.

[114] O. Petit, M. Kuper, E. López-Gunn, J.D. Rinaudo, A. Daoudi, C. Lejars, Can agricultural groundwater economies collapse? An inquiry into the pathways of four groundwater economies under threat, Hydrogeol. J. 25 (6) (2017) 1549–1564.
[115] F. Mukhtarov, K.A. Daniell, Transfer diffusion adaptation and translation of water policy models, in: K. Conca, E. Weinthal (Eds.), The Oxford Handbook of Water Politics and Policy, Oxford University Press, 2016.
[116] O. Tanouti, La Gestion Intégrée des Ressources en eau à l'épreuve du bassin versant—Le cas du bassin du Tensift au Maroc, Thèse de Doctorat, Université de Paris Nanterre, 2017.
[117] D. Sims, Egypt's desert dreams: development or disaster? American University in Cairo Press, Cairo, 2015.
[118] A. Houdret, The water connection: irrigation water grabbing and politics in southern Morocco, Water Altern. 5 (2) (2012) 284–303.
[119] C. Sanchis-Ibor, R. Boelens, M. García-Mollá, Collective irrigation reloaded, re-collection and re-moralization of water management after privatization in Spain, Geoforum 87 (2017) 38–47.
[120] M. García-Mollá, M. Ortega-Reig, R. Boelens, C. Sanchis-Ibor, Hybridizing the Commons. Privatizing and Outsourcing Collective Irrigation Management After Technological Change in Spain, 2020 (under preparation).

CHAPTER 5

Water reuse: A resource for Mediterranean agriculture

Nassim Ait-Mouheb[a], Pierre-Louis Mayaux[b], Javier Mateo-Sagasta[c], Tarik Hartani[d], Bruno Molle[a]
[a]INRAE, UMR GEAU, University of Montpellier, Montpellier cedex 5, France
[b]CIRAD, UMR GEAU, University of Montpellier, Montpellier cedex 5, France
[c]International Water Management Institute (IWMI), Colombo, Sri Lanka
[d]Centre Universitaire de Tipaza, Tipaza, Algeria

5.1 Introduction

The Mediterranean region is particularly vulnerable to the adverse consequences of multiple and mutually reinforcing structural changes that include global warming, biodiversity loss, soil erosion, and population growth. Many of these changes contribute to water scarcity, one the most pressing challenges the region is facing in the very near term. The southern and western shores of the Mediterranean, the Middle East, and North Africa (MENA) are already the most water-stressed regions in the world. But the European shore is not spared, as Greece, Cyprus, Macedonia, and Spain also top the list of the world's most water-stressed countries.[a] In many places, absolute water scarcity is compounded by a marked degradation in water quality due to pollution and salinization. The heightened competition over an increasingly scarce resource raises fears of social unrest and uncontrolled population flows.

If all economic sectors consume ever larger quantities of water, irrigated agriculture is still by far the largest user. Between 80% and 90% of all available water is consumed by irrigation in the southern shore, although that ratio is halved at the other end of the spectrum, in the case of France. Therefore, the widening gap between water supply and demand is primarily a challenge for food production, especially as all governments appear anxious to prioritize other uses (i.e., domestic and industrial) should a water shortage erupt.

To what extent, then, can water reuse be part of a broader policy mix aiming at providing food production with a larger and more reliable supply of water when the resource is limited and irregular? And, in addition to its strictly quantitative potential, under what hydrological, agronomic, institutional conditions can water be reused safely and at reasonable cost? Before giving an overall assessment, we should emphasize that the term

[a] For a regular updating of this list, and bearing in mind the shortcomings related to national averaging, see: https://www.wri.org/applications/aqueduct/country-rankings/?indicator=bws.

Water Resources in the Mediterranean Region
https://doi.org/10.1016/B978-0-12-818086-0.00005-4

"water reuse" can be applied to a wide variety of situations. To begin with, the extent to which reused water is treated can vary considerably. Today, it has been calculated that a total of 36 million hectares worldwide (some 12% of all irrigated land) receive some urban wastewater; however, only around 15% of this water is deemed adequately treated [1]. When it is not, those who use wastewater typically disregard specific practices that might protect them or their consumers from health risks. Second, irrespective of whether it is adequately treated or not, wastewater can either be tapped at the point of rejection (direct reuse) or withdrawn a little farther away, mixed to varying degrees with surface or groundwater (indirect reuse). The latter is by far the most common, although it is not adequately captured by national statistics that define "reuse" only as direct reuse (Box 5.1). Finally, some projects are formally planned and regulated by state authorities while other types of uses occur informally, although in the latter case state officials may still exercise some measure of informal control. Again, and by definition, unplanned uses are not well captured by national statistics that focus on planned reuse. Thus, although examples of unplanned water reuse are common on the Southern shore of the Mediterranean, total wastewater reuse is only few volume of all officially quantified reclaimed wastewater.

Against this complex background, it should be noted that, over the past few decades, the development of more efficient treatment systems for domestic and industrial wastewater has made water reuse increasingly safe for human health, and an increasingly viable option in economic terms. Many projects now have an attractive cost-benefit ratio compared to the construction of dams that would provide a comparable volume of water.

BOX 5.1 Main stages and types of wastewater treatment processes [2].

- Before treatment, sewage usually goes through pretreatment to remove grit, grease, and gross solids that could hinder subsequent treatment stages.
- Next, *primary treatment* aims to settle and remove both organic and inorganic suspended solids. The most common primary treatments are primary settlers and septic tanks.
- During *secondary treatment,* soluble biodegradable organics are degraded and removed by bacteria and protozoa through (aerobic or anaerobic) biological processes. This process may include the removal of nutrients (mainly nitrogen and phosphorous). Typical secondary treatments include aerated lagoons, activated sludge, trickling filters, oxidation ditches, and other extensive processes such as constructed wetlands.
- *Tertiary treatment* aims at refining the effluent before it is discharged or reused and can include the removal of some nutrients and residual suspended matter by filtration, nutrients and residual suspended matter, or microorganisms (disinfection with chlorine, ozone, ultraviolet radiation, or others). Tertiary treatment can include membrane filtration (micro-, nano-, ultra- and reverse osmosis), infiltration/percolation, and activated carbon. This third stage/level seldom exists in low-income countries.

The potential for water reuse, and the fact that it could provide some good quality water in a reliable way throughout the year, is now widely recognized. However, water reuse can also be justified on grounds other than purely quantitative ones. At least three additional justifications are routinely given. First, wastewater supplies valuable additional nutriments (mainly nitrogen and phosphorus) to crops. Second, when officially organized, it can provide an additional source of revenue for water utilities. Third, it can be considered a supplementary treatment stage that improves the quality of water ultimately discharged into ecosystems. Given these rationales, it is important to take a careful look at how this resource is currently managed, to assess whether it can live up to its promises, and to think about possible ways to better harness it by combining legal reforms, policy incentives, technological improvements, and better management practices.

Despite the ancient and well-recognized use of wastewater around the Mediterranean, scientific knowledge is still in need of improvement, especially with respect to real practices on the ground. The aim of this chapter is to provide an overview of the current situation across the region, including the types of treatment processes, legal frameworks, policy orientations, irrigation practices, and agronomical performances. Some local cases are analyzed in greater detail. Finally, an integrated approach is recommended for future studies that includes environmental, economic, and equity considerations.

5.2 Water's second life for agriculture

5.2.1 Definitions and qualities of water reuse

Water reuse is probably as old as cities themselves. According to Angelakis et al. [3], domestic wastewater (sewage) has been used for irrigation by a number of civilizations since the beginning of the Bronze Age (around 3200–1100 BC), including cities in Egypt, Mesopotamia, and Crete. Later (around 1000 BC–AD 330), wastewater was used for irrigation and fertilization by both Greek and Roman civilizations, especially around major cities such as Athens and Rome. Throughout human history, the term "reuse" has referred to the use of raw or only partially treated wastewater for beneficial purposes, mostly in agriculture.

In recent decades, a fundamental change has taken place, as the diffusion of treatment plants has allowed treated water reuse to become an object of public policies. Many countries now consider water reuse as an important component of their overall water mix, mainly for the agricultural sector as irrigated area continues to expand [4, 5]. Countries like Jordan, where 87% of the safely treated wastewater is now reused directly or with little dilution, Israel, where treated wastewater accounted for 40% of all water used for irrigation in 2011, and to a lesser degree Tunisia, where about 25% of treated water is now reused, have become recognized leaders in this rapidly expanding field [6]. Israel, in particular, uses more than 300 Mm^3/year of tertiary-treated effluents or secondary chlorinated effluents in agricultural irrigation, mainly for industrial crops, cotton and fodder, citrus trees, cooked food, and unrestricted irrigation crops [7]. To put this volume into perspective, it represents about 40% of the total amount of treated water used in the whole European Union.

However, despite recent improvements, wastewater treatment is still far from being universal in the MENA region. For example, it is estimated that only 8% of the wastewater generated in Lebanon is treated. About 11% of the population benefits from safely managed wastewater systems in the northern and southern governorates, compared to only 7% and 3% in Greater Beirut and Bekaa, respectively [7]. This calls into question the possibility of planned water reuse under national regulations [7]. As a consequence, we are currently witnessing a wide array of modes of reuse that range from unplanned, small-scale use of raw wastewater for vegetable production in peri-urban areas of Morocco, Algeria, or Egypt, to the sophisticated use of ultra-purified water for sensitive crops such as vegetables in Italy. Box 5.1 below summarizes the main stages of standard wastewater treatment. Across the Mediterranean there are examples of all types of reuse levels, from no treatment at all to tertiary treatment. It should also be noted that a systemic evaluation of real performance of many treatment plants or disinfection technologies has not been undertaken and relevant information such as maintenance, costs, and safety is not available.

In the Mediterranean region, when urban areas expand without a concomitant increase in the wastewater treatment capacity, a large quantity of wastewater is discharged into surface water bodies with little or no treatment. Downstream irrigated areas then depend on these same water sources. The diverse modes of water reuse are summarized in Box 5.2. It should again be emphasized that the site-specific applicability of many disinfection technologies has not been evaluated and relevant information about maintenance, costs, safety, etc. is not available.

5.2.2 Key regional trends and figures

In the following section, we have gathered a few key facts and figures regarding water reuse in four contrasted countries (Morocco, Jordan, France, and Algeria) to highlight the wide diversity of situations, issues, and current challenges. It is important to recognize the diversity because even though it is always possible for best practices to travel from one country to the next, local conditions should be duly analyzed: when it comes to water reuse, there can be no one size fits all.

5.2.2.1 Morocco

Up until the beginning of the 21st century, public investments in urban water in Morocco focused mainly on universalizing access to drinking water, leaving aside the disposal and treatment of wastewater. However, informal reuse of raw water was common in many city peripheries including Meknès, Marrakesh, and Settat [12–14]. This

BOX 5.2 Diversity of wastewater reuse on the Mediterranean rim.

- Direct use of untreated wastewater occurs in areas where collective treatment plants are not in place and where alternatives are limited, usually in arid or semiarid areas. For example, these practices may involve direct pumping of raw wastewater into the pipe before it arrives at the treatment plant. The reasons for such uses may also be linked with the poor quality of conventional water resources (e.g., groundwater salinity) or the exorbitant cost of access to fresh water (e.g., pumping costs). Although officially banned in most countries, direct use of untreated wastewater occurs in many urban and peri-urban areas [8].
- Indirect or "de facto" use of untreated or treated wastewater is by far the most common type of use [2, 9]. In this case, untreated wastewater or wastewater from treatment plants is discharged into surface water streams and/or percolates to groundwater, where it dilutes and is subsequently used—most often unintentionally—by downstream farmers. For example, in the South of France, the estimated impact of wastewater effluents in the Montpellier river basin was estimated to vary between 1.5% and 51% on average, depending on flow conditions and seasons [10].
- The planned use of reclaimed water (treated wastewater) is typically driven by water scarcity, but where the environment is highly degraded, the reclaimed wastewater is also used to save fresh water and ecosystems. Reclaimed water can be used directly for many purposes such as irrigation, city landscaping, and golf courses. Along the Mediterranean rim, wastewater is rarely used for groundwater recharge or indirectly used for drinking purposes, as is the case of Windhoek in Namibia [11]. The planned use of reclaimed water is now one of the priorities in the Middle East and North African countries, as well as in the Eastern and Mediterranean countries.

changed markedly with the launching of the national sanitation plan in 2006 (French acronym PNA). The plan aimed to increase the overall treatment rate from a mere 8%–60% by 2020. This required building—and rehabilitating—some 187 plants by 2020, a target later increased to 330 by 2030, 145 of which were effectively operational in 2015 [15]. This opened up new opportunities for water reuse. The national water plan (French acronym PNE), released in 2009, forecasts a total volume of reusable wastewater of 424 Mm^3 by 2020 and 935 Mm^3 by 2030. The plan aimed to achieve a reuse rate of 19% by 2020 and 31% by 2030; at that time, this amounted to a total volume of reused water of 300 Mm^3, later increased to 325 Mm^3.

In 2017, 47.5 Mm^3 were effectively being reused in 24 different projects [16]; by far the most for golf resorts. The Cherifian Office of Phosphates also reuses treated water to wash its mineral rocks, while some cities, including Ouarzazate, use their treated water to irrigate their green spaces and green belts. After a number of small-scale experiments dating back to the early 1990s [17], three projects for agricultural use of treated water are now underway with treated water in Tiznit, Settat, and Oujda.

Aside from the national water plan, three different planning documents in 2017 dealt with wastewater reuse. The National Sanitation program for rural areas (French acronym PNAR) was prepared by the Ministry of the Interior. The National Plan for treated wastewater reuse (French acronym PNRUE) was drafted by the Water Department of the Ministry of Energy, Mines and Sustainable Development, and the National Plan for wastewater reuse in Agriculture (French acronym PDREUTI) issued by the Ministry of Agriculture. In 2018, these plans were merged into a single, comprehensive National Plan for Integrated Sanitation (French acronym PNAM). The new plan aims to reuse some 428 Mm^3 of water by 2025, 474 Mm^3 by 2030, and 573 Mm^3 by 2040. This includes an objective of 134 Mm^3 for agriculture to eventually irrigate a surface area of 15,363 ha.

Promising though these figures may seem on chapter, they overlook the fact that a significant amount of water is already being reused informally. Therefore, the claim that wastewater represents a previously "untapped" resource should be examined critically. As mentioned above, part of the difficulty involved stems from the lack of official, aggregate data on illegal uses. However, case studies in Settat, Marrakesh, and Tiznit have all shown that at least part of the raw wastewater is already used downstream. The new projects (for golfing resorts in Marrakesh, for agriculture in Settat and Tiznit) are undoubtedly safer, as they are formally organized around treatment plants. But they all entail some reallocation from previous users to new beneficiaries, and in one case at least (i.e., Marrakech) this reallocation clearly shifted upward, targeting wealthy international consumers. However, to go beyond such anecdotal evidence, we would need better aggregate data on *all* current uses, including information on the types and numbers of beneficiaries, and the incomes they derive from wastewater, whether directly (through productive uses) or indirectly (through the rental value of their land).

5.2.2.2 Jordan

Jordan suffers from severe water scarcity, intensified by climate change and the continuous flow of refugees from Syria since 2011. Today the per capita supply of water is 140 m^3/capita/year while the threshold of severe water scarcity is 500 m^3/capita/year [18].

Water reuse in Jordan is not merely an option but a real necessity and most of the treated wastewater generated in the country is already used, directly or indirectly, for productive purposes, mainly in agriculture. The government is working to increase wastewater treatment capacity and the share of treated wastewater in its water budget. To this end, the government has created policies that allow better water (re)allocation across sectors in order to save more fresh water for domestic use [19].

There are 27 wastewater treatment plants in Jordan that currently serve 63% of Jordan's population and produce 121 Mm^3/year of treated wastewater [19]. The Al-Samra treatment plant alone generates about 80 Mm^3/year, most of which is reused indirectly in the Jordan Valley for unrestricted irrigation, after "dilution"[b] along the Al Zarqa Wadi and storage in the King Talal Dam. Much of the production in the Jordan Valley supplies international markets and preserving food safety in the valley is a government priority.

On the other hand, about 45 Mm^3/year of treated wastewater is used in restricted irrigation to produce fodder crops and fruit trees in projects that experience low-cost recovery and whose long-term financial sustainability is questionable. There are opportunities to shift toward more economically efficient and productive reuse if economic barriers as well as formal constraints (rules and regulations) and informal constraints (behaviors and human capacity) are addressed [20].

5.2.2.3 France

Currently, France only faces local and seasonal episodes of water resource deficits. Therefore, the reuse of wastewater is restricted to certain regions, and only about 40 treated wastewater (TWW) reuse projects were identified, most involving golf courses, production of turf and garden or agricultural irrigation. In 2014, the average daily volume of TWW reused in France was estimated at 19,200 m^3 (about 7 Mm^3 per year), representing about 0.1% of the total TWW produced and less than 0.3% of the all the water used for irrigation. The biggest agricultural projects are located in Clermont-Ferrand (central France), where 700 ha of seed maize, maize, red beet, and wheat are irrigated with 0.9 Mm^3 of TWW per year; and in Noirmoutier Island, where 320 ha of quality potatoes are irrigated with TWW, representing an average yearly volume of 0.38 Mm^3. Actually, 20 golf courses are irrigated with TWW in France. The majority are located on the Atlantic coast where the main driver of these projects is the need to limit TWW discharge into the sea and the environment. The golf courses have access to this water resource all year round, whereas water restrictions often apply to conventional water resources in summer [21].

The use of TWW from waste water treatment plants (WWTPs) for agricultural irrigation, turf production, and garden irrigation has been regulated since 2010, but few projects emerged from the new water reuse regulation. The 2010 regulation introduced quality standards for TWW based on four quality levels (from A—high quality—to D).[c] The higher the risk of human exposure (valorization of the crops, type of irrigation), the higher the level of treatment required. The regulation was updated in 2014 to better account for the case of sprinkler irrigation techniques. It provides wind velocity limits at which irrigation should be stopped depending on the operating pressure in use, and stipulates safety distances that account for the maximum sprinkler range. Many new water reuse projects are planned for the coming years mainly concentrated close to the sea. In 2016, the Rhone-Mediterranean Water Agency launched call for projects to encourage local authorities

[b] Although dilution is actually negligible, particularly in the dry season, this practice enables the production of vegetables, which is forbidden by law in direct reuse schemes.

[c] French Law No. *2010*-788 of 12 July 2010.

and industry to study and carry out projects to promote reuse. More than 47 projects where cofinanced. Many of the projects in this call remain experimental and research oriented. For example, setting up an experimental platform studies the impacts of water reuse on the agro-system in Murviel-Lès-Montpellier [22] or Saint Martin de Castillon.

5.2.2.4 Algeria

In June 2018, total production of domestic wastewater in Algeria was 1596 Mm^3. Table 5.1 lists forecast indicators for the 2035 horizon based on the National Water Plan, which is used as a tool for water resources management in Algeria [24].

In 2018, the total volume of wastewater produced and managed by the Algerian wastewater treatment office (ONA) was more than 577 Mm^3. The number of treatment plants also monitored by ONA was 148 for a treated volume of 123 Mm^3.

The first legislative framework for the treatment of wastewater, named "Concession for the use of treated wastewaters for irrigation purposes," dates from May 23rd, 2007 and provides a general framework for using these waters (definition of wastewater, sanitary controls, and financial agreement). An interministerial decree published in January 2012 lists the required properties of treated wastewater for irrigation purposes. It defines thresholds for microbiological parameters, physical parameters, chemical parameters, and heavy metals. These thresholds are inspired by WHO (World Health Organization) standards and are probably too inflexible for the Algerian context. Today, the number of treatment plants officially concerned by wastewater reuse is 17 and the volume reused is 8 Mm^3. The formal reuse of wastewater for agricultural purposes started in 2007, meaning it is a relatively new practice. The total irrigated area covers more than 11,000 ha, particularly fruit trees (palm dates, olive, etc.) and some cereals. The most emblematic cases for controlled water reuse are Hennaya-Tlemcen in the North West and Guelma-Bouchegouf in the north east, with 912 ha and 6980 ha of irrigated land, respectively. In both cases, water is released from an activated sludge plant into a river and then pumped to the plots.

Today, two major constraints on the expansion of reuse are the limited volume available for irrigation and the limitation to a secondary level of treatment in all the plants.

Table 5.1 Predicted increase in the quantity of wastewater in Algeria [23].

Horizon	Total wastewater (Mm^3/year)	Plant capacity (Mm^3/year)	Treated wastewater (Mm^3/year)
2020	1190[a]	1060	650
2025	1360	1320	820
2030	1570	1510	1020
2035	1780	–	1625

[a] The expected volume was reached in 2012.

Unplanned reuse has not been officially quantified but a previous survey reported that 10% of the land in the high plains of Algeria received untreated wastewater [25]. This practice continued in many areas of the country until August 2018 when the use of wastewater was strictly prohibited due to an outbreak of cholera in the districts of Blida and Tipaza. Indeed, before this date, some drip irrigation systems used reclaimed wastewater to irrigate cucumber, tomato, and watermelon in Tipaza, particularly in the summer season (see Box 5.4). Much progress has been made in the last decade toward the safe use of treated wastewater to meet farmers' demand for irrigation water. However, further efforts are still required to bridge the gap between unplanned and planned practices.

5.3 The governance of wastewater reuse: Regulations, policies, and politics

5.3.1 A comparative overview of regulatory frameworks

The 1970s were an important international landmark with respect to national regulations and guidelines for water reuse.[d] This critical juncture matters as it was also a period of increased awareness of both environmental and public health risks [26, 27]. The 1970s informed the way wastewater reuse has been framed up to now in international arenas of expertise, where concerns for possible effects on public health have fostered a cautious approach.

In 1973, the World Health Organization (WHO) released its first water reuse guidelines, which at that time, mostly targeted developing countries. In 1978, the State of California issued its own criteria for wastewater irrigation, which would soon prove to be very influential worldwide. By the end of the decade, the first national regulations specifically tailored for water reuse had been adopted in Tunisia, Italy, and Israel. These regulations subsequently increased in geographical scope and sophistication, with diffusion patterns in the Mediterranean displaying two notable features. First, temporally, government outputs intensified markedly following the second generation of WHO guidelines in 1989, and even more markedly from the beginning of the 2000s on. Second, spatially, no Mediterranean subregion appeared to advance more rapidly than the others. Thus, although scholars have found that at similar level of water stress, the adoption of *tangible projects* in wastewater reuse tended to be faster in countries with stronger economies [6], this does not appear to be the case for wastewater *regulations* and *policies*. Regarding the latter, national idiosyncrasies have mattered most, under the key influence of international standards and guidelines. Table 5.2 provides a timeline of selected water reuse regulations and national strategies in the Mediterranean region.

[d] Although the State of California, a pioneer regarding water reuse, originally introduced regulations in 1918.

Table 5.2 Timeline of selected water reuse regulations and strategies in the Mediterranean region.

1973	**WHO releases water reuse guidelines that include quality thresholds (100 FC/100 mL) and treatment requirements for several uses, including crop irrigation and potable reuse (*Reuse of effluents: Methods of wastewater treatment and public health safeguards*)**
1975	The Tunisian Water Act (n° 75-16): the way for water reuse
1977	Italy regulates water reuse for irrigation and describes extensive treatment processes (CITAI, February 4th and February 21st 1977, n° 48)
1978	Following older regulations, the State of California issues its influential criteria for wastewater reclamation for reuse (*Wastewater reclamation criteria, title 22, division 4: Environmental Health*)
1978	Israel issues regulations for water reuse in irrigation that define treatment requirements, quality limits (unrestricted irrigation: 12 FC/100 mL in 80% of samples: 2.2 FC/100 mL in 50% of samples), crops, and additional barriers
1985	FAO releases its "Water Quality for Agriculture" guidelines [28]
1987–1992	The fourth European environmental action program encourages water reuse
1989	First WHO revision of water reuse guidelines (unrestricted irrigation: 1000 FC/100 mL; <1 nematode egg/L) based on the conclusions of the previous reports [29]
1989	Tunisia issues standards for water reused in irrigation based on FAO (1985) and WHO (1989) guidelines for restricted irrigation (<1 nematode egg/l) and a multibarrier approach
1991	French recommendations for water reuse based on WHO guidelines (Circular n° 51 dated July 22, 1991, issued by the Ministry of Health)
1999	Revised Israel regulations: Unrestricted irrigation <1 FC/100 mL and a multibarrier approach [30]
2002	Moroccan regulations on water quality for irrigation (both for natural water and wastewater)
2003	WHO State of the Art Report on Artificial Recharge of Groundwater with Recycled Water [31]
2003	Revised Italian regulations for water reuse (Ministerial Decree n° 185/2003)
2005	Cyprus issues criteria for water reuse in agriculture (Decree n° 296/03.06.05)
2006	WHO releases its second revision of water reuse guidelines on Treated Wastewater in Agriculture: Risk analysis and management, which adopt a quantitative risk assessment methodology [32]
2006	Portugal releases regulations for water reuse (Portuguese Standard NP 4434)
2007	Spain issues water reuse regulations (Royal Decree 1620/2007)
2010	France sets out water reuse criteria (*arrêté du 2 août* 2010)
2011	The National Health and Medical Research Council and National Resource Management Ministerial Council publish Australian Drinking Water Guidelines
2011	Greece issues water reuse regulations (the 145,116 Common Ministerial Decision, March 2011)

Table 5.2 Timeline of selected water reuse regulations and strategies in the Mediterranean region—cont'd

1973	**WHO releases water reuse guidelines that include quality thresholds (100 FC/ 100 mL) and treatment requirements for several uses, including crop irrigation and potable reuse (*Reuse of effluents: Methods of wastewater treatment and public health safeguards*)**
2012	National Strategy for wastewater, including wastewater reuse, in Lebanon (*Stratégie nationale du secteur de l'eau*)
2014	Revised French water reuse regulation (*arrêté du 25 juin* 2014)
2017	World Health Organization, Geneva, Potable Reuse: Guidance for Producing Safe Drinking Water
2018	EU minimum water quality requirements for irrigation and aquifer recharge. An EU legislative proposal will be published in the near future

Based on A.N. Angelakis, T. Asano, A. Bahri, B.E. Jimenez, G. Tchobanoglous, Water reuse: from ancient to modern times and the future, Front. Environ. Sci. 6 (May) 2018; J. Fawell, K. Le Corre, and P. Jeffrey, Common or independent? The debate over regulations and standards for water reuse in Europe, Int. J. Water Resour. Dev. 32 (2016) 559–572; WWAP (United Nations World Water Assessment Programme), The United Nations World Water Development Report 2017: Wastewater, The Untapped Resource, UNESCO, Paris, 2017.

Although the timeline may give the impression that national regulations are broadly homogeneous, this is far from true. Rather, from the outset, the two benchmark guidelines, the California guidelines [33] and the WHO guidelines [29], have represented two contrasting approaches to water reuse. The California guidelines are stricter. They are based on the premise that pathogenic microorganisms can survive for weeks in the soil and on crops, implying that the detection of any of these organisms is sufficient to constitute a public health problem (see Section 5.4.1). This leads to stringent requirements regarding, for example, chlorine disinfection of any reused water that may come into contact with the edible portion of the crop [34]. This "zero risk" approach therefore calls for the "best available technology" [35]. By contrast, WHO guidelines are more flexible, following a "calculated risk" approach. They consider irrigation as an additional treatment stage in itself, while farming, harvesting, and hygiene practices are also considered as key components of risk mitigation [36]. This "multibarrier" approach involves a wider array of guidelines applying to all stages of the reuse process. It also tolerates up to 1000 fecal coliforms (FCs) per 100 mL for unrestricted irrigation of any crop, and for *Escherichia coli*, a limit of 10^4 colony forming units (CFUs) per 100 mL for wastewater used to irrigate vegetables consumed as fresh food [32, 37]. According to WHO guidelines, and different from Californian standards, this can be achieved through a series of stabilization ponds, for example [37].

National standards in the Mediterranean region derived either from the Californian model (e.g., Greece, Cyprus, and Italy), from the WHO criteria (e.g., France, Morocco, Tunisia), or from a combination of the two (e.g., Spain and Portugal) [38]. Italy is perhaps the best example of the "zero risk" approach. It defines limits for 54 different parameters, 37% of which are not considered for drinking water analysis. This is in sharp contrast to Israel, the leading country in water reuse, which monitors less than a dozen parameters [39] but requires the construction of barriers to contamination based on a range of technical practices (level of treatment, type and contact with the crop/produce, type of irrigation technique). Some limits defined by the Italian regulations are very strict, such as the parameters for BOD5 (< 20 mg/L), total suspended solids (< 10 mg/L), and *E. coli* (< 10 CFU/100 mL for conventional treatment systems). Not surprisingly, the number of parameters to be monitored is said to have negatively affected the total volume of reused water in the country [40]. It can also push actors toward informal practices, as is apparently the case in Milan. The current proposal for a future European Directive on the "Regulation on minimum requirements for water reuse" [41] leans more on the stringent side for pathogens, beyond the bathing water quality regulation, albeit less so than the Italian case. At least 11 out of 22 Arab States have adopted legislation permitting the use of reclaimed wastewater, issued by the national institutions responsible for the use and discharge of wastewater, the ministries responsible for the environment in Kuwait, Lebanon and Oman, health in Iraq, agriculture in Tunisia, housing in Egypt, and the institutes responsible for standards in Jordan and Yemen [32].

National regulatory frameworks do not only vary according to their stringency. They also differ in the types of use authorized, as illustrated by Table 5.3.

For some observers, the high level of discrepancy between national standards may well be the single biggest impediment to the expansion of treated water reuse across the Mediterranean [39, 42]. As a consequence, the same authors commonly call for more international harmonization. In their opinion, harmonization should preferably occur around flexible and realistic regulatory frameworks that avoid unnecessarily restrictive standards [43]. More recently, a "one water" concept has been gaining momentum among sanitation engineers. It asserts that standards should be established as a function of the category of water use, independently of the water source or origin [44]. However, disagreements over regulations are only one type of institutional challenge that, in some cases, may hinder the expansion of treated wastewater reuse. The lack of clarity and/or dispersion of formal prerogatives, for example over who is in charge of setting rates, are another. They test the institutional capacity to design integrated policies around water reuse, beyond the accumulation of specific regulations.

5.3.2 From regulations to policies

Wastewater reuse is, by definition, an issue that straddles many policy sectors including sanitation and plant management, water resources at a broader scale, public works, agriculture, the Interior, Public health and the Environment. The exact distribution of prerogatives among these sectors is far from straightforward. Rather, it tends to vary greatly, not only

Table 5.3 Authorized uses for wastewater in the Mediterranean countries.

	Algeria	Cyprus	Egypt	France	Greece	Israel	Italy	Jordan	Lebanon	Morocco	Portugal	Spain	Syria	Tunisia	West bank
Agricultural irrigation	Authorized	Authorized	Forbidden: R+ P+E / Authorized	Authorized	Authorized	Authorized	Authorized	Authorized		Authorized	Authorized	Authorized	Forbidden: R+ P / Authorized	Forbidden: R / Authorized	Forbidden: R / Authorized
Landscape irrigation/ golf courses		Authorized	Authorized	Authorized	Authorized	Authorized	Authorized	Authorized		Authorized	Authorized	Authorized			
Aquifer recharge[a]	Forbidden				Authorized					Authorized	Authorized	Authorized	Forbidden: *		
Environment					Authorized						Authorized	Authorized			
Industrial recycling					Authorized						Authorized	Authorized			
Urban use		Authorized		Authorized	Authorized	Authorized					Authorized	Authorized			
Domestic use					Authorized							Authorized			
Potable water															

Legend: Authorized (green); Forbidden (orange); Unregulated (white)

[a] Aquifers exploited for drinking water.

R, products eaten raw; F, fruit without pericarp; *E*, crops for export; *P*, pasture.
Based on N. Condom, M. Lefebvre, L. Vandome, "La réutilisation des eaux usées traitées en Méditerranée: retour d'expériences et aide à l'élaboration de projets". Plan Bleu, Valbonne (Les Cahiers du Plan Bleu 11), 2012.

from country to country but also from one project to the next in the same country. For the sake of clarity, at least five key areas of prerogatives should be distinguished: (i) who proposes and designs the basic sociophysical parameters of the project (i.e., location, surface area, beneficiaries, mode of water reuse)? (ii) who authorizes the project after assessing its social desirability and whether it complies with existing regulations? (iii) who decides who pays what in upfront investments, operation, and maintenance? (iv) who operates and maintains the project? (v) who subsequently monitors water quality and its impacts on soil, crops, and human health? (vi) who assumes the responsibility of any unintended dispersion of contaminant?

In many countries, at least some of these areas of prerogatives lack clear lines of responsibility and accountability. In Morocco, for example, there is no formally agreed-upon process for formulating and designing new projects. Thus, in the case of Settat, near Casablanca, the governor and the river basin agency happened to play a central role in defining the basic components of the project, while in Tiznit, both the municipality and the agricultural administration happened to play a more prominent role [14]. This matters because even with the most up-to-date regulatory framework, insufficient guidelines (and "ownership" of their various steps) can impede projects. When responsibilities and criteria for decision-making are blurred, risk-averse public officials are more likely to reject a water reuse project due to concern for their own position and unwillingness to breach perceived water management norms on quality and safety [45].

Conversely, countries in which treated water reuse is more widely used are also those with clearer lines of authority.[e] In Tunisia, for instance, regional departments of agriculture (French acronym CRDA) built and have maintained their authority concerning water reuse for agriculture (Box 5.3).

BOX 5.3 Wastewater policy in Tunisia.

Tunisia has been developing planned water reuse since the 1960s. In the aftermath of its 1975 water act, it launched an ambitious national program. More specific standards for the treatment, distribution, and quality control of the recycled water were set up in 1989 [46]. As these dates indicate, Tunisia closely followed the WHO guidelines and their focus on public health. It forbids the use of wastewater that is only treated to secondary level (which includes most Tunisian wastewater, which is treated at this level through activated sludge) for irrigating vegetables. Regional departments of agriculture (CRDA) supervise the treated water reuse projects for agriculture in their jurisdiction and collect water fees from water users' associations (GDA) based on the volume of water delivered.

As a consequence, in 2016, about 30% of reclaimed wastewater was being recycled, most of it to irrigate 8415 ha of farmland (industrial, fodder and tree crops, and cereals) and 1040 ha of golf courses. Many other projects are planned, with the objective of covering some 30,000 ha over the next few years. However, mainly because of restrictions on vegetables, many farmers continue to prefer to irrigate using well water. Farmers in the southern driest part of the country also express concern regarding the long-term effects of using salty wastewater on their soils. To address these problems, Tunisian policy-makers are currently rethinking their national regulations and strategy.

[e] Authority, here, should be taken in its original, Weberian sense as combining legitimacy to exercise decision-making power and a capacity to do so.

Aside from fuzziness in institutional prerogatives, lack of proper coordination can also be a serious impediment and, as such, has been widely criticized. It is linked to a lack of institutional arrangements concerning water cycle management in general, without which the potential for water reuse cannot be adequately assessed. At the national level, overcoming this shortcoming requires creating a functional interministerial committee, with a permanent secretary, regular follow-up meetings, and clear decision-making rules. Second-order coordination is also easier once an overarching policy has been adopted. This is why many Mediterranean countries are seeking to transition from mere regulatory frameworks to integrated national strategies. Lebanon is a case in point. Its national strategy for sanitation, passed in 2012, combined a program for building treatment plants (as only 8% of wastewater is currently safely treated) with a detailed list of priority reuse projects and financial mechanisms. Expenditures were expected to total US$ 3.1 billion. However, program implementation was interrupted by political uncertainties and the fallout of the Syrian conflict. In February 2016, Jordan also passed a "national policy for water substitution and reuse." It provides detailed guidelines for water rates, both for "pure" wastewater and for wastewater mixed with more conventional waters [47]. It also sets up specific decentralized arrangements for small cities.

At the local level, the efficiency of a reuse project can only be improved by incorporating water reuse projects in sanitation plans and master plans for water management at the basin level right from the outset. A number of reuse projects have been hampered or rendered impossible by the failure to plan wastewater treatment, reuse, and disposal together in a consistent way. Administrative reorganization may be necessary but in itself is not enough to prevent such situations arising. Rationalizing integrated water management and planning requires new tools: integrated technical-economic models and decision-making tools to work out and compare different water management scenarios, for example, technical-economic evaluation and comparing wastewater reuse options with other alternatives [48]. However, we should be fully aware that these optimization strategies have their limits: in so far as recycling projects allocate costs and benefits, while redistributing decision-making power and social legitimacy, they are inescapably political. The politics are managed differently depending on local, national, and even international political-institutional environments in Mediterranean countries.

5.3.3 Beyond optimization: The politics of treated water reuse projects

So far, most studies of wastewater reuse have prioritized a technical focus, that is, the performance of treatment processes and, more recently, impacts on human health. As for the—more limited—social science literature, it has mostly operated under the assumption that treated water reuse was self-evidently beneficial, and therefore centered on explaining why it had not yet developed further. In so doing, it has largely framed its issue of concern as an acceptance problem (for a detailed review, see Ref. [49]).

For example, a recent public consultation report compiled by Deloitte for the European Commission [50] highlighted that, while around 70% of respondents thought water reuse was valuable for adaptation to climate change and resource efficiency, only 30% were in favor of water reuse involving contact with food and people. To overcome this social "barrier," Po et al. [51] show that the worries harbored by the European population diminish notably when awareness and knowledge increase and residents can make an "informed choice," with awareness campaigns and information seen as instrumental in moving away from seemingly emotional [52] public responses. However, this public acceptance literature has two limits.

First, it has focused mostly on advanced, especially European countries, leaving aside countries located on the southern and eastern shores of the Mediterranean. This matters because very often, in these countries, treated wastewater reuse replaces previous irrigation with untreated wastewater. Trajectories are therefore unmistakably positive, which can only increase the level of social acceptance. However, anecdotal evidence in Settat, Morocco, point to a 10%–15% discount for parsley irrigated with treated wastewater on the local market. Some farmers, however, reported increased prices because the parsley looked greener. In any case, the lack of traceability prevents any observation of consumers' acceptance beyond local markets, as consumers are seldom aware of the mode of irrigation used for the product they buy.

Second, the social dimension of water reuse cannot be reduced to an acceptance problem, for it is, just like any infrastructure project, an inherently political issue. The first sensitive issue concerns the selection of beneficiaries: how to justify the inclusion of some and the exclusion of others, invoking a greater good in a way that might be broadly considered acceptable? This question arises especially in agriculture, where the production of new water resources with potentially important fertilizing properties is bound to generate strategies of appropriation and control by interested actors. In Settat, for example, the selection of beneficiaries caused considerable controversy, as before the treatment plant was built, the raw wastewater was collected by farmers who flanked the river. However, both the plant and the new perimeter were located downstream from this former territory, thereby depriving these farmers of irrigation water, and in the process cutting by half the rental value of their land. Financial compensations were discussed but not implemented. A similar process happened in Marrakech where wastewater used to irrigate golf courses no longer benefited famers who used it to grow wheat and olive trees downstream [13].

Another issue relates to political accountability. Who should take the blame, and the related economic costs, if a project malfunctions, be it due to reduced water quality, a decrease in farm outputs due to an increase in salinity, an outbreak of cholera or complaints related to bad smells or even visual pollution (as happened in Marrakesh where clients complained about algae development on the ponds)? Given the difficulty of properly assessing risks and proposing adequate remedies, different stakeholders can be expected to engage in blame avoidance strategies [53]. These strategies may include scapegoating (blaming others), passing the buck (forcing somebody else to take the blame), wholesale defection (starting to criticize the whole project as misguided), and circling the wagon (co-opting dissent to contain blame, even at the cost of not solving the problem).

However, the opposite dynamics, namely credit claiming, is also pervasive, in so far as treated water reuse projects deliver socially valued benefits. In this respect, initiating a project can be a way to enhance one's organizational standing, as may be the case for river basin agencies in Morocco, whose position within the water subsystem is structurally fragile. Likewise, leading a wastewater users' association can strengthen one's social visibility and status. Here the logic tends to be self-reinforcing: leaders tend to already be notables, as members trust them more to represent their interests vis-a-vis State agencies; but assuming these responsibilities leads them, in turn, to accumulate more social legitimacy. This legitimacy might be later used for other pursuits, including a local political career.

Finally, the politics of treated water reuse project revolves a lot around issues of political economy, as the way costs are distributed can be hotly debated (see Section 5.4.3).

5.4 Field issues related to sustainable water reuse

The sustainability of water reuse in agriculture is strongly dependent on the farmer and irrigation practices, and more generally on the capacity of the irrigation system manager to control unintended impacts. Therefore it mainly relies on the irrigation technology selected, its design management, and maintenance that will result in homogeneous water distribution, in space (uniformity of water application) and over time (proper irrigation scheduling). The impacts on soil, plants, and the environment will then be easier to manage. These impacts need to be identified clearly, considered and managed transparently to improve public awareness and perception of WWR.

5.4.1 Irrigation technologies and reuse practices

In Mediterranean countries, voluntarist policies for modernizing irrigation technologies used by farmers are being implemented, and may even provide comprehensive financial support for the purchase of said water-saving irrigation equipment. However, reality is never idyllic and performances are often far below expectations [54]. The same applies to old equipment whose distribution performance is so bad that water use efficiency can fall to under 50% (irrigation efficiency can be defined as the proportion of water applied that is effectively used by the crop for its growth). This poses environmental as well as agronomical problems in the case of salty waters that will be reinforced when wastewater is reused in semiarid regions (i.e., where winter rainfall is not sufficient to leach out excess salt).

It is thus crucial to maintain, or achieve, a high level of water use efficiency in agriculture by implementing consistent resource management through: (i) well understood and applied scheduling methods, (ii) properly designed and managed irrigation systems, and (iii) adapted maintenance methods, preferably preventive rather than curative. When reusing wastewater, any drop in irrigation efficiency may result in pollution or contamination through unintended or prolonged contact with the product, dispersion or runoff of contaminated water, bad use of nutrients, and accumulation of salt or micropollutants that may later be leached to the groundwater.

The efficiency concept can apply at different spatial scales (from the irrigation block or perimeter to the river basin); here, we focus on the plot scale. It should be noted that the question of irrigation efficiency is still the subject of debate, as a high level of efficiency does not necessarily imply significant water savings. This will depend on the spatial scale addressed, for example in gravity irrigation, efficiency may be poor when up to 50% of the water flows out of the plot (5%–10% in drip irrigation), but is much better at a larger scale, given that the outflow will be used in plots downstream [55].

5.4.1.1 Design and management of irrigation systems

On the farm, the use of wastewater involves additional constraints. For example, depending on the treatment process, the irrigation systems need to be designed to take into account the rich nutrient concentrations and suspended solids [56], or the unintentional dispersion of pollutants (pathogens, bacteria and antibiotic resistance genes, etc.). In several of the Mediterranean countries, regulations for the use of wastewater were established based on an integrated approach to irrigation technologies (Fig. 5.1). This is the case in France where constraints inherent in irrigation technology were taken into account in the regulation (the law of August 2, 2010, revised on June 25, 2014); the same applies for Israel where the irrigation technology neither or nor plays a role of barrier. Three distinct irrigation methods are available for water application in agriculture:

(A) (B) (C)

Fig. 5.1 (A) Surface irrigation on olive trees, Morocco, (B) sprinkler irrigation on potatoes, France, and (C) drip irrigation on vineyard with reclaimed wastewater, France.

surface/gravity irrigation, pressurized sprinkler irrigation, and drip irrigation. These three methods are able to reach good level of efficiencies (typically up to 90% for drip and sprinkler and up to 75% for surface irrigation). Cost is also highly variable, mainly human time for surface irrigation, mostly capital for pressurized. The observed life duration varies also greatly depending on maintenance, water quality, and sensitivity of the technique. Each has benefits and risks in the wastewater reuse cases. These sources of variability in technological performance of WWR projects can be effectively controlled applying suitable management and maintenance practices designed to decrease the potential level of hazards.

Gravity or surface irrigation (immersion, furrow, etc.) is still the most widely used technique in Mediterranean agriculture and in the world, despite billions of dollars of government subsidies for pressurized irrigation systems [55]. Gravity systems take advantage of the potential energy of water to flood the plots as homogeneously as possible through soil leveling, regulating plot inlet flow, and the length of the irrigation turn. This technology is the lowest cost technology, requires filed practices that are well known traditionally, and restricts any pathogens or pollutants aerial dispersion. This technique is insensitive to any clogging related to water quality. In gravity irrigation, the main problems are avoiding (i) both under-irrigated and over-irrigated zones, to prevent salinization and water logging, which may result in progressive soil clogging and accumulation of different pollutants (this will be obtained through a consistent field or furrows leveling); (ii) direct contact with the product: root crops mainly and the farmer/operator. Consequently, surface irrigation with wastewater is usually limited to unplanned or indirect reuse.

Sprinkler irrigation, which tends to reproduce homogeneous rainfall over the entire plot (field crops) or restricted area explored by the roots (trees), limits potential contact between the irrigator and contaminated water. But hazards are related to the unintentional release of effluents in fine droplets that may contain pathogens or other harmful contaminants when sprinklers operate under different wind regimes [57]. National regulations generally require high water quality (pathogen reduction) and advanced treatments before permitting reuse of wastewater for agricultural purposes. In sprinkler irrigation, the point of concerns relate to (i) the quantification of airborne dissemination of fine droplets through drift and transport that may contain pathogens and pollutant [57], (ii) the management of instantaneous application rate (rainfall intensity and kinetic energy) that may result in runoff, (iii) the range of sprinkler and spacing between them to achieve a good uniformity, (iv) the setting of pressure to limit drift volume potential under windy conditions, (v) the position of sprinklers (over/under crop canopy) and resulting contact/contamination of leaves and production.

Drip irrigation is widely promoted as a way to improve plot-level water efficiency [58], which is true when the equipment is new but also depends on farmer practices

[54]. Drip irrigation has the potential for much greater efficiency than other irrigation techniques because water is applied directly in the immediate vicinity of the plant, thereby increasing water productivity [59]. Drip irrigation systems can also improve nutrient use efficiency (through fertigation), reduce energy requirements, and improve cultivation practices [60]. In water reuse, this technology is considered the best option in terms of protection against contact with and dispersion of pathogens. Nevertheless, the long-term sustainability of drip irrigation is not yet ensured. The presence of nutrients and ions in wastewater increases the possibility of chemical precipitation and of the development of biofilm in the drippers [61–65]. In addition, clogging due to sedimentation of suspended solids can rapidly lead to a significant decrease in distribution performance [66, 67]. At the field level, the additional concentration of solids in wastewater compared to conventional water is not necessarily taken into account in the design and installation of the equipment (high-flow rate drippers, filtration, etc.) or in the maintenance protocols for drip irrigation systems (by chlorination and acidification). This can mean rapid loss of investment and subsidies for drip installation or the massive use of disposable dripper systems (3–6 months of shelf life?), which is not environmentally satisfactory. A better understanding of clogging phenomena and maintenance solutions requires further research into flow topology within narrow labyrinth channels, biofilm characteristics, and response mechanisms that account for both nutrient transport and hydraulic shear forces, and into chemical precipitation. In addition, research is also required into the barrier or assumed "positive" effects of biological fouling of pipes and drippers by reducing pathogenic or chemical pollution [22].

5.4.2 Environmental, agro-systems, sanitary and risk aspects

The impact of the quality of irrigation water is the most important issue to address in the case of water reuse projects. The management of risks may determine the perception and hence the acceptability of these practices. Depending on the category of wastewater, impacts on health, the environment, agricultural yields, and economic-technological criteria can be positive or negative. The success and sustainability of water reuse projects in Mediterranean countries will depend to a great extent on controlling these factors.

5.4.2.1 Environmental impacts

From an environmental point of view too, the impacts of water reuse in agriculture can be positive and negative. Virtuous agricultural practices, consistent irrigation management, and efficient treatments will limit the impact of water reuse on the environment [5] and make it possible to reserve good quality water for more sensitive uses such as drinking water. Regarding impacts on the soil, reusing wastewater can progressively increase the salt content in the soil, jeopardizing soil stability and changing its osmotic properties. Excessive salt content will lead to dispersion of soil aggregates, reduce hydraulic conductivity and aeration of the soil matrix [68]. Further, groundwater pollution is highly dependent on biogeochemical processes in the unsaturated zone. These points are critical for the protection of soil fertility against salinization, alkalinization, and degradation of the soil structure. The influence of water reuse on agricultural soils properties is mainly due to its content: nutrients (mainly nitrogen and phosphorus), dissolved salts (sodium, boron, etc.), and other constituents such as heavy metals that may accumulate in the soil over time. Regarding salts, in arid or semiarid conditions, they may accumulate in the root zone because of the absence of leaching which occurs when winter rainfall exceeds the soil water holding capacity. Over time, the absence of leaching may have harmful impacts on soil biological and chemical activity and hence on crop yields. The problem of soil salinity and sodicity can be solved by increasing the soil mineral retention capacity through the application of natural (manure, compost) or artificial soil conditioners (e.g., polyacrylamide compounds [69]). However, such soil reclamation measures are costly, adding an economic burden that reduces in crop financial productivity.

At the other end of the scale, over-irrigation using wastewater can damage the quality of groundwater resources in the long run through the delivery of excess nutrients and salts caused by wastewater leaching below the plant root zone [70]. Particularly if the farmers' fertilization practices do not take the nutrient concentrations in wastewater into account. Pollution effects will also depend on factors including the depth of the groundwater table, soil hydraulic conductivity, and the quality of the wastewater used for irrigation. Thus, the potential for groundwater contamination needs to be evaluated before starting on a major wastewater irrigation program. In addition to the accretion of salts and nitrates, wastewater irrigation can, under certain conditions, transfer pathogenic bacteria and viruses to groundwater, that is, in karst systems. However, it should be noted that wastewater can be also reused for artificial groundwater recharge, as was the case of the coastal groundwater of Korba, Cap-Bon, Tunisia [71] or as a coastal barrier against seawater intrusion.

5.4.2.2 Impacts on plants

Concerning agronomic issues, several studies of water reuse in agriculture have been performed on different crops including some eaten raw, such as lettuce [72], lemon [73], eggplant, and tomato [40]. The authors of these studies concluded that, beyond the benefits of using reclaimed wastewater for irrigation, such as increased soil nutrients, increased crop yield and reduced doses of fertilizer, depending on the level of wastewater treatments, some drawbacks including soil salinization, contamination of sensitive crops (vegetable crops) by pathogens and loss of soil infiltration capacity remain a concern [73]. Regarding nutrients, in their study, Qadir et al. [74] reported that for every $1000\,m^3$ of reclaimed wastewater, from 4 to 24 kg of phosphorus, 16 to 62 kg of nitrogen, 2 to 69 kg of potassium, plus 18 to 208 kg of calcium, 9 to 110 kg of magnesium, and 27 to 182 kg of sodium may have positive or negative effects when accumulated. If using wastewater is well managed, it is a form of a fertigation that can reduce the need for chemical fertilizer and save money. From an agronomical point of view, in a controlled experimental system under semiarid conditions in Sicily, Italy, Cirelli et al. [40] demonstrated that reclaimed wastewater can increase tomato yield by 20% in comparison with crops using conventional water. In contrast, eggplants were shown to be very sensitive (with a reduction in yield of around 22%) due to clogging of the drippers caused by calcium carbonate precipitation and high pH. Clogging resulted in plant water stress and a decrease in the quality of the eggplant produced. However, studies that account for farmers' adaptations and practices at the scale of large fields irrigated using wastewater in Mediterranean context and assess crop yield are rare, and farmers' practices and their effects on productivity call for further investigations involving extension services along with researchers.

5.4.2.3 Health impacts

Concerning the impact of irrigation with wastewater on health, as shown by the examples in different Mediterranean countries reported in this chapter, knowledge of wastewater reuse and management of health risk varies with the regional context and with the country concerned. In the many water reuse projects in Jordan or Tunisia with adequate treatment levels and appropriate irrigation practices, no diseases associated with wastewater reuse have been reported. In addition, current treatment technologies are able to reduce the concentration of pathogens in water below the level of detection used by the monitoring methods. However, national regulations require careful monitoring of parameters, usually based on WHO recommendations, but these can be difficult to apply in practice (too many parameters to monitor at reasonable cost, very few long-term experiments) and are not enforced in the case of unplanned or indirect water reuse. Where the wastewater treatment process is not effective or is absent, it is suspected that many pathogenic microorganisms (bacteria, parasites, viruses) are transported and disseminated by wastewater, which can lead to diseases or to the contamination of farmers and consumers. This is rendered even more complicated by the huge variability of wastewater microbiological composition depending on the season, the origin of the water collected, the health status of the population, and on the wastewater treatment process itself and its efficacy (see Box 5.4). In addition, the irrigation technologies and type of crop (e.g., if the edible part is in contact with the soil) will play a key role in the level of transmission of these pathogenic microorganisms. When distributed by sprinklers, dispersion on the soil and possibly on fruits and the aerosolization process are taken into consideration, while in the case of drip irrigation, contact with the soil is the point of vigilance. The detection and exhaustive quantification of these pathogenic microorganisms is very difficult. This is the reason why only indicators are to be monitored [36, 75]. In water reuse regulations or guidelines this mainly concerns [36]:

BOX 5.4 Cholera crisis in Algeria summer 2018.

In summer 2018, a severe cholera outbreak occurred in the small village of Sidi Kebir located in the district of Tipaza, and rapidly spread to the city of Blida not far from Algiers. The unplanned use of wastewater in agriculture was immediately suspected of being responsible for the death of a number of people. The Algerian authorities rapidly prohibited the use of wastewater for irrigation throughout the country but the consumption of summer crops particularly watermelon also decreased drastically. The halt to the unplanned reuse of wastewater did not account for the level of wastewater treatment, dilution with surface water, or after a sanitary analysis of real crop contamination by vibrio cholera.

Today, only state planned cases of wastewater reuse are authorized and the public policy is focused on developing the tertiary treatment stage in wastewater treatment plants. In conclusion, the general perception is that cholera is closely linked to poor wastewater reuse practices in agriculture and that farmers should find another resource when safe waters are not available. Yet, the validity of crop contamination remains to be analyzed.

- Bacteriophage viruses (Somatic Coliphages, RNA-F Coliphages),
- Bacteria (thermotolerant coliforms, *Escherichia coli*, intestinal Enterococci),
- Protozoa (*Clostridium perfringens*)
- Helminthes (Ascaris eggs).

The main factors involved in exposure to pathogens are ingestion of wastewater, consumption of food that has been in contact with wastewater, or inhalation of bioaerosols contaminated with pathogens. Several issues remain to be addressed, including the multiple barrier effects after treatment or storage, the dynamics of contaminants during water transport and irrigation [22, 76] and effects on soil microbiota, which have also been ignored in most water reuse irrigation studies. In a context of reusing untreated or nonoptimal wastewater, the effects of these biological contaminants (human and animal pathogens, phytopathogens, antibiotic-resistant bacteria, and antibiotic resistance genes) are rarely studied at large field scale.

In addition to microorganisms, it should be noted that despite national regulations, industrial wastewaters are rarely pretreated or separated from urban wastewater in the Mediterranean region. Microcontaminants are also widely found in urban wastewater, including pharmaceuticals, antibiotics, chemicals from personal care products, endocrine-disrupting compounds, plasticizers, flame retardants, surfactants, illicit drugs, transformation products, and pesticides. Although the concentrations are usually very low (in the nanomolar range), their impacts on ecosystems and human health have not been extensively studied to date. Of course, industrial wastewater may present a more serious threat because certain chemicals may be present at higher concentrations. In the case of endocrine disrupting compounds, for example, these can mimic the activity of certain hormones and interfere with endocrine system functioning. Concerning micropollutants, in Europe, many discussions are underway to decide if it is necessary to add this list of pollutants in the water reuse regulations. Many of these substances are resistant to conventional wastewater treatment and persist in the environment. They are spread during irrigation on fields and can then reach surface water rivers or groundwater. Antibiotics and their metabolites are mainly detected in hospital effluents. The characteristics of this wastewater, which is also rich in microorganisms, can provide ideal conditions for ecosystem evolution, transfer, and the development of resistance genes [77]. Persistence of these compounds accelerates the development of resistance genes and of bacterial resistance that may increase the selection pressure on environmental microorganisms.

5.4.3 Cost-effective risk management in wastewater use systems

Different approaches have been proposed to mitigate health risks. Traditionally, the approaches used have placed the emphasis on water quality and strict regulations at the place of use, making the treatment of wastewater essential for water reuse [29].

However, in low-income areas, stringent water quality standards for reuse are often perceived as unaffordable and consequently fail. The most recent WHO guidelines [35] for the safe use of wastewater acknowledge that reducing health risks is necessary and possible even in scenarios where wastewater is not treated at all or is treated improperly. In places where there are no viable alternatives to irrigation with contaminated water, the guidelines also enable the adoption of risk mitigation options on-farm and in postharvest operations [78]. This broader security concept follows the principles of the Hazard Analysis and Critical Control Point (HACCP) approach, which is widely used in the food industry and in drinking water safety management, and is in accordance with the Convention on Stockholm for the development of harmonized guidelines [35].

The WHO [35] guidelines propose the use of the multiple barrier approach along the food and sanitation chain, from the generation of wastewater to the consumption of food, instead of focusing only on the quality of the waters at the place of use (Fig. 5.2). For example, barriers can be placed at wastewater generation (low-cost treatment), on agricultural land (sedimentation lagoons, crop restriction, or safer irrigation techniques), in markets (washing crops with cleaner water), and even at the consumer level (disinfection, cooking, or dehulling). These practices must go hand in hand with people's exposure control (e.g., gloves and boots for farmers) and may require the participation of a wide range of stakeholders, beyond those responsible for wastewater treatment facilities. This approach is more cost-effective and feasible in countries or cities where the treatment and exhaustive disinfection of wastewater are failing, due to the high costs of construction, operation, and maintenance of the treatment plants, which also require high technical and management skills. The multiple barrier approach allows for a more flexible regulatory system, in tune with the socioeconomic realities of the country or locality. The

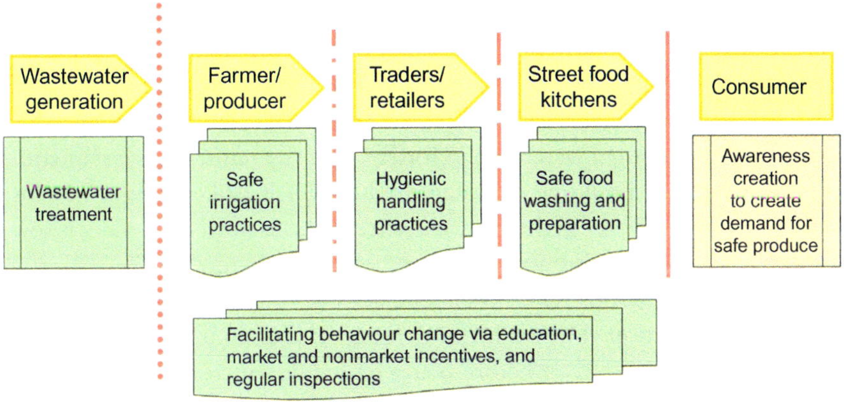

Fig. 5.2 The multiple-barrier approach as applied in wastewater irrigation [79].

reader will find examples of its application in countries with different levels of development in the guides published by the WHO [37].

5.5 Main conclusions

As the effects of climate change intensify, the levels of rainfall steadily decline and aquifers become systematically overexploited across the Mediterranean region, planned water reuse for agriculture is increasingly put forward by national governments as a valuable solution within a wider menu of options. There is little doubt that it costs less to produce and that it generates less environmental externalities than the other mainstream type of "unconventional water," desalination. Additionally, the nitrogen and phosphorus contained in wastewater reduce the need for fertilizers by at least 50%. Also, by providing a regular and predictable volume throughout the year, water reuse can help maintain peri-urban agriculture, thereby containing urban sprawl.

However, many major challenges still lie ahead that prevent a massive upscaling. First, the water produced is expensive for most agricultural use. If the full operation and maintenance costs were to be recovered, wastewater could generally not compete with wells, especially when energy costs are subsidized. Water reuse for agriculture (as opposed to green spaces, golf courses, or water-intensive industrial processes) only makes strict economic sense for high-value crops and where there is no alternative resource readily available at a reasonable cost. Even then, operation and management may still need to be partly subsidized. Second, the risks for human health, however limited, have important implications: the procedures for monitoring water quality are expensive and often cumbersome, whereas State officials and agencies might be reluctant to authorize projects and assume formal responsibilities for fear of a public backlash if a health crisis ever erupted. The potential for barrier effects between treatment and use points (along the irrigation systems and supply chains until final consumption) is a promising avenue but remains to be assessed. Third, as a crosscutting issue spanning many established policy domains, the governance of water reuse for agriculture is complex. However, this complexity can be reduced when wastewater is reused indirectly, for instance by being stored back in dams and mixed with conventional water like in Jordan, thereby reintegrating standard policy subsystems. This option should be considered whenever it makes economic sense.

Even then, it is useful to recall the obvious: water reuse is not a new water resource, or only secondarily so (when wastewater was previously discharged in the sea or when it evaporated in treatment ponds with no beneficial use). In this sense, it is fundamentally different from desalinization. The reallocation it entails is often innocuous, when wastewater was highly diluted in a river before being reused downstream. But it can be more contentious when wastewater was either already reused directly (typically informally), or when it played a major support role during periods of low river flow. In such cases, reuse projects can still be justified on an economic basis, but former users need to be compensated for their loss, either economically or by being allowed to access other sources of water.

Finally, much like the water cycle in general, the water reuse chain for agriculture has chemical, hydrological, agronomical, economic, social, and political dimensions that cannot be viewed in isolation from each other when it comes to tackling real-world problems and situations. Interdisciplinary approaches and interdisciplinary teams of researchers should be encouraged. For instance, many fields of expertise should be called for to conduct sensible cost-benefit analysis where all types of relevant costs and benefits are appropriately considered and weighted. Different backgrounds are also appropriate to assess the many social, economic, and agronomical implications of technological innovations such as small-scale, decentralized treatment facilities.

References

[1] A.L. Thebo, P. Drechsel, E.F. Lambin, Global assessment of urban and peri-urban agriculture: irrigated and rainfed croplands, Environ. Res. Lett. 9 (11) (2014) 114002.

[2] J. Mateo-Sagasta, L. Raschid-Sally, A. Thebo, Global wastewater and sludge production, treatment and use, in: Drechsel, et al., (Ed.), Wastewater: Economic Asset in an Urbanizing World, Springer, 2015.

[3] A.N. Angelakis, T. Asano, A. Bahri, B.E. Jimenez, G. Tchobanoglous, Water reuse: from ancient to modern times and the future, Front. Environ. Sci. 6 (2018).

[4] S. Chittoor Jhansi, S. Kumar Mishra, Wastewater treatment and reuse: sustainability options, J. Sustain. Dev. 10 (1) (2013) 1–15.

[5] V. Lazarova, B. Akica, Water Reuse for Irrigation: Agriculture, Landscapes, and Turf Grass, CRC Press, 2004.

[6] M. Kellis, I.K. Kalavrouziotis, P. Gikas, Review of wastewater reuse in the Mediterranean countries, focusing on regulations and policies for municipal and industrial applications, Global NEST J. 15 (3) (2013) 330–350.

[7] A. Karnib, Research paper assessing population coverage of safely managed wastewater systems: a case study of Lebanon, J. Water Sanit. Hyg. Dev. (2016) 313–319.

[8] T. Asano, Water re-use, Chem. Eng. News 44 (12) (2010) 90–102.
[9] A.L. Thebo, P. Drechsel, E.F. Lambin, K.L. Nelson, A global, spatially-explicit assessment of irrigated croplands influenced by urban wastewater flows, Environ. Res. Lett. 12 (7) (2017).
[10] J.E. Drewes, U. Hübner, V. Zhiteneva, S. Karakurt, Characterization of unplanned water reuse in the EU, Garching: Technical Univ. of Munich (prepared for the European Commission DG Environment), 2017, pp. 1–61.
[11] J. Lahnsteiner, P. van Rensburg, J. Esterhuizen, Direct potable reuse—a feasible water management option, J. Water Reuse Desalin. 8 (1) (2018) 14–28.
[12] P. Dugué, E. Valette, Des agriculteurs marginalisés au coeur des villes: le cas de Meknès (Maroc), GREP, Pour (2015) 61–67.
[13] O. Tanouti, F. Molle, Réappropriations de l'eau dans les Bassins Versants Surexploités. Le Cas du Bassin du Tensift (Maroc), Etud. Rurales 192 (2) (2013) 79–96.
[14] P.L. Mayaux, A. Massot, Entre décharge participative et Etat développeur: des élites locales marocaines en quête de légitimité, Gouvernance 7 (1) (2019) 84–109.
[15] CREM, Ministère de l'Équipement, du Transport, de la Logistique et de l'Eau-Département de l'Eau-Deutsche Gesellschaft für Internationale Zusammenarbeit (GIZ) Gmb H, Etat des lieux du secteur de l'eau au Maroc, 2018. Projet "CREM" Coopération Régionale pour une Gestion Durable des Ressources en Eau au Maghreb.
[16] L'Economiste, Réutilisation des eaux usées: Un plan en cours de finalisation, 2017, March, 23).
[17] M. Belghiti, Eau usée en agriculture Expériences et perspectives, Présentation, Direction de l'Irrigation et de l'Aménagement de l'Espace Agricole, atelier AGIRE, GIZ, 4 et 5 février, 2013.
[18] UNWWDR, UN-World Water Development Report, Wastewater the Untapped resource, http://www.unesco.org/new/en/natural-sciences/environment/water/wwap/wwdr/, 2017. (Accessed 16 January 2018).
[19] A.A. Ulimat, Wastewater production, treatment and use in Jordan, in: Water Authority Jordan. Ministry of Water & Irrigation. Second Regional Workshop 'safe use of wastewater in Agriculture', 16–18 May 2012, New Delhi, India, 2012.
[20] Water Governance in Jordan, OECD, 2014.
[21] N. Ait-Mouheb, et al., The reuse of reclaimed water for irrigation around the Mediterranean Rim: a step towards a more virtuous cycle? Reg. Environ. Chang. 18 (3) (2018).
[22] K. Lequette, N. Ait-Mouheb, N. Wéry, Drip irrigation biofouling with treated wastewater: bacterial selection revealed by high-throughput sequencing, Biofouling 35 (2) (2019) 1–13.
[23] Ministère des ressources en eaux, Nouveau modèle de croissance économique. Stratégie du secteur de l'eau à l'horizon 2035, (2017), p.77.
[24] Ministère des ressources en eau, Plan national de l'eau, (2010), p.64.
[25] T. Hartani, La réutilisation des eaux usées en irrigation: cas de la Mitidja en Algérie, La Mod. l'agriculture irriguée. Actes du séminaire Euro-Méditerranéen, 2005, April.
[26] J. Radkau, The Age of Ecology, Polity Press, 2014.
[27] U. Beck, Risk Society, Sage Publications, Londres, 1992.
[28] FAO, Water quality for agriculture, in: R.S. Ayers, D.W. Westcot (Eds.), FAO Irrigation and Drainage Paper 29 (Rev. 1), Food and Agriculture Organization of the United Nations, Rome, 1985.
[29] W.H.O. Health, Guidelines for the Use of Wastewater in Agriculture and Aquaculture. Report of a WHO Scientific Group, Technical Report Series 778, World Health Organization, Geneva, 1989.
[30] P. Fine, R. Halperin, E. Hadas, Economic considerations for wastewater upgrading alternatives: an Israeli test case, J. Environ. Manag. 78 (2006) 163–169.
[31] R. Aertgeerts, A.N. Angelakis, State of the Art Report: Health Risks in Aquifer Recharge Using Reclaimed Water, WHO, Water, Sanitation and Health Protection and the Human Environment, Copenhagen, WHO, Geneva and WHO Regional Office for Europe, 2003.
[32] WHO, Guidelines of the Safe Use of Wastewater, Excreta and Grey Water; Vol. 2: Wastewater Use in Agriculture, World Health Organization, Geneva, Switzerland, 2006.

[33] State of California, Wastewater Reclamation Criteria, an Excerpt From the California Code of Regulations, Title 22, division 4, environmental health. Sacramento: Dept. of Health Services, 1978.
[34] State of California, "The Purple Book" Excerpts from the Health and Safety Code, Water Code, and Titles 22 and 17 of the California Code of Regulations, Department of Health Services, Sacramento, California, USA, 2001.
[35] F. Licciardello, M. Milani, S. Consoli, N. Pappalardo, S. Barbagallo, L.G. Cirelli, Wastewater tertiary treatment options to match reuse standards in agriculture, Agric. Water Manag. 210 (2018) 232–242.
[36] WHO, Water Quality: Guidelines, Standards and Health—Assessment of Risk and Risk Management for Water-Related Diseases, 2001.
[37] WHO, Guidelines for the safe use of wastewater, excreta and grey water in agriculture and aquaculture, third ed., 2010. (Second information kit). Guidance Note for national programme managers and engineers: applying the guidelines along the sanitation ladder, IWMI.
[38] S. Barbagallo, L.G. Cirelli, S. Consoli, F. Somma, Wastewater quality improvement through storage: a case study in sicily, Water Sci. Technol. 47 (2003) 169–176.
[39] A.N. Angelakis, P. Gikas, Water reuse: overview of current practices and trends in the world with emphasis on EU states, Water Util. J. 8 (2014) 67–78.
[40] G.L. Cirelli, S. Consoli, F. Licciardello, R. Aiello, F. Giuffrida, C. Leonardi, Treated municipal wastewater reuse in vegetable production, Agric. Water Manag. 104 (Feb. 2012) 163–170.
[41] S. Lavrnic, M. Zapater-Pereyra, M.L. Mancini, Water scarcity and wastewater reuse standards in southern Europe: focus on agriculture, Water Air Soil Pollut. (2017) 228–251.
[42] F. Brissaud, Criteria for water recycling and reuse in the Mediterranean countries, Desalination 218 (2008) 24–33.
[43] J. Fawell, K. Le Corre, P. Jeffrey, Common or independent? The debate over regulations and standards for water reuse in Europe, Int. J. Water Resour. Dev. 32 (2016) 559–572.
[44] N.V. Paranychianakis, M. Salgot, S.A. Snyder, A.N. Angelakis, Water reuse in EU states: necessity for uniform criteria to mitigate human and environmental risks, Crit. Rev. Environ. Sci. Technol. 45 (13) (2015) 1409–1468.
[45] A.N. Angelakis, M.H.F.M. do Monte, L. Bontoux, T. Asano, The status of wastewater reuse practice in the Mediterranean basin: need for guidelines, Water Res. 33 (10) (1999) 2201–2217.
[46] A. Bahri, Wastewater reclamation and reuse in Tunisia, in: T. Asano (Ed.), Wastewater Reclamation and Reuse, Water Quality Management Library, vol. 10, Technomic Publishing Co., Inc, 1998, pp. 877–916.
[47] MWI (Ministry of Water and Irrigation), National water strategy of Jordan, 2016–2025, Hashemite Kingdom of Jordan (In text as: MWI-NWS, 2015), 2015.
[48] G. Zeng, R. Jianga, G. Huanga, M. Xua, J. Lia, Optimization of wastewater treatment alternative selection by hierarchy grey relational analysis, J. Environ. Manag. 82 (2007) 250–259.
[49] R. Beveridge, T. Moss, M. Naumann, A socio-spatial understanding of water politics: tracing topologies of water reuse, Water Altern. 10 (1) (2017) 22–40.
[50] BIO by Deloitte, Optimising water reuse in the EU–Final report prepared for the European Commission (DG ENV), Part I, in: Collaboration with ICF and Cranfield University. Luxembourg, Publications Office of the European Union, 2015.
[51] M. Po, J.D. Kaercher, B.E. Nancarrow, Literature Review of Factors Influencing Public Perceptions of Water Reuse, http://www.clw.csiro.au/publications/technical2003/tr54-03.pdf, 2004.
[52] D. Bixio, C. Thoeye, J. de Koning, D. Joksimovic, D. Savic, T. Wintgens, T. Melin, Wastewater reuse in Europe, Desalination 187 (2006) 89–101.
[53] R.K. Weaver, The Politics of Blame Avoidance, J. Publ. Policy 6 (4) (1986) 371.
[54] M. Benouniche, M. Kuper, A. Hammani, H. Boesveld, Making the user visible: analysing irrigation practices and farmers' logic to explain actual drip irrigation performance, Irrig. Sci. 32 (6) (2014) 405–420.
[55] P. Steduto, et al., The paradox of irrigation efficiency, Science 361 (6404) (2018) 748–750.
[56] A. Adin, M. Sacks, Dripper-clogging factors in wastewater irrigation, J. Irrig. Drain. Eng. 117 (6) (1991) 813–827.

[57] B. Molle, S. Tomas, L. Huet, M. Audouard, Y. Olivier, J. Granier, Experimental approach to assessing aerosol dispersion of treated wastewater distributed via sprinkler irrigation, J. Irrig. Drain. Eng. 142 (9) (2016) 04016031.

[58] D. Luquet, A. Vidal, M. Smith, J. Dauzat, 'More crop per drop': how to make it acceptable for farmers? Agric. Water Manag. 76 (2) (Aug. 2005) 108–119.

[59] F.R. Lamm, et al., Using Livestock Wastewater With SDI. A Status Report After Four Seasons, (2002), pp. 8–15.

[60] J.E. Ayars, et al., Subsurface drip irrigation of row crops: a review of 15 years of research at the water management research laboratory, Agric. Water Manag. 42 (1) (1999) 1–27.

[61] W. Niu, L. Liu, X. Chen, Influence of fine particle size and concentration on the clogging of labyrinth emitters, Irrig. Sci. 31 (4) (2013) 545–555.

[62] A. Capra, B. Scicolone, Emitter and filter tests for wastewater reuse by drip irrigation, Agric. Water Manag. 68 (2) (2004) 135–149.

[63] S. Gamri, A. Soric, S. Tomas, B. Molle, N. Roche, Biofilm development in micro-irrigation emitters for wastewater reuse, Irrig. Sci. 32 (1) (2014) 77–85.

[64] N. Ait-Mouheb, et al., Impact of hydrodynamics on clay particle deposition and biofilm development in a labyrinth-channel dripper, Irrig. Sci. 37 (1) (2019).

[65] N. Rizk, N. Ait-Mouheb, G. Bourrié, B. Molle, N. Roche, Parameters controlling chemical deposits in micro-irrigation with treated wastewater, J. Water Supply: Res. Technol.—AQUA 66 (8) (2017) 587–597.

[66] S. Bounoua, S. Tomas, J. Labille, B. Molle, J. Granier, P. Haldenwang, S.N. Izzati, Understanding physical clogging in drip irrigation: in situ, in lab and numerical approaches, Irrig. Sci. 34 (2016) 327.

[67] F.C. Oliveira, R. Lavanholi, A.P. Camargo, A. Frizzone, N. Ait-Mouheb, S. Tomas, B. Molle, Influence of concentration and type of clay particles on dripper clogging, Irrig. Drain. Syst. Eng. 6 (1) (2017) 1–5.

[68] V.R. Urbano, T.G. Mendonça, R.G. Bastos, C.F. Souza, Effects of treated wastewater irrigation on soil properties and lettuce yield, Agric. Water Manag. 181 (2017) 108–115.

[69] V.S. Green, D.E. Stott, Polyacrylamide: A review of the use, effectiveness and cost of a soil erosion control amendment, in: D.E. Stott, R.H. Mohtar, G.C. Steinhardt (Eds.), Proceedings of the 10th International Soil Conservation Organization Meeting, Purdue University and the USDA-ARS National Soil Erosion Research Laboratory, West Lafayette, IN, USA, 2001, pp. 384–389.

[70] G.J. Levy, P. Fine, A. Bar-tal, Treated Wastewater in Agriculture: Use and Impacts on the Soil Environment and Crops, Wiley-Blackwell, 2011.

[71] F. El Ayni, S. Cherif, A. Jrad, M. Trabelsi-ayadi, Impact of Treated Wastewater Reuse on Agriculture and Aquifer Recharge in a Coastal Area : Korba Case Study, (2011), pp. 2251–2265.

[72] R.D.S. Cuba, et al., Potencial de efluente de esgoto doméstico tratado como fonte de água e nutrientes no cultivo hidropônico de alface (Ambient. e Agua-An Interdiscip), J. Appl. Sci. 10 (3) (2015) 574–586.

[73] F. Pedrero, I. Kalavrouziotis, J.J. Alarcón, P. Koukoulakis, T. Asano, Use of treated municipal wastewater in irrigated agriculture—review of some practices in Spain and Greece, Agric. Water Manag. 97 (9) (Sep. 2010) 1233–1241.

[74] M. Qadir, B.R. Sharma, A. Bruggeman, R. Choukr-Allah, F. Karajeh, Non-conventional water resources and opportunities for water augmentation to achieve food security in water scarce countries, Agric. Water Manag. 87 (1) (Jan. 2007) 2–22.

[75] P. Drechsel, C.A. Scott, L. Raschid-Sally, M. Redwood, A. Bahri, Wastewater Irrigation and Health: Assessing and Mitigating Risk in Low-Income Countries, Earthscan-IDRC-IWMI, UK, 2010.

[76] Y. Yao, O. Habimana, Biofilm research within irrigation water distribution systems: trends, knowledge gaps, and future perspectives, Sci. Total Environ. 673 (2019) 254–265.

[77] F. Cerqueira, V. Matamoros, J.M. Bayona, T.U. Berendonk, L.M. Hornstra, B. Piña, Antibiotic resistance gene distribution in agricultural fields and crops. A soil-to-food analysis, Environ. Res. 177 (2019) 18–26.

[78] B. Keraita, J. Mateo-Sagasta, M. Qadir, K. Medlicott, P. Drechsel, B. Lamizana, Chapter 4: Risk mitigation in wastewater irrigation systems and examples in Latin America and the Caribbean, in: J. Mateo-Sagasta (Ed.), Safe and Productive Use of Wastewater in Latin America and the Caribbean: Principles, Status and Needs, Food and Agriculture Organization of the United Nations, Santiago de Chile, 2017.

[79] P. Amoah, B. Keraita, M. Akple, P. Drechsel, R.C. Abaidoo, F. Konradsen, "Low-Cost Options for Reducing Consumer Health Risks from Farm to Fork where Crops Are Irrigated with Polluted Water in West Africa" IWMI Research Report 141, Colombo, Sri Lanka, 2011.

Further reading

WWAP (United Nations World Water Assessment Programme), The United Nations World Water Development Report 2017: Wastewater, The Untapped Resource, UNESCO, Paris, 2017.

N. Condom, M. Lefebvre, L. Vandome, La réutilisation des eaux usées traitées en Méditerranée: retour d'expériences et aide à l'élaboration de projets, Plan Bleu, Valbonne (Les Cahiers du Plan Bleu 11), 2012.

CHAPTER 6

Remote sensing

Jean-Christophe Calvet[a], Clément Albergel[a], Nicolas Baghdadi[b], Delphine Leroux[a], Simon Munier[a]
[a]CNRM, University of Toulouse, Meteo-France, CNRS, Toulouse, France
[b]IRSTEA, University of Montpellier, TETIS, Montpellier, France

6.1 Introduction

Human-induced climate change tends to increase the frequency of extreme events such as heat waves [1], droughts [2], and intense precipitation events [3, 4]. In this rapidly changing world, monitoring severe conditions over land is necessary. Land surface variable products derived from Earth Observation (EO) can be used together with land surface models for this purpose [5, 6]. Accurate characterization and simulation of hydrological and biophysical variables at the land surface are not easy tasks given that terrestrial surfaces generally present large spatial heterogeneities. Because of anthropogenic factors, the land use can change over time, sometimes very rapidly. This is particularly true for water resources as they can be rapidly mobilized for agriculture through irrigation systems. Observations and simulations of the response of land surface conditions to extreme events are needed in order to manage adaptation to climate change impacts.

The role of land surface models has evolved over the years, from providing boundary conditions to atmospheric models to monitoring and forecasting land surface conditions [7–9]. Moreover, modeling of terrestrial variables can be improved through the dynamical integration of observations [10–12]. There is a growing awareness that constraining land surface simulations with past and near-real-time observations is necessary, as well as coupling them with other models of the Earth system [8, 13–15]. Land surface models can be implemented over a large range of spatial resolutions, ranging from $0.25° \times 0.25°$ to $1° \times 1°$ for global climate modeling applications and from about $1\,\text{km} \times 1\,\text{km}$ to $10\,\text{km} \times 10\,\text{km}$ for numerical weather predictions (NWP). It must be noted that there is a trend toward the increase in spatial resolution of NWP models at the sub-kilometric scale [16].

Remote sensing observations are particularly useful in this context as they are now available at the global scale with high spatial resolution and with long-term records. Many satellite-derived products relevant to the hydrological (e.g., soil moisture, snow depth/cover, terrestrial water storage), vegetation (e.g., leaf area index, biomass), and energy (e.g., land surface temperature, albedo) cycles are readily available [17].

Water Resources in the Mediterranean Region
https://doi.org/10.1016/B978-0-12-818086-0.00006-6

6.2 Observing terrestrial variables from space

A key variable observable from space for water resource monitoring is surface soil moisture. However, for most applications, the most relevant variable is root-zone soil moisture. The latter can be derived from surface soil moisture by statistical techniques (e.g., [18]) or by data assimilation into a land surface model [19]. Monitoring other variables (including vegetation variables) is needed to monitor water resources. For example, the Leaf Area Index (LAI) is a key driver of evapotranspiration. The LAI observations can also be assimilated in order to analyze the root-zone soil moisture [20]. At a global scale, low spatial resolution products can be used to detect the onset of severe conditions. On the other hand, monitoring such events requires a more detailed description of the soil–plant system and high-resolution products are needed. Mature methods originally developed for low-resolution EO data (~1–50 km) are now being adapted to high spatial resolution (sub-kilometric scale to field scale).

Linking the signal observed from a spaceborne sensor to end-user applications is a complex and difficult task involving several levels of data processing. A top-down approach consists of producing so-called "Level 2" (or "Level 3," "Level 4") products that can be directly used for monitoring purposes through simple statistical analyses or more complex data fusion techniques or integration into numerical models. A bottom-up approach consists for users in simulating the "Level 1" radiance or brightness, after possible geometric or atmospheric corrections. The ability to simulate Level 1 observations permits their integration into numerical models using data assimilation techniques in an end-to-end processing chain. This technique is mainly used for atmospheric variables in numerical weather prediction models to initialize weather forecast models. For land applications, most users now use Level 2 products, either directly or through their assimilation into land surface models.

Land surface models are components of Earth system models representing the soil plant system. For example, the ISBA land surface model [21–24] was recently upgraded within the SURFEX modeling platform of Meteo-France [25] and is now able to assimilate Level 2 surface soil moisture and LAI products. This is called a Land Data Assimilation System (LDAS). When coupled with an atmospheric model, SURFEX is able to represent the soil–plant system and also the impact of urban areas and of water bodies on atmospheric simulations. The need for geographic information and for surface state variables in modeling platforms used in Earth system models is detailed below for vegetation and soils. It must be noted that other variables related to the water and energy budgets can be monitored from space and used to validate or force land surface models: the fractional coverage of water bodies, the snow fractional coverage and water equivalent, the land surface temperature, and the surface albedo. In addition to these surface variables, the terrestrial water storage can be derived from Gravity Recovery and Climate Experiment (GRACE) satellites (e.g., [26]), but at a very low spatial resolution of about 300 km × 300 km.

6.2.1 Vegetation

Together with atmospheric variables, land cover is probably the most critical forcing for land surface models. Numerical land cover maps are used to specify the model parameters at a given grid-cell, together with the fraction of model surface types such as bare soil, permanent snow, water bodies, urban areas, and several types of forests, crops, and grasslands. Since most land cover maps do not distinguish various types of crops or grasslands, ancillary information such as agricultural statistics is used [27]. The representation of agricultural practices is particularly challenging in land surface models, and additional information such as irrigation maps [28] must be integrated in the forcing files.

The most popular satellite-derived vegetation index is still the Normalized Difference Vegetation Index (NDVI). This product was first proposed in the 1970s [29] and has a long heritage of uptake in various applications. A more recent vegetation product is the Fraction of Absorbed Photosynthetically Active Radiation (FAPAR). The definition of FAPAR is related to the biophysical functioning of the plant canopies, and FAPAR can be simulated by state-of-the-art land surface models. This quantity is considered as an Essential Climate Variable (ECV) by the Global Climate Observing System (GCOS). Another key land ECV is LAI, which can either be used to force land surface models or be simulated. Contrary to FAPAR, a representation of LAI is needed in any land surface model. Several LAI and FAPAR products have been provided in near-real-time from either the Moderate Resolution Imaging Spectroradiometer (MODIS) [30] or European satellites [31].

Vegetation products can also be derived from microwave observations, either passive or active. At L-band, the Vegetation Optical Depth (VOD) can be related to both LAI and to the wood biomass [32]. At C-band, the VOD is more directly related to LAI for all vegetation types at mid-latitudes [12], but negative correlation coefficients were reported for Australian forests [33].

6.2.2 Soil

Soil surface characteristics play a key role in different hydrological processes (floods, runoff, evapotranspiration, infiltration). They also drive other processes such as soil erosion, soil contamination by toxic elements, and imbalances in the water and carbon cycles. Over the last three decades, remote sensing has been shown to have considerable potential for the retrieval of soil characteristics. Various methodologies, based on the use of different remote sensing sensors and techniques (passive and active), have been proposed for the estimation of soil parameters, of which the four most frequently considered techniques are: (i) multispectral optical remote sensing (e.g., Sentinel-2, Landsat, Spot, Aster, etc.); (ii) hyperspectral optical remote sensing, which is particularly well adapted to the description of soil texture; (iii) passive microwave remote sensing, which is used for the estimation of surface soil moisture; and (iv) active microwave remote sensing, which is used for the estimation of soil moisture and soil roughness.

Today, most surface soil property products are generated from microwave observations. Passive microwave research missions such as the L-band ESA Soil Moisture and Ocean Salinity (SMOS) [34] and the NASA Soil Moisture Active and Passive (SMAP) [35] have produced low spatial resolution surface soil moisture products for several years. The EUMETSAT meteorological space agency has produced low spatial resolution surface soil moisture products from the C-band ASCAT radar. Time series start in 2010 and 2007, respectively. Going back in time is possible by harmonizing these time series with data from pre-existing sensors [36].

A key soil property for land surface models is the soil maximum available water content. This quantity is related to soil water holding capacity and to plant rooting depth, and is not directly observable from space. Dewaele et al. [37] showed that this parameter can be retrieved over wheat croplands using a land surface model together with satellite-derived LAI observations.

Soil texture of the topsoil layer tends to correlate with soil reflectances in bare soil conditions provided soil roughness does not vary much [38]. This quantity can also be derived from radar observations. Zribi et al. [39] analyzed the difference in X-band radar backscatter before and after a rain over a semiarid area. They showed that this difference correlates with the fractional content of clay.

6.2.3 Evapotranspiration and fluorescence

Evapotranspiration (ET) is critical for monitoring water resources in semi-arid regions. While this quantity is produced on a sub-daily basis by land surface models, model-free daily estimates can also be generated from EO data [40]. An intermediate approach consists of integrating a variety of EO-derived products into a simplified model working at a low spatial resolution and at a daily time step [41]. Major efforts have been and are currently being made by the EO community to improve ET products, especially using the Sentinels missions.

At the same time, Sun Induced Fluorescence (SIF) has emerged as a new independent source of information to validate land surface models. Now, low spatial resolution SIF products from satellite instruments dedicated to atmospheric chemistry are available. For example, data from the GOME-2 instrument on the METOP European meteorological satellites are used to produce SIF estimates that can be compared to existing Gross Primary Production (GPP) estimates [42, 43], at a spatial resolution of about $0.5° \times 0.5°$. While the GPP variable corresponds to photosynthesis, it was also shown that SIF can be used to validate plant transpiration simulations [44]. The Fluorescence Explorer (FLEX) will be the first satellite mission dedicated to the observation of SIF. FLEX is scheduled for launch in 2023.

Both ET and SIF products can be used to validate land surface models and to assess the added-value of data assimilation.

6.2.4 Data assimilation

Land surface models are not perfect and simulations of land surface variables are affected by uncertainties. Four main kinds of uncertainties can be identified: (i) assumptions and simplifications in the model equations, in relation to a lack of knowledge; (ii) errors in forcing variables (e.g., atmospheric variables such as precipitation or incoming solar radiation, land cover maps); (iii) uncertainties in model parameter values; and (iv) uncertainties in initial conditions (e.g., root-zone soil moisture, vegetation biomass). Integrating observations into land surface models can enhance the realism of the simulations. In particular, key land surface variables such as the root-zone soil moisture or the vegetation biomass change relatively slowly over time compared to other variables (e.g., surface temperature). This property tends to trigger a memory effect. Improving the representation of these variables at a given time can drastically improve the capacity of the model to predict the evolution of the whole soil–plant system. Satellite remote sensing can be used to analyze slow-varying variables by using data assimilation techniques. Variables observable from space such as LAI and surface soil moisture are used to simulate land surface states, including for variables that are not remotely observable, such as the root-zone soil moisture. For example, Kalman filters can be used to correct the model trajectory in the observable variable space. When observations are available, the assimilation generates increments in simulated LAI and root-zone soil moisture. The succession of increments through time gradually improves the consistency of the model simulations with the observations.

The latter method is a sequential data assimilation approach. The advantage of a sequential data assimilation technique is that near-real-time monitoring of extreme events is possible. Another way to improve the model simulations is to tune the values of key parameters of the model. In certain modeling configurations based on highly parameterized phenology schemes, this is the only way to assimilate vegetation products. For example, the CCDAS approach [45] is based on a variational data assimilation approach using a time window of several months. It must be noted that key model parameters can also be tuned a posteriori by using a sequential approach. For example, Dewaele et al. [37] showed that rooting depth of straw cereals can be retrieved by minimizing LAI increments during the senescence phase.

Another aspect of data assimilation is the level of pre-processing of the assimilated data. The use of Level 1 data is generally recommended because Level 1 is closer to the radiance or brightness observed by the spaceborne sensor. Level 2 products may involve empirical, physical, or statistical models, and can be considered as an interpretation of the observations. For example, LAI and surface soil moisture are Level 2 products.

In this case, the observation operator consists of the identity matrix or a simple linear rescaling equation. Assimilating Level 1 products is more complex because the observation operator must be able to connect the simulated land surface variables to the sensed physical signal. In this case, reflectance and radiative transfer models must be coupled to the land surface model. Alternatively, statistical methods based on machine learning techniques may be used. For example, Forman and Xue [46] compared the use of an artificial neural network and of a support vector machine to simulate microwave brightness temperatures over snow-covered areas. They found that the support vector machine technique is particularly good at representing the spatial and temporal variability of the signal.

6.3 Applications

6.3.1 Integration of satellite-derived terrestrial variables into models

Examples given in this section make use of low (or moderate) spatial resolution products. The same methods could be applied at a higher spatial resolution if consistent atmospheric forcings are available.

6.3.1.1 Joint assimilation of LAI and surface soil moisture

Although assimilating Level 2 soil moisture products is generally simpler than assimilating Level 1 products, the integration of Level 2 products in land surface models is not always straightforward. For example, the volumetric surface soil moisture expressed in $m^3 m^{-3}$ is model-dependent. Its value depends on components and on forcing environmental data of the model such as pedotransfer functions and soil parameter maps. Moreover, retrieved Level 2 product values may be affected by perturbing factors poorly accounted for in the retrieval algorithm. In the case of surface soil moisture, undetected changes in vegetation and soil roughness can be considered as perturbing factors [47]. Because of the possible mismatch between Level 2 and simulated surface soil moisture, the former have to be rescaled to be consistent with the model. A simpler linear rescaling may be sufficient as shown by Calvet and Noilhan [48] for in situ soil moisture observations. In the case of satellite-derived products, a seasonal rescaling is needed in many areas [49]. An example is shown in Fig. 6.1 for surface soil moisture derived from the ASCAT C-band radar instrument over the Euro-Mediterranean area, and from the ISBA land surface model.

The seasonal cycle of the raw soil wetness index (dimensionless) differs markedly from the modeled cycle. In particular, the relatively wet conditions found by the ASCAT product at summertime are not consistent with the model. Assimilating this product applying a simple linear rescaling would be detrimental to the model simulation of the root-zone soil moisture. The seasonally rescaled observations are much more consistent with the model and can be assimilated.

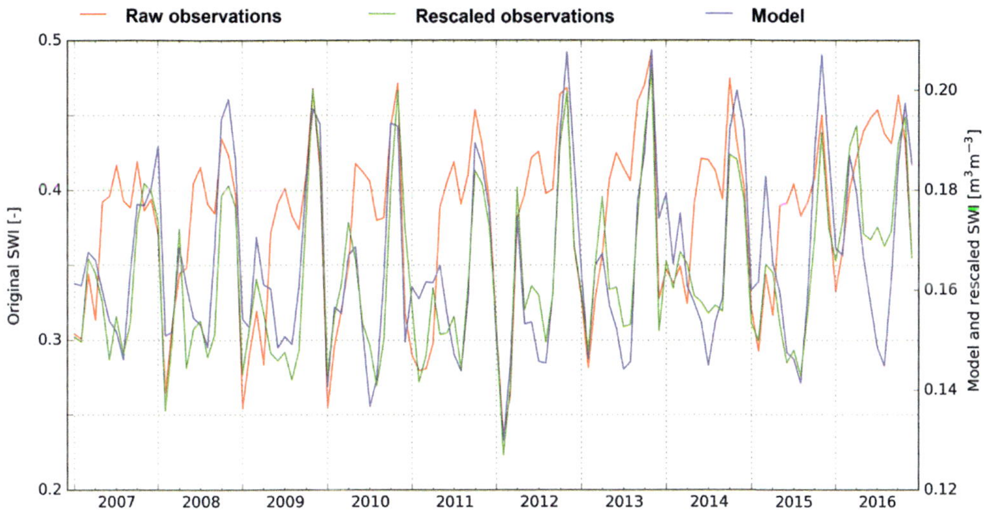

Fig. 6.1 Observed and modeled surface soil moisture over the Euro-Mediterranean area (25°N–75.5°N, 11.75°W–62.5°E) from 2007 to 2016. A seasonal statistical rescaling of the ASCAT-derived observations is needed before integrating them into the model.

Another limitation of the assimilation of surface soil moisture products is that simulations of this quantity are decoupled from the root-zone soil moisture in dry conditions [50]. In such conditions, the observation of vegetation variables such as LAI or the microwave vegetation optical depth (VOD) can be more informative than surface soil moisture [20]. Vegetation observations can be assimilated in order to analyze the root-zone soil moisture. This was shown by [20] for LAI using the LDAS implemented within the SURFEX (Surface Externalisée, Version 8.1) modeling platform of Meteo-France [25].

The LDAS [5, 19, 51, 52] developed in the research department of Meteo-France, the CNRM, permits integrating satellite products into the ISBA land surface model [21–24] using a sequential data assimilation scheme. In ISBA, phenology is completely driven by photosynthesis and no growing degree-day sub model is used. As a result, the simulated LAI can be updated whenever an observation is available. The LDAS is able to assimilate LAI and surface soil moisture products jointly. It was extended to the global scale by [20] (LDAS-Monde). At the same time, the coupling to hydrological models (ISBA-CTRIP for ISBA-CNRM, Total Runoff Integrating Pathways) was consolidated [53]. The obtained land surface reanalyses from LDAS-Monde account for the synergies of various upstream products (e.g., model- and satellite-derived observations) and are able to provide an improved representation of the land surface variables. LDAS-Monde can also provide improved initial conditions to initialize forecasts of the land surface variables, because its impacts on key land surface variables can persist over time [6, 12].

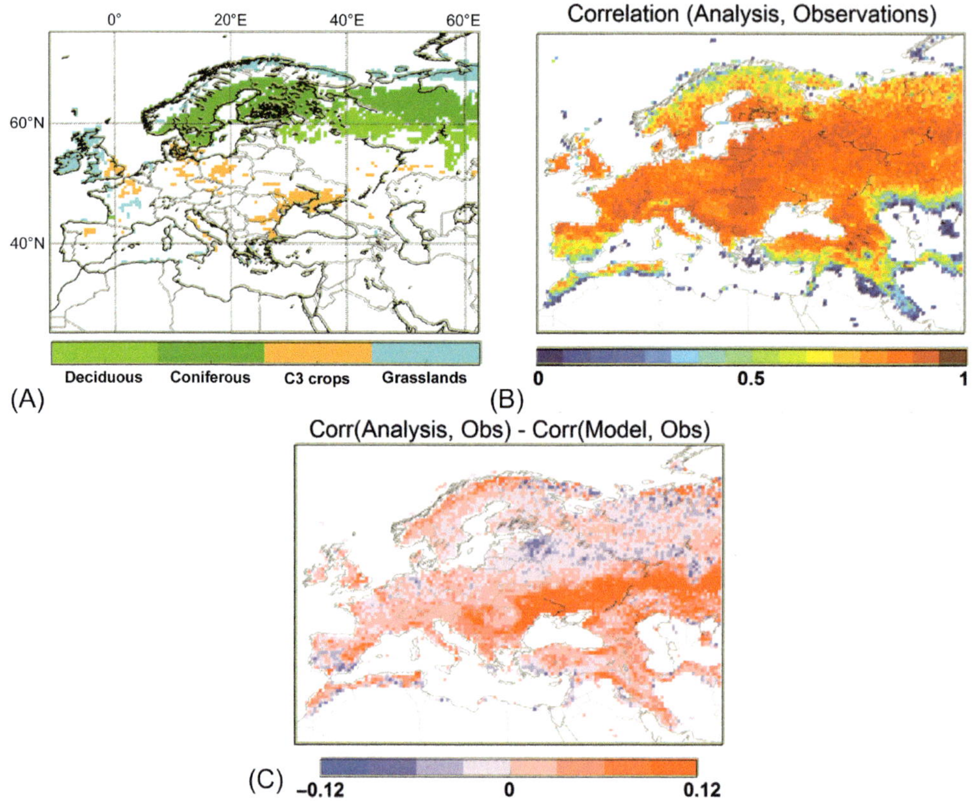

Fig. 6.2 Dominant land cover classes over the Euro-Mediterranean area (25°N–75.5°N, 11.75°W–62.5°E) (A) at a spatial resolution of 0.5° × 0.5° with grid cells where at least 50% is covered by a single vegetation type (deciduous forests, coniferous forests, C3 crops, grassland). Correlation maps between the SIF observations and the GPP analysis for the 2007–2015 period (B) and the difference map (C), where red means an improvement after assimilation. *(Adapted from D. Leroux, J.-C. Calvet, S. Munier, C. Albergel, Using satellite-derived vegetation products to evaluate LDAS-Monde over the Euro-Mediterranean area, Remote Sens. (Basel) 10 (2018) 1199 1–21.)*

As illustrated by Fig. 6.2, SIF can be used to perform an independent validation of the assimilation. The SIF observations were found to correlate with GPP by a number of authors (e.g., [54]). Therefore, a positive impact on the correlation between the simulated GPP and SIF observations is an indirect way to validate upgrades of a land surface model. Fig. 6.2 shows that assimilating satellite products (other than SIF) in the ISBA land surface model generally increases the correlation between the simulated GPP and SIF. However, the correlation can decrease at the edge of boreal forest ecosystems, at the interface between semi-arid and desert areas, and in areas corresponding to large sub-grid heterogeneities. This illustrates the limit of global simulations of land surface variables at a low spatial resolution (0.5° × 0.5° in the case of [43]). Implementing a LDAS at a better spatial resolution could be recommended.

6.3.1.2 Impact on river discharge simulations

A key variable for water resource monitoring is river discharge. A dedicated space mission called Surface Water and Ocean Topography (SWOT) is scheduled for launch in 2021 [55]. The main SWOT instrument consists of a Ka-band radar interferometer. Over land, SWOT would provide river discharge estimates every 11 days on average. Water storage in lakes and reservoirs would be estimated at a spatial resolution of about 100 m × 100 m.

River discharge can be estimated over large basins by using global land surface models. Fig. 6.3 shows the impact on river discharge of assimilating LAI and surface soil moisture observations into the ISBA land surface model coupled to the CTRIP hydrological model [53].

Over the Euro-Mediterranean area, the impact of the assimilation on the mean water flux values is relatively small. More meaningful results were found by [18] over

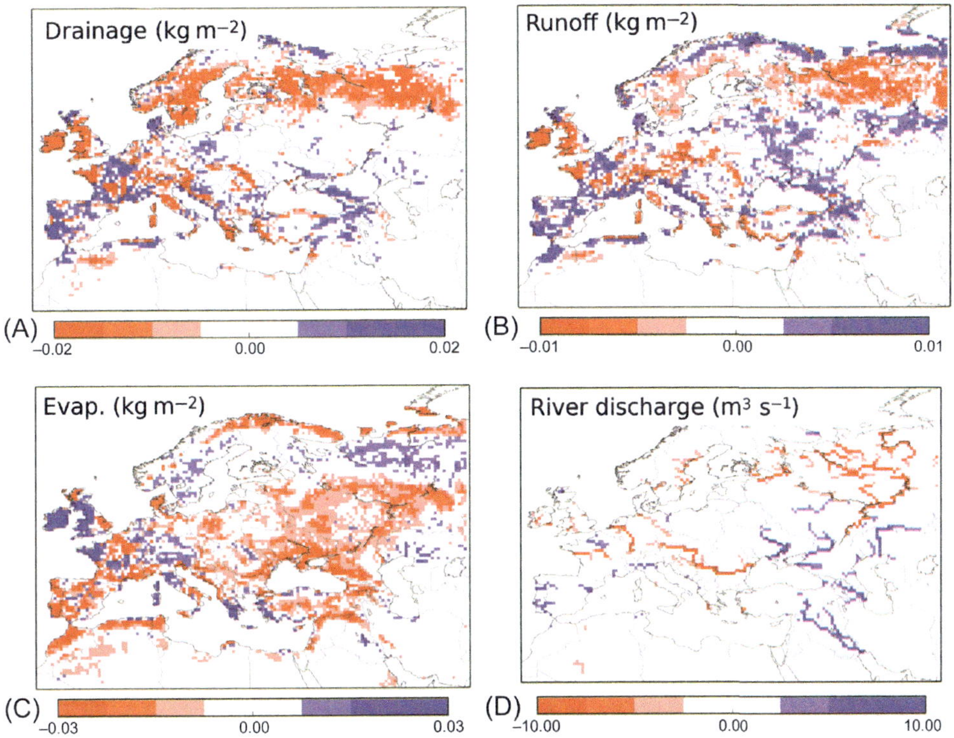

Fig. 6.3 LDAS-Monde mean analysis impact on land water flux variables over the period 2000–2012 over the Euro-Mediterranean area (25°N–75.5°N, 11.75°W–62.5°E): (A) drainage, (B) runoff, (C) evapotranspiration, and (D) river discharge. *(Adapted from C. Albergel, S. Munier, D.J. Leroux, H. Dewaele, D. Fairbairn, A.L. Barbu, E. Gelati, W. Dorigo, S. Faroux, C. Meurey, P. Le Moigne, B. Decharme, J.-F. Mahfouf, J.-C. Calvet, Sequential assimilation of satellite-derived vegetation and soil moisture products using SURFEX_v8.0: LDAS-Monde assessment over the Euro-Mediterranean area, Geosci. Model Dev. 10 (2017) 3889–3912.)*

agricultural regions of the USA using the same LDAS-Monde tool. This difference of the hydrological response to land surface variables could be related to the larger spatial heterogeneity of the land cover over the Euro-Mediterranean area. Implementing a LDAS at a better spatial resolution is needed over this area.

6.3.1.3 Toward the assessment of severe conditions over the Mediterranean basin

Water resources in the Mediterranean basin are particularly vulnerable to climate change [56]. In this context, developing observational, monitoring, and forecasting capabilities able to favor the adaptation to the risk of drought is necessary. Given the large land cover heterogeneities in this area, the access to high spatial resolution land surface remote sensing products is critical. Land surface models forced by high-resolution atmospheric variables produced by operational weather forecast models (e.g., HRES at ECMWF, AROME over western Europe, at 9 km × 9 km and 1.3 km × 1.3 km, respectively) can already produce very good soil moisture estimates. However, land use and agricultural practices such as irrigation are not represented or represented with difficulty in these models so far. From this point of view, mapping and quantifying irrigation constitutes a major added value of soil moisture products derived from microwave sensors over these areas. In addition to surface soil moisture, vegetation products such as LAI can be integrated into land surface models. In the LDAS-Monde platform, for example, the assimilation of LAI can be used to analyze LAI and vegetation biomass, and also the root-zone soil moisture. Routinely producing analyses of land surface variables in near-real-time over the whole Mediterranean is technically feasible at a low spatial resolution (e.g., 0.25° × 0.25°). Such analyses could then be used to initialize on-alert forecasts of land surface variables at a better spatial resolution (e.g., 1 km × 1 km) over areas presenting negative LAI and soil moisture anomalies, while accounting for sub-grid heterogeneities. Short- and medium-range (1–10 days) land surface forecasts could be complemented with seasonal forecasts (1–3 months). Proper initialization of such forecasts would be critical as root-zone soil moisture and vegetation variables such as LAI present a memory effect. Their initial values can condition their future values up to several months in advance [18].

6.3.2 High spatial resolution products

6.3.2.1 Surface soil moisture

Active microwave remote sensing is generally based on the use of two types of radars: high spatial resolution, synthetic aperture radars (SAR), which are very well suited to local and regional studies, and low-resolution scatterometer sensors, which are better adapted to the estimation of land surface variables at a global scale. The SAR measurements can be made under all meteorological conditions (clouds, fog, etc.) and at any time of the day or night, using wavelengths ranging from 1 mm to 1 m. In the case of bare soils, the SAR signal, which is affected by three different radar parameters (polarization, incidence angle, and frequency), is also correlated with the surface roughness and the moisture content of the topsoil layer. Speckle noise, due to the coherent interference of waves reflected from many elementary scatterers, is present on SAR images and makes the pixel-by-pixel interpretation of SAR images extremely difficult. This is why soil surface characteristics are generally estimated over homogeneous areas including several pixels, or at the field scale (which helps to reduce the speckle effect).

The first generation of products from spaceborne radars, in the 1990s, was based on a single polarization. The ERS-1/2 ESA Remote Sensing Satellites operated in the VV polarization (vertical transmit and receive) mode, at an incidence angle of 23 degrees. The RADARSAT-1 operated in the HH polarization (horizontal transmit and receive) mode, at incidence angles ranging from 20 to 50 degrees. Both of these sensors were operated in the C-band (5.3 GHz), and were widely used to retrieve surface soil moisture and soil roughness. In March 2002, the launch of the European Environmental Satellite (ENVISAT), equipped with a C-band Advanced Synthetic Aperture Radar (ASAR), allowed the scientific community to use images in dual-polarization mode (two simultaneous polarizations selected from four available polarizations: HH, HV, VH, and VV) and at various incidence angles ranging from 15 to 45 degrees. The nominal swath widths for ASAR and ERS-2 were about 100 km or more, with 12.5 m × 12.5 m pixels but with a spatial resolution no finer than 25 m. With RADARSAT-1, the nominal swath width was about 50 km × 50 km or more, and the spatial resolution was about 10 m × 10 m at best. In 2006 and 2007, two polarimetric SAR sensors were launched: ALOS (Japanese Space Agency, NASDA), operated in the L-band (1.2 GHz), and RADARSAT-2 (Canadian Space Agency, CSA), operated in the C-band. For the first time in 2007, very high-resolution radars with a spatial resolution of about 1 m were launched: TerraSAR-X (German Space Agency, DLR) and the COSMO-SkyMed constellation (Italian Space Agency), both of which operated in the X-band (9.65 GHz). All these SAR sensors did not have a high revisit time nor a high spatial resolution. Additionally, images were not freely available. All this was detrimental to the development of operational approaches for monitoring soil characteristics.

With the advent of Sentinel-1 (S1), routine operational mapping of surface soil moisture has become possible. The S1 SAR satellites within the Copernicus program of the European Union has provided users with free, open access data at high spatial resolution (10 m × 10 m) and high sampling time (every 6 days over Europe). The current S1 mission from the European Space Agency (ESA) consists of a constellation of two polar orbiting SAR satellites (Sentinel-1A and Sentinel-1B, launched in 2014 and in 2016, respectively) operating at C-band (~5.4 GHz).

In the last years, different algorithms based mainly on two approaches were proposed to estimate surface soil moisture using SAR remote sensing. The first approach is based on neural network techniques, and usually involves the combined use of one or two radar signal polarizations and the Normalized Difference Vegetation Index (NDVI) from optical data. The second approach uses the multi-temporal change detection technique. For example, Paloscia et al. [57] used the neural network technique to invert S1 C-band data and estimate surface soil moisture values. Mattia et al. [58] estimated soil moisture at a spatial resolution of 1 km × 1 km using the change detection technique applied to S1 SAR time series. El Hajj et al. [59] developed an operational method to map soil moisture at the plot scale over agricultural areas based on the fusion of S1 SAR data and Sentinel-2 (S2) optical data using a neural network technique called S2MP. It combined the WCM (Water Cloud Model) with the IEM (Integral Equation Model). The latter is a surface backscattering model used for bare soils. The S2MP algorithm provided surface soil moisture products at the field scale for several reference Mediterranean sites located in France, Greece, Italy, Germany, and Lebanon. Gao et al. [60] combined S1 data and the NDVI derived from S2 images to retrieve a soil moisture index ranging from 0 to 100 (from driest to wettest conditions) at a spatial resolution of 100 m × 100 m. More recently, the Copernicus Global Land Service has started providing surface soil moisture of the soil's topmost 5 cm layer over the European continent from S1 sensors at a spatial resolution of 1 km × 1 km using the change detection approach [61].

For dense vegetation canopies, retrieving surface soil moisture at C-band is difficult because the radar signal is markedly attenuated by vegetation. This limitation is related to the low penetration depth of the C-band signal into well-developed canopy covers such as dense crop types or forests. Namely, the emitted SAR wave cannot reach the underlying soil or the soil contribution to the backscattered wave is too small. In these conditions, the low penetration makes the C-band backscattered SAR signal insensitive to soil moisture, and estimations of this quantity become unreliable. El Hajj et al. [59, 62] reported that the penetration depth of the C-band SAR signals into wheat and grassland canopies is low for NDVI values larger than 0.7. This means that there is no sensitivity to surface soil moisture for NDVI values larger than 0.7. For vegetation covers presenting NDVI values smaller than 0.7, results showed that volumetric surface soil moisture estimates can be obtained with relatively good accuracy: root mean square errors ranging from 0.04 to 0.06 m^3m^{-3}.

The S2MP approach for surface soil moisture mapping was applied for 2 years over a study site located in the French Occitanie administrative unit in southern France. The S1-derived surface soil moisture maps are produced at a high spatial resolution, compatible with monitoring applications at the field scale. These products are freely available via the French Theia Land Data Center (http://www.theia-land.fr/en/thematic-products).

Fig. 6.4 shows a comparison between S1-derived surface soil moisture maps and precipitation estimates derived from GPM (Global Precipitation Measurement) data at four dates. For each date, a cumulative rainfall map for the 6 days prior to the soil moisture map date is presented. On 21 December 2016, the eastern and central parts of the region present high soil moisture values larger than 0.20 $m^3 m^{-3}$ corresponding to accumulated

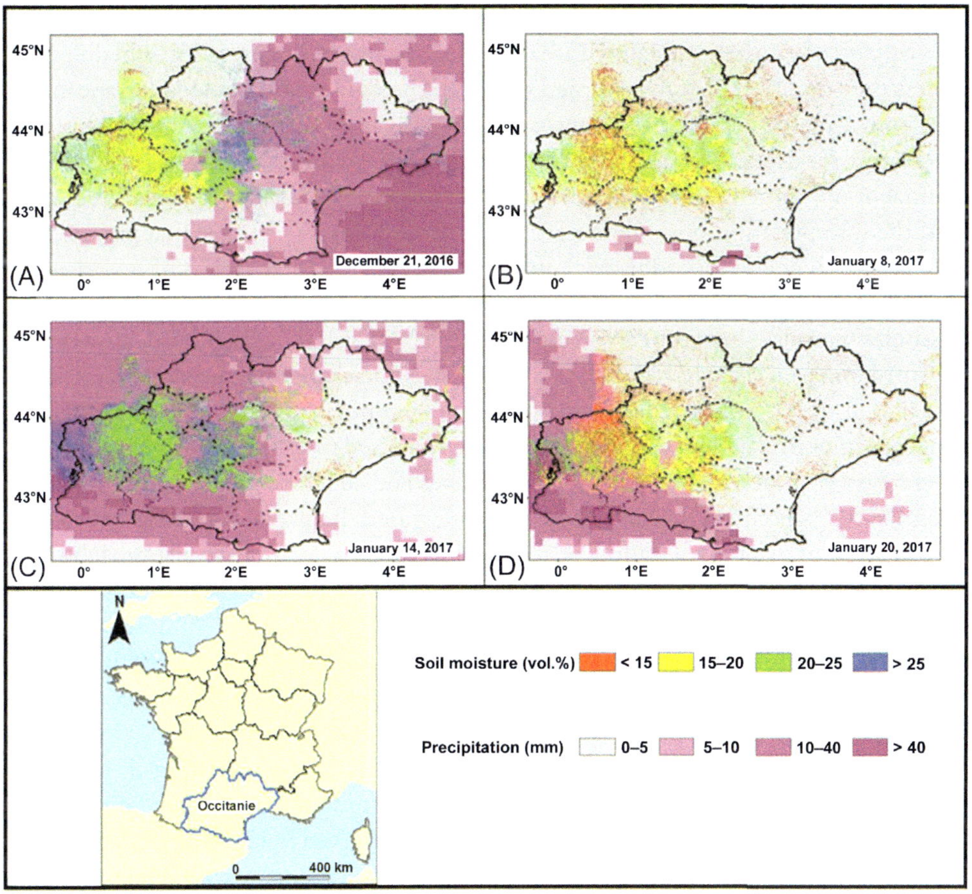

Fig. 6.4 S1-derived surface soil moisture maps over the Occitanie region in France overlaid with GPM cumulative precipitation data: (A) 21 December 2016, (B) 8 January 2017, (C) 14 January 2017, (D) 20 January 2017. Dotted lines indicate the department administrative units.

precipitation of 10–20 mm during the past 6 days. On the other hand, the western part of the region presents small soil moisture values (less than 0.20 $m^3\ m^{-3}$) corresponding to the absence of precipitation in the area for the same period of time. No rain is observed from 27 December 2016 to 8 January 2017, and surface soil moisture values drop to values smaller than 0.15 $m^3\ m^{-3}$ on 8 January 2017 over the whole region. Six days later, on 14 January 2017, surface soil moisture values remain small over the eastern part of the region as no precipitation is recorded. A moderate increase in soil moisture values is observed in the central and western parts of the region corresponding to accumulated precipitation of 5–20 mm between 8 January 2017 and 14 January 2017. Twenty-four days with no precipitation recorded on the eastern part of the region from 27 December 2016 to 20 January 2017 trigger dry surface soil moisture values of less than 0.15 $m^3\ m^{-3}$.

6.3.2.2 Mapping and monitoring irrigation

The mapping of irrigated and rainfed areas can be used to prepare irrigation schedules and to assess the condition of water-stressed crops. In the context of the Mediterranean climate, with a marked scarcity of water resources, this information is essential for stakeholders and for water managers. Several studies have demonstrated the considerable potential of optical remote sensing for irrigation mapping. Thiruvengadachari [63] proposed the interpretation of single-date Landsat optical imagery for the mapping of irrigated areas, by including additional information related to major surface irrigation projects, canal network maps, drainage patterns, and recorded groundwater use. Although single-date imagery can be used to identify irrigated fields, it is not always reliable, since single-date analysis does not take the dynamic changes of the vegetation cover into account over a full agricultural season. Multi-temporal analysis is better adapted to the detection of irrigated fields. Thenkabail et al. [64] used time-series from MODIS data to generate a map of irrigated areas for the Ganges and Indus river basins. Studies combining the two aforementioned methods have increased the spatial and temporal resolution of irrigation mapping. Gumma et al. [65] developed a decision-tree approach using Landsat 30 m × 30 m single-date data, combined with MODIS 250 m × 250 m time series. They achieved a precision ranging from 67% to 93% for the identification of irrigated fields, with different classes of vegetation. Despite the strong potential of these techniques, the use of optical data is heavily dependent on the prevailing weather conditions, and cloud cover in particular.

Over areas presenting large cloud coverage, these methods can be difficult to implement. During wet years, under low water stress conditions with vegetation cover, it can be difficult to distinguish between irrigated and non-irrigated fields using optical indices only. The use of soil moisture estimations can improve the accuracy with which irrigated and non-irrigated areas are classified.

Gao et al. [66] proposed an analysis of Sentinel-1 radar time series over an irrigated area close to Urgell in Spain. They analyzed different indices such as the mean temporal value of radar signals, the variance, the fractal dimension, and the temporal correlation length. Using Support Vector Machine and Random Forest approaches, they demonstrated the capacity of the radar signal to separate between rainfed fields and crop or tree irrigated fields with a precision higher than 85%. Fig. 6.5 illustrates results of the proposed mapping of irrigated and rainfed field over this area.

Bousbih et al. [67] analyzed the use of two satellite products over central Tunisia: (i) soil moisture at the field scale from the synergistic use of S1 and S2; and (ii) a NDVI product from S2. They used the temporal mean value and variance of the soil moisture product, and the mean value of NDVI, respectively. They showed that the highest potential for mapping irrigation is achieved using the soil moisture indices. The lowest potential is observed with NDVI, which seems insufficient to identify irrigation, particularly in wet conditions. This approach is applied using a Support Vector Machine approach.

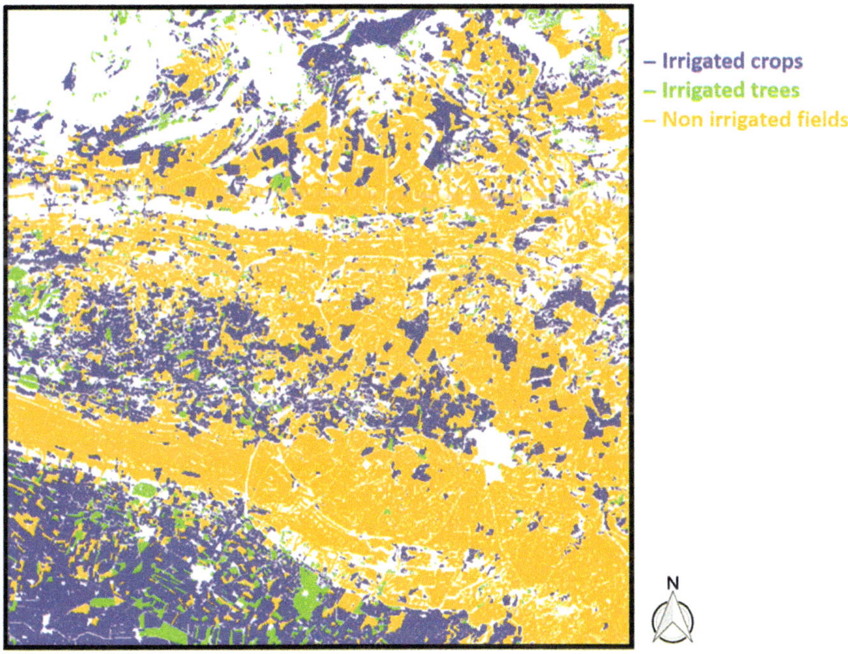

Fig. 6.5 Irrigation mapping close to Urgell in Spain over a 20 km × 20 km area after masking out forests, urban areas, and water bodies. *(Adapted from Q. Gao, M. Zribi, M. Escorihuela, N. Baghdadi, P. Segui, Irrigation mapping using Sentinel-1 time series at field scale, Remote Sens. (Basel) 10 (2018) 1495 1–18.)*

Because both S1 and S2 images can now be combined at high sampling time, different approaches can be proposed to estimate the irrigation water use in addition to irrigation mapping. This can be achieved through the assimilation of surface soil moisture products into land surface models able to work at the field scale. These methods can also be implemented using low-resolution surface soil moisture products [68].

6.4 Main conclusions

New satellite-derived land surface products are being produced at increasingly high spatial resolution. New variables and surface characteristics could be available in the next years in relation to the launch of new satellites and to the development of new algorithms. Preparing the uptake of these new products into existing applications and into new applications is key.

The European Copernicus program, initiated in 2014 after a research preparation phase of about 10 years, has led to the operational production of new freely available products over Europe and at a global scale: high-resolution land cover, soil moisture, and vegetation products such as LAI. At the same time, techniques designed to integrate these products into land surface models were developed. The use of satellite-derived products is particularly relevant to monitor water resources in the Mediterranean. Monitoring irrigation practices is notably important. High spatial resolution soil moisture and LAI products can be used for this purpose. Land surface models based on initial land surface conditions constrained by remote sensing observations can be used to monitor and forecast the evolution of drought events.

References

[1] P.B. Duffy, C. Tebaldi, Increasing prevalence of extreme summer temperatures in the U.S., Clim. Change 111 (2012) 487–495.
[2] B. Mueller, S.I. Seneviratne, Hot days induced by precipitation deficits at the global scale, Proc. Natl. Acad. Sci. U. S. A. 109 (2012) 12398–12403.
[3] E. Scoccimarro, S. Gualdi, Heavy precipitation events in a warmer climate: Results from CMIP5 models, J. Climate 26 (2012) 7902–7911.
[4] Z. Feng, L.R. Leung, S. Hagos, R.A. Houze, C.D. Burleyson, K. Balaguru, More frequent intense and long-lived storms dominate the springtime trend in central US rainfall, Nat. Commun. 7 (2016) 134291–8.
[5] A.L. Barbu, J.-C. Calvet, J.-F. Mahfouf, S. Lafont, Integrating ASCAT surface soil moisture and GEOV1 leaf area index into the SURFEX modelling platform: a land data assimilation application over France, Hydrol. Earth Syst. Sci. 18 (2014) 173–192.
[6] C. Albergel, E. Dutra, B. Bonan, Y. Zheng, S. Munier, G. Balsamo, P. de Rosnay, J. Munoz-Sabater, J.-C. Calvet, Monitoring and forecasting the impact of the 2018 summer heatwave on vegetation, Remote Sens. (Basel) 11 (2019) 520 1–22.
[7] P.A. Dirmeyer, X. Gao, M. Zhao, Z. Guo, T. Oki, N. Hanasaki, The second global soil wetness project (GSWP-2): Multi-model analysis and implications for our perception of the land surface, B. Am. Meteorol. Soc. 87 (2006) 1381–1397.
[8] G. Balsamo, C. Albergel, A. Beljaars, S. Boussetta, E. Brun, H. Cloke, D. Dee, E. Dutra, J. Muñoz-Sabater, F. Pappenberger, P. de Rosnay, T. Stockdale, F. Vitart, ERA-interim/land: A global land surface reanalysis data set, Hydrol. Earth Syst. Sci. 19 (2015) 389–407.
[9] J. Schellekens, E. Dutra, A. Martínez-de la Torre, G. Balsamo, A. van Dijk, F. Sperna Weiland, M. Minvielle, J.-C. Calvet, B. Decharme, S. Eisner, G. Fink, M. Flörke, S. Peßenteiner, R. van Beek, J. Polcher, H. Beck, R. Orth, B. Calton, S. Burke, W. Dorigo, G. P. Weedon, A global water resources ensemble of hydrological models: The eartH2Observe Tier-1 dataset, Earth Syst. Sci. Data 9 (2017) 389–413.
[10] R.H. Reichle, R.D. Koster, P. Liu, S.P.P. Mahanama, E.G. Njoku, M. Owe, Comparison and assimilation of global soil moisture retrievals from the Advanced Microwave Scanning Radiometer for the Earth Observing System (AMSR-E) and the Scanning Multichannel Microwave Radiometer (SMMR), J. Geophys. Res. 112 (2007) D09108, 1–14.
[11] W.A. Lahoz, G.J.M. De Lannoy, Closing the gaps in our knowledge of the hydrological cycle over land: Conceptual problems, Surv. Geophys. 35 (2014) 623–660.
[12] C. Albergel, S. Munier, A. Bocher, B. Bonan, Y. Zheng, C. Draper, D.J. Leroux, J.-C. Calvet, LDAS-monde sequential assimilation of satellite derived observations applied to the contiguous US: An ERA5 driven reanalysis of the land surface variables, Remote Sens. (Basel) 10 (2018) (1627) 1–24.
[13] P. de Rosnay, M. Drusch, D. Vasiljevic, G. Balsamo, C. Albergel, L. Isaksen, A simplified extended Kalman filter for the global operational soil moisture analysis at ECMWF, Q. J. Roy. Meteorol. Soc. 139 (2013) 1199–1213.
[14] P. de Rosnay, G. Balsamo, C. Albergel, J. Muñoz-Sabater, L. Isaksen, Initialisation of land surface variables for numerical weather prediction, Surv. Geophys. 35 (2014) 607–621.

[15] S.V. Kumar, M. Jasinski, D. Mocko, M. Rodell, J. Borak, B. Li, H. Kato Beaudoing, C.D. Peters-Lidard, NCA-LDAS land analysis: Development and performance of a multisensor, multi-variate land data assimilation system for the National Climate Assessment, J. Hydrometeorol. (2018) (in press).

[16] R. Honnert, F. Couvreux, V. Masson, D. Lancz, Sampling the structure of convective turbulence and implications for grey-zone parametrizations, Bound.-Lay. Meteorol. 160 (2016) 133–156.

[17] D.P. Lettenmaier, D. Alsdorf, J. Dozier, G.J. Huffman, M. Pan, E.F. Wood, Inroads of remote sensing into hydrologic science during the WRR era, Water Resour. Res. 51 (2015) 7309–7342.

[18] C. Albergel, C. Rüdiger, T. Pellarin, J.-C. Calvet, N. Fritz, F. Froissard, D. Suquia, A. Petitpa, B. Piguet, E. Martin, From near-surface to root-zone soil moisture using an exponential filter: an assessment of the method based on in-situ observations and model simulations, Hydrol. Earth Syst. Sci. 12 (2018) 1323–1337.

[19] A. Barbu, J.-C. Calvet, J.-F. Mahfouf, C. Albergel, S. Lafont, Assimilation of soil wetness index and leaf area index into the ISBA-A-gs land surface model: Grassland case study, Biogeosciences 8 (2011) 1971–1986.

[20] C. Albergel, S. Munier, D.J. Leroux, H. Dewaele, D. Fairbairn, A.L. Barbu, E. Gelati, W. Dorigo, S. Faroux, C. Meurey, P. Le Moigne, B. Decharme, J.-F. Mahfouf, J.-C. Calvet, Sequential assimilation of satellite-derived vegetation and soil moisture products using SURFEX_v8.0: LDAS-Monde assessment over the Euro-Mediterranean area, Geosci. Model Dev. 10 (2017) 3889–3912.

[21] J. Noilhan, J.-F. Mahfouf, The ISBA land surface parameterisation scheme, Global Planet. Change 13 (1996) 145–159.

[22] J.-C. Calvet, J. Noilhan, J.-L. Roujean, P. Bessemoulin, M. Cabelguenne, A. Olioso, J.-P. Wigneron, An interactive vegetation SVAT model tested against data from six contrasting sites, Agric. For. Meteorol. 92 (1998) 73–95.

[23] J.-C. Calvet, V. Rivalland, C. Picon-Cochard, J.-M. Guehl, Modelling forest transpiration and CO_2 fluxes—response to soil moisture stress, Agric. For. Meteorol. 124 (2004) 143–156.

[24] A.-L. Gibelin, J.-C. Calvet, J.-L. Roujean, L. Jarlan, S. Los, Ability of the land surface model ISBA-A-gs to simulate leaf area index at the global scale: comparison with satellites products, J. Geophys. Res. 111 (2006) D18102 1–16.

[25] V. Masson, P. Le Moigne, E. Martin, S. Faroux, A. Alias, R. Alkama, S. Belamari, A. Barbu, A. Boone, F. Bouyssel, P. Brousseau, E. Brun, J.-C. Calvet, D. Carrer, B. Decharme, C. Delire, S. Donier, K. Essaouini, A.-L. Gibelin, H. Giordani, F. Habets, M. Jidane, G. Kerdraon, E. Kourzeneva, M. Lafaysse, S. Lafont, C.L. Brossier, A. Lemonsu, J.-F. Mahfouf, P. Marguinaud, M. Mokhtari, S. Morin, G. Pigeon, R. Salgado, Y. Seity, F. Taillefer, G. Tanguy, P. Tulet, B. Vincendon, V. Vionnet, A. Voldoire, The SURFEXv7.2 land and ocean surface platform for coupled or offline simulation of earth surface variables and fluxes, Geosci. Model Dev. 6 (2013) 929–960.

[26] V. Humphrey, L. Gudmundsson, S.I. Seneviratne, Assessing global water storage variability from GRACE: trends, seasonal cycle, subseasonal anomalies and extremes, Surv. Geophys. 37 (2016) 357–395.

[27] S. Faroux, A.T. Kaptué Tchuenté, J.-L. Roujean, V. Masson, E. Martin, P. Le Moigne, ECOCLIMAP-II/Europe: a twofold database of ecosystems and surface parameters at 1 km resolution based on satellite information for use in land surface, meteorological and climate models, Geosci. Model Dev. 6 (2013) 563–582.

[28] J. Meier, F. Zabel, W. Mauser, A global approach to estimate irrigated areas—a comparison, between different data and statistics, Hydrol. Earth Syst. Sci. 22 (2018) 1119–1133.

[29] J.W. Rouse, R.H. Haas, J.A. Scheel, D.W. Deering, Monitoring vegetation systems in the Great Plains with ERTS, Proceedings, in: 3rd Earth Resource Technology Satellite (ERTS) Symposium, 1 1974, pp. 48–62.

[30] K. Yan, T. Park, G. Yan, C. Chen, B. Yang, Z. Liu, R.R. Nemani, Y. Knyazikhin, R. B. Myneni, Evaluation of MODIS LAI/FPAR Product Collection 6. Part 1: Consistency and Improvements, Remote Sens. (Basel) 8 (2016) 359 2–16.

[31] F. Baret, M. Weiss, R. Lacaze, F. Camacho, H. Makhmared, P. Pacholczyk, B. Smets, GEOV1: LAI and FAPAR essential climate variables and FCOVER global time series capitalizing over existing products. Part 1: Principles of development and production, Remote Sens. Environ. 137 (2013) 299–309.

[32] N.J. Rodríguez-Fernández, A. Mialon, S. Mermoz, A. Bouvet, P. Richaume, A. Al Bitar, A. Al-Yaari, M. Brandt, T. Kaminski, T. Le Toan, Y.H. Kerr, J.-P. Wigneron, An evaluation of SMOS L-band vegetation optical depth (L-VOD) data sets: high sensitivity of L-VOD to above-ground biomass in Africa, Biogeosciences 15 (2018) 4627–4645.

[33] M. Vreugdenhil, S. Hahn, T. Melzer, B. Bauer-Marschallinger, C. Reimer, W.A. Dorigo, W. Wagner, Assessing vegetation dynamics over mainland Australia with Metop ASCAT, IEEE J. Sel. Top. Appl. Earth Obs. Remote Sens. 10 (2016) 2240–2248.

[34] Y.H. Kerr, A. Al-Yaari, N. Rodriguez-Fernandez, M. Parrens, B. Molero, D. Leroux, S. Bircher, A. Mahmoodi, A. Mialon, P. Richaume, S. Delwart, A. Al Bitar, T. Pellarin, R. Bindlish, T.J. Jackson, C. Rüdiger, P. Waldteufel, S. Mecklenburg, J.-P. Wigneron, Overview of SMOS performance in terms of global soil moisture monitoring after six years in operation, Remote Sens. Environ. 180 (2016) 40–63.

[35] S.K. Chan, R. Bindlish, P. O'Neill, T. Jackson, E. Njoku, S. Dunbar, J. Chaubell, J. Piepmeier, S. Yueh, D. Entekhabi, A. Colliander, F. Chen, M.H. Cosh, T. Caldwell, J. Walker, A. Berg, H. McNairn, M. Thibeault, J. Martínez-Fernández, F. Uldall, M. Seyfried, D. Bosch, P. Starks, C. H. Collins, J. Prueger, R. van der Velde, J. Asanuma, M. Palecki, E.E. Small, M. Zreda, J. Calvet, W.T. Crow, Y. Kerr, Development and assessment of the SMAP enhanced passive soil moisture product, Remote Sens. Environ. 204 (2018) 931–941.

[36] W. Dorigo, W. Wagner, C. Albergel, F. Albrecht, G. Balsamo, L. Brocca, D. Chung, M. Ertl, M. Forkel, A. Gruber, E. Haas, P.D. Hamer, M. Hirschi, J. Ikonen, R. de Jeu, R. Kidd, W. Lahoz, Y.Y. Liu, D. Miralles, T. Mistelbauer, N. Nicolai-Shaw, R. Parinussa, C. Pratola, C. Reimer, R. van der Schalie, S.I. Seneviratne, T. Smolander, P. Lecomte, ESA CCI soil moisture for improved earth system understanding: state-of-the art and future directions, Remote Sens. Environ. 203 (2017) 185–215.

[37] H. Dewaele, S. Munier, C. Albergel, C. Planque, N. Laanaia, D. Carrer, J.-C. Calvet, Parameter optimisation for a better representation of drought by LSMs: inverse modelling vs. sequential data assimilation, Hydrol. Earth Syst. Sci. 21 (2017) 4861–4878.

[38] P. Lagacherie, F. Baret, J.B. Feret, J. Madeira Netto, J.M. Robbez-Masson, Estimation of soil clay and calcium carbonate using laboratory, field, and airborne hyperspectral measurements, Remote Sens. Environ. 112 (2008) 825–835.

[39] M. Zribi, F. Kotti, Z. Lili-Chabaane, N. Baghdadi, N. Ben Issa, R. Amri, B. Duchemin, A. Chehbouni, Soil texture estimation over a semiarid area using TerraSAR-X radar data, IEEE Geosci. Remote Sens. Lett. 9 (2012) 353–357.

[40] B. Gallego-Elvira, A. Olioso, M. Mira, S. Reyes-Castillo, G. Boulet, O. Marloie, S. Garrigues, D. Courault, M. Weiss, P. Chauvelon, O. Boutron, EVASPA (EVapotranspiration assessment from SPAce) tool: An overview, four decades of progress in monitoring and modeling of processes in the soil-plant-atmosphere system: Applications and challenges, Procedia Environ. Sci. 19 (2013) 303–310.

[41] B. Martens, D.G. Miralles, H. Lievens, R. van der Schalie, R.A.M. de Jeu, D. Fernández-Prieto, H. E. Beck, W.A. Dorigo, N.E.C. Verhoest, GLEAM v3: Satellite-based land evaporation and root-zone soil moisture, Geosci. Model Dev. 10 (2017) 1903–1925.

[42] J. Joiner, L. Guanter, R. Lindstrot, M. Voigt, A.P. Vasilkov, E.M. Middleton, K.F. Huemmrich, Y. Yoshida, C. Frankenberg, Global monitoring of terrestrial chlorophyll fluorescence from moderate-spectral resolution near-infrared satellite measurements: methodology, simulations, and application to GOME-2, Atmos. Meas. Tech. 6 (2013) 2803–2823.

[43] D. Leroux, J.-C. Calvet, S. Munier, C. Albergel, Using satellite-derived vegetation products to evaluate LDAS-Monde over the Euro-Mediterranean area, Remote Sens. (Basel) 10 (2018) 1199 1–21.

[44] B.R. Pagán, W.H. Maes, P. Gentine, B. Martens, D.G. Miralles, Exploring the potential of satellite Solar-Induced Fluorescence to constrain global transpiration estimates, Remote Sens. (Basel) 11 (2019) 413 2–15.

[45] T. Kaminski, W. Knorr, G. Schürmann, M. Scholze, P.J. Rayner, S. Zaehle, S. Blessing, W. Dorigo, V. Gayler, R. Giering, N. Gobron, J.P. Grant, M. Heimann, A. Hooker-Stroud,

S. Houweling, T. Kato, J. Kattge, D. Kelley, S. Kemp, E.N. Koffi, C. Köstler, P.P. Mathieu, B. Pinty, C.H. Reick, C. Rödenbeck, R. Schnur, K. Scipal, C. Sebald, T. Stacke, A.T. van Scheltinga, M. Vossbeck, H. Widmann, T. Ziehn, The BETHY/JSBACH Carbon Cycle Data Assimilation System: experiences and challenges, Eur. J. Vasc. Endovasc. Surg. 118 (2013) 1414–1426.

[46] B.A. Forman, Y. Xue, Machine learning predictions of passive microwave brightness temperature over snow-covered land using the special sensor microwave imager (SSM/I), Phys. Geogr. 38 (2017) 176–196.

[47] S.-B. Kim, H. Huang, T.-H. Liao, A. Colliander, Estimating vegetation water content and soil surface roughness using physical models of L-band radar scattering for soil moisture retrieval, Remote Sens. (Basel) 10 (2018) 556 1–16.

[48] J.-C. Calvet, J. Noilhan, From near-surface to root-zone soil moisture using year-round data, J. Hydrometeorol. 1 (2000) 393–411.

[49] A. Stoffelen, S. Aaboe, J.-C. Calvet, J. Cotton, G. De Chiara, J. Figua-Saldana, A.A. Mouche, M. Portabella, K. Scipal, W. Wagner, Scientific developments and the EPS-SG scatterometer, IEEE J. Sel. Topics Appl. Earth Obs. Remote Sens. 10 (2017) 2086–2097.

[50] M. Parrens, J.-F. Mahfouf, A. Barbu, J.-C. Calvet, Assimilation of surface soil moisture into a multilayer soil model: design and evaluation at local scale, Hydrol. Earth Syst. Sci. 18 (2014) 673–689.

[51] D. Fairbairn, A.L. Barbu, J.-F. Mahfouf, J.-C. Calvet, E. Gelati, Comparing the ensemble and extended Kalman filters for in situ soil moisture assimilation with contrasting conditions, Hydrol. Earth Syst. Sci. 19 (2015) 4811–4830.

[52] D. Fairbairn, A.L. Barbu, A. Napoly, C. Albergel, J.-F. Mahfouf, J.-C. Calvet, The effect of satellite-derived surface soil moisture and leaf area index land data assimilation on streamflow simulations over France, Hydrol. Earth Syst. Sci. 21 (2017) 2015–2033.

[53] B. Decharme, C. Delire, M. Minvielle, J. Colin, J.-P. Vergnes, A. Alias, D. Saint-Martin, R. Séférian, S. Sénési, A. Voldoire, Recent changes in the ISBA-CTRIP land surface system for use in the CNRM-CM6 climate model and in global off-line hydrological applications, J. Adv. Model. Earth Syst. 11 (2019) 1207–1252.

[54] Y. Sun, C. Frankenberg, J.D. Wood, D.S. Schimel, M. Jung, L. Guanter, D.T. Drewry, M. Verma, A. Porcar-Castell, T.J. Griffis, L. Gu, T.S. Magney, P. Köhler, B. Evans, K. Yuen, OCO-2 advances photosynthesis observation from space via solar-induced chlorophyll fluorescence, Science 358 (2017) eaam5747 1–6.

[55] S. Biancamaria, D.P. Lettenmaier, T.M. Pavelsky, The SWOT mission and its capabilities for land hydrology, Surv. Geophys. 37 (2016) 307–337.

[56] W. Cramer, J. Guiot, M. Fader, J. Garrabou, J.-P. Gattuso, A. Iglesias, M. Lange, P. Lionello, M. Llasat, S. Paz, J. Peñuelas, M. Snoussi, A. Toreti, M.N. Tsimplis, E. Xoplaki, Climate change and interconnected risks to sustainable development in the Mediterranean, Nat. Clim. Chang. 8 (2018) 972–980.

[57] S. Paloscia, S. Pettinato, E. Santi, C. Notarnicola, L. Pasolli, A. Reppucci, Soil moisture mapping using Sentinel-1 images: Algorithm and preliminary validation, Remote Sens. Environ. 134 (2013) 234–248.

[58] F. Mattia, A. Balenzano, G. Satalino, F.P. Lovergine, A. Loew, J. Peng, U. Wegmüller, M. Santoro, O. Cartus, K. Dabrowska-Zielinska, J. Musial, M. Davidson, S.H. Yueh, S.-B. Kim, N.N. Das, A. Colliander, J.T. Johnson, J.D. Ouellette, J.P. Walker, X. Wu, H. McNairn, A. Merzouki, J. Powers, T. Caldwell, D. Entekhabi, M.H. Cosh, T.J. Jackson, Sentinel-1 high resolution soil moisture, IEEE Int. Geosci. Remote Sens. Symposium (2017) 5533–5536.

[59] M. El Hajj, N. Baghdadi, M. Zribi, H. Bazzi, Synergic use of Sentinel-1 and Sentinel-2 images for operational soil moisture mapping at high spatial resolution over agricultural areas, Remote Sens. (Basel) 9 (2017) 1292 1–28.

[60] Q. Gao, M. Zribi, M.J. Escorihuela, N. Baghdadi, Synergetic use of Sentinel-1 and Sentinel-2 for soil moisture mapping at 100 m resolution, Sensors 17 (2017) (1966) 1–21.

[61] B. Bauer-Marschallinger, V. Freeman, S. Cao, C. Paulik, S. Schaufler, T. Stachl, S. Modanesi, C. Massari, L. Ciabatta, L. Brocca, W. Wagner, Toward global soil moisture monitoring with Sentinel-1: Harnessing assets and overcoming obstacles, IEEE Trans. Geosci. Remote Sens. 57 (2019) 520–539.

[62] M. El Hajj, N. Baghdadi, H. Bazzi, M. Zribi, Penetration analysis of SAR signals in the C and L bands for wheat, maize, and grasslands, Remote Sens. (Basel) 11 (2019) 31 1–14.
[63] S. Thiruvengadachari, Satellite sensing of irrigation pattern in semiarid areas: An Indian study, Photogramm. Eng. Remote. Sens. 47 (1981) 1493–1499.
[64] P.S. Thenkabail, M. Schull, H. Turral, Ganges and Indus river basin land use/land cover (LULC) and irrigated area mapping using continuous streams of MODIS data, Remote Sens. Environ. 95 (2004) 317–341.
[65] M.K. Gumma, P.S. Thenkabail, F. Hideto, A. Nelson, V. Dheeravath, D. Busia, A. Rala, Mapping irrigated areas of Ghana using fusion of 30 m and 250 m resolution remote-sensing data, Remote Sens. (Basel) 3 (2011) 816–835.
[66] Q. Gao, M. Zribi, M. Escorihuela, N. Baghdadi, P. Segui, Irrigation mapping using Sentinel-1 time series at field scale, Remote Sens. (Basel) 10 (2018) 1495 1–18.
[67] S. Bousbih, M. Zribi, M. El Hajj, N. Baghdadi, Z. Lili-Chabaane, Q. Gao, P. Fanise, Soil moisture and irrigation mapping in a semi-arid region based on Sentinel-1 and Sentinel-2 synergy, Remote Sens. (Basel) 10 (2018) (1953) 1–22.
[68] F. Zaussinger, W. Dorigo, A. Gruber, A.P. Tarpanelli Filippucci, L. Brocca, Estimating irrigation water use over the contiguous United States by combining satellite and reanalysis soil moisture data, Hydrol. Earth Syst. Sci. 23 (2019) 897–923.

CHAPTER 7

Modeling for integrated water resources management in the Mediterranean region

Michel Le Page[a], Younes Fakir[b,c], Jalel Aouissi[d]
[a]CESBIO (UPS/CNES/CNRS/INRAE/IRD), Toulouse, France
[b]TREMA/GEOHYD, Cadi Ayyad University, Marrakech, Morocco
[c]CRSA (Center for Remote Sensing Application), UM6P, Benguerir, Morocco
[d]Institut National Agronomique de Tunisie, Laboratoire GREEN-TEAM, Université de Carthage, Tunis, Tunisia

7.1 State of the art

7.1.1 Integrated water resources management

According to the Global Water Partnership [1], the water resources are under increasing pressure due to the demographic pressure and improved standard of living. The global population increased by a factor of three during the 20th century and will reach 10 billion people by 2050. Agriculture needs to satisfy food necessities. Climate change is progressively modifying both plant water demand and the water available from rainfall. The factors affecting the water availability and demand are very well illustrated in Ref. [2] and shown here in Fig. 7.1.

Several countries have already entered an era of water shortage [3] or can be classified as water scarce or water stressed [2, 4–6]. The competition and conflicts over freshwater resources are intensifying. Human activities provoke domestic, agricultural, and industrial pollutions that are transported by the water fluxes. The pollution is concentrated in rivers, threatening human health and ecosystems. Conversely, if water resources decrease, the concentration of pollutants will increase, increasing their impact on water quality. By the turn of the 21st century, the sector-based approach of water resources was also perceived to be deficient because of the fragmented and uncoordinated management of water resources.

Watershed management has existed for millennia [7]; nevertheless, a holistic and bottom-up approach [8] has only emerged since the 1970s. Darghouth et al. [9] consider that the shift from an "engineering-led" approach to a participatory approach actually occurred during the 1990s. Various authors agree to place the birth of Integrated Water Resource Management (IWRM) in 1992 with the Dublin world conference and the World Summit in Rio [1, 8–10]. Chapter 18 of Agenda 21 [11] states that all states could establish the institutional basis for IWRM by the year 2000 and have completed parts of

Water Resources in the Mediterranean Region
https://doi.org/10.1016/B978-0-12-818086-0.00007-8

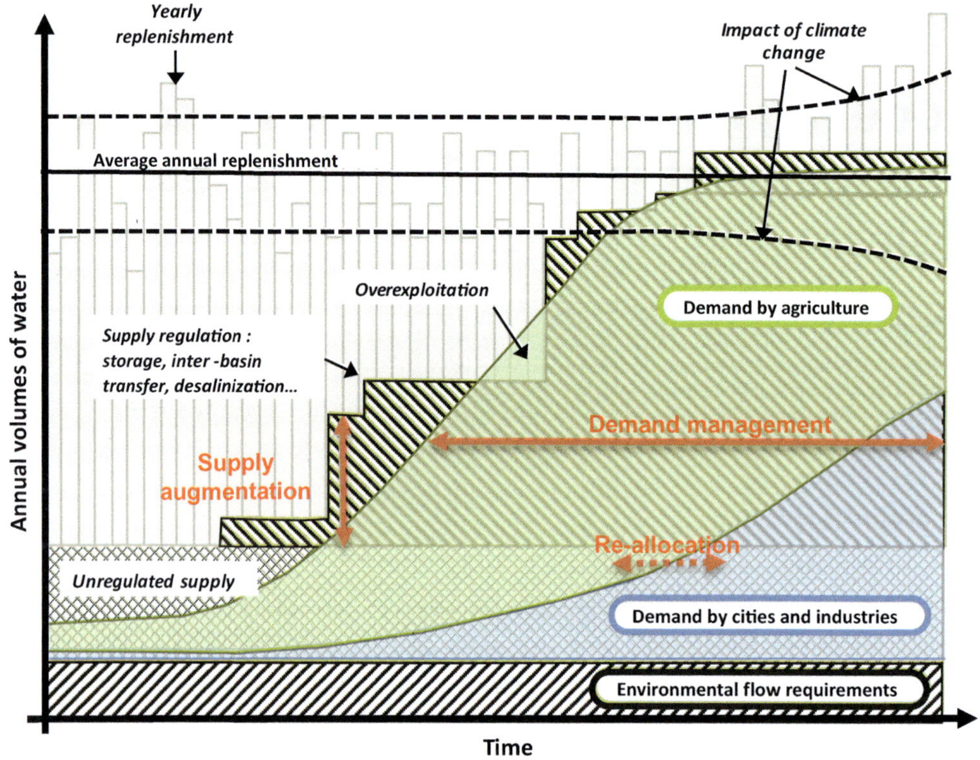

Fig. 7.1 The factors affecting water availability and demand. The water available to the system is mobilized by supply regulation in order to satisfy the different water demands. Water management operates on the supply, the demand, and allocation of water. *Adapted from FAO (Ed.), Coping With Water Scarcity: An Action Framework for Agriculture and Food Security. Food and Agriculture Organization of the United Nations, Rome, 2012.*

all freshwater programs by 2025. The Dublin Statement on Water and Sustainable Development set out four guiding principles:

1. Freshwater is a finite and vulnerable resource, essential to sustain life, development, and the environment.
2. Water development and management should be based on a participatory approach, involving users, planners, and policy-makers at all levels.
3. Women play a central part in the provision, management, and safeguarding of water.
4. Water has an economic value in all its competing uses and should be recognized as an economic good.

At the turn of the 21st century, many countries adopted the Agenda 21 guidelines and IWRM—a 2008 survey showed that 22% of the world countries had implemented IWRM plans, while 37% partially implemented it [12]—which has been defined as "a process which promotes the coordinated development and management of water, land

and related resources, in order to maximize the resultant economic and social welfare in an equitable manner without compromising the sustainability of vital ecosystems" [1]. Nevertheless, after a few years of experimentation of the process, skepticism arose. Biswas [13] criticized a blurry concept, wondering what aspect should be integrated and underpinning the difficulties of implementation. Molle [14] qualified IWRM as a nirvana concept, an ideal image of what the world should tend toward, but which is unreachable. After the candid years and skeptic years, IWRM has evolved from theory to practice. The Australian Water Reform extensively related in Ref. [15] is an excellent example of this evolution in the last 30 years. It includes the transformation of water allocation, the improvement of environmental management, the reform of pricing of water services, modernization of institutions, community and stakeholder engagement, and the improvement of water information and knowledge.

Interestingly, Borchardt and Ibisch [16] argue that the strong links and important trade-offs between water security, food security, and energy security mean that IWRM should be considered as a pathfinder process for the implementation of integrated resource management.

IWRM is, therefore, a holistic approach to water management involving technical, economic, financial, and institutional aspects [8]. IWRM is included in a virtuous circle of knowledge fueled by the participatory approach so that institutions can play their decision-making role in the most transparent way possible.

IWRM has management instruments which are "tools and methods that enable and help decision-makers to make rational and informed choices between alternative actions" [1]. A wide range of quantitative and qualitative methods are available and combine disciplines such as economics, hydrology, hydraulics, environmental sciences, and sociology.

The first aspect of IWRM is a hydrological assessment, which characterizes the properties and interactions of water resources within the watershed. The assessment includes land use, rainfall, runoff of streams and rivers, and groundwater. The hydrological assessment crosses the sectors of the economy, energy, agriculture, and environment [16] and looks for water quantity, quality, and demand [10]. IWRM should also deal with the variability of water in time and space, and manage risks associated with extreme variations of the resource. Drought and flood events can have dramatic effects on the population and economic activities. Pollution can be aggravated by those extremes. However, the reverse is also true, to some degree. Hydraulic works may increase the impact of floods, especially the larger ones (e.g., river channeling).

Within the participatory approach, the assessment is also an opportunity to gather stakeholders together and build a shared vision planning (SVP) of the watershed, i.e., "computer simulation models of water systems built, reviewed, and tested collaboratively with all stakeholders" [17]. It should be noted that for other authors [18], SVP has a broader meaning but we retain the original meaning here. A shared vision is a preamble to bottom-up management like that shown in the Matarraña catchment in Spain [19]. Darghouth et al. [9] note that the most successful applied research and knowledge sharing in watershed management programs worldwide occur when stakeholders are full partners in the research process at all the stages of identification, design, implementation, and evaluation, and technologies are offered as a range of choices to be adapted. However, they also remark that it requires careful institutional organization upfront to coordinate research agencies at various levels and to factor in other stakeholders, including farmers and the private sector. As a preliminary to SVP, a framework like the Transboundary Water Opportunity Analysis [20] can be useful to identify environment and ecosystem services, and to analyze the opportunities in the basin.

7.1.2 Issues of integrated modeling

Watershed management involves important trade-offs at the decision-making level, yet the technical aspect of integrated modeling of water resources is also exposed to interlinked trade-off issues. The modeler will deal with data availability and quality issues, the spatial, temporal, and physical resolution of the model, and the complexity of the model. Apart from being validated by peers and stakeholders, the model must be calibrated and the uncertainties should be evaluated at different levels. The model should obviously be able to respond to the major questions identified inside the basin, which generally includes dealing with extremes and the short- and long-term future. Finally, the model should rely on software that is not too complex to use and that is well maintained.

7.1.2.1 Spatial, temporal, and physical scales

Hydrological processes occur at a wide range of temporal and spatial scales [21]. Many hydrological processes operate in response to precipitation. The effect of precipitation can be taken at a very fine temporal and spatial scale—for example, a submetric spatial resolution and subminute time step allow analysis of water infiltration, soil saturation, and soil accommodation. At a larger time step (e.g., 5 min), different techniques or modeling approaches allow studying hydrograph distribution of flood events. At the daily time step, the flood event cannot be studied in detail, but the amount of runoff produced by a rainfall-runoff model can still be valuable information for basin analysis. Going to a monthly time step is still interesting to compare the evolution of basin production over the long term, but the runoff produced by a monthly model is not able to provide information about flood intensity. The resolution of the studied process is therefore important to take into account for integrated modeling.

Spatial resolution involves the concept of heterogeneity and variability of the studied objects. For example, when studying evapotranspiration of crops, a 1-m^2 plot can usually be considered homogeneous regarding the soil, the weather, the crop cover and its development, and the management practices. When going to the farmer plot scale, there is an inner heterogeneity that in some cases may be a blocking factor for modeling.

The modeler relies on different techniques to reach the desired temporal and spatial scale of information. Distributing involves interpolation techniques like Spline or Kriging, for example. Aggregation and disaggregation involve a spatial or temporal average of values. Upscaling is the process of distributing and aggregating, while downscaling is the process of disaggregating and singling out the values. Dynamical and statistical downscaling are the two principal ways to combine the information on local conditions with large-scale climate projection. The statistical downscaling however rely heavily on historical observations [22].

Liu et al. [23] have proposed SAHRA integrated modeling, which is a multidisciplinary integrated approach with three resolutions (fine, medium, and coarse). Another option is to couple models through a computational framework [24]. However, these complex multiresolution approaches should ensure that the different models are consistent with each other.

7.1.2.2 Data issue

The assessment of water resources is based on local water resources knowledge, in particular concerning the description of the watershed: land cover and land use, hydrological network, canals, aquifer, reservoirs, and the measurement of water fluxes: rainfall, river gauging, irrigation withdrawal, drinking water needs. The connection between resources and demand sites should be well-known. In addition, environmental and nonconsuming services should be identified (recreational, hydro-electricity) in order to cap their needs during the year. As monitoring and gauging systems have been deteriorating during the last 30 years [25], some of the data needed for IWRM are lacking or of poor quality. IWRM is, therefore, not only an opportunity to emphasize the utter importance of measurements but also a complex exercise of analysis, interpolation, extrapolation, and spatialization of data, which add to the inherent uncertainty of modeling. Prior to modeling, the data should be analyzed in order to check for the stationarity of time series, the spatial coherence, the accuracy of the spatial locations, and consistency between data coming from different agencies [26]. A whole arsenal of tools and techniques are available for the practitioner. Plots allow a rapid and visual assessment of data (e.g., time series plots, scatter plots). Statistical tests are useful to analyze the data, such as the Mann Kendall and Pettitt tests for trend and rupture analysis. The detection of errors can be done systematically in many cases with range tests, time consistency tests, and internal and spatial consistency tests [27]. Gap-filling techniques can then be applied to the data [28]. With systematic Earth Observation and global or regional meteorological modeling, new data sets are now available for integrated modeling. In particular, data sets such as the evolution of land cover, irrigated areas, and crop vitality are very relevant.

7.1.2.3 Model complexity

IWRM models are generally based on entities (objects, cells, etc.) which simulate something related to the water cycle (watershed, dam, aquifer, irrigation district, etc.). One task of the modeler is to find an adequate number of entities regarding the actual problem. If the upper streamflow is of no interest in terms of decision-making or if the available data will not allow forcing or calibrating the hydrologic model upstream, it is advised to have a minimum number of upstream entities. On the other hand, if the chosen modeling approach (e.g., distributed modeling) or the issue to be resolved (e.g., upstream hillside reservoirs) compels toward a large number of entities, the tasks regarding each entity should be well-programmed (input data, calibration, and error propagation). The modeler can also choose to detail the functioning of its entities: for example, a conceptual approach of a particular watershed may not be sufficient; it may be necessary to include an approach of the vadose zone with dynamic crop growth to study the effect of nitrate on the hydrosystem. While the detailed functioning responds to some questions and may also enhance accuracy, an increase in the number of parameters and input variables may prohibit the generalization over larger areas or along time. The number of connections between the entities of the model is directly linked to complexity. As the numbers of input and output fluxes are interconnected, the complexity of the model, and also its uncertainty, increases. In general, the IWRM model should be as simple and parsimonious as possible. More simplicity reduces the potential problems of maintenance (input data, calibrations) and model manipulation (scenario). Based on his long experience of modeling, [29] also considers that the model complexity can be determined adaptively in order to respond to different concerns.

7.1.2.4 Uncertainty

Water resource managers are aware that their decisions must be made in an uncertain and changing environment. Taking this uncertainty into account makes decisions very complex. Models are an imperfect abstraction of reality [8]. As stated earlier, the data carry their own part of uncertainty and gaps, and the scaling of the model forcing data introduce another level of uncertainty. The model parameterization and the numerical implementation are also possible sources of error. Walker et al. [30] proposed to distinguish three dimensions of uncertainty: the location, the nature, and the level of uncertainty. The location corresponds to the source of uncertainties mentioned earlier (context, model, inputs, and parameters). The nature of uncertainty is due to either a lack of knowledge (epistemic) or inherent variability. The levels of uncertainty range from deterministic to undetermined and can be separated into three classes. The statistical uncertainty is usually referred as to "uncertainty" earth sciences, which also suppose that the model fairly simulates the system and that the error can be measured through sensitivity analysis [31]. At the scenario level, a plausible future of the system and its driving forces is described, but because the scenario assumptions are not verifiable, the uncertainty level is beyond statistical uncertainty. Finally, at the level of recognized ignorance, the knowledge is too weak to propose a scenario. Refsgaard et al. [32] reviewed 14 different methods of uncertainty assessment and proposed a guide to select an appropriate methodology for uncertainty assessment. The communication about uncertainty increases the transparency of the scientific assessment process. However, strategies and caution must be taken into consideration in order to provide stakeholders with a clear and unambiguous report on the strengths and weaknesses of the assessment [33].

7.1.2.5 Sensitivity, calibration, and validation

Following the assessment of uncertainty, a sensitivity analysis aims at ranking the importance of input variables or parameters. Over-parametrization is a well-known problem of hydrological models [34]. Four families of methods can be separated: the local methods, which evaluate sensitivity at one point of the model hyperspace; the First Order Second Moment (FOSM), which estimates the mean and variance in a single point; the integration of local method to a global method (for example, by the sum of squared error); and finally the global sampling methods, which include Monte Carlo methods and Latin-Hypercube methods.

A calibration is the best estimate of the parameters by comparing the model output to reference data, by splitting the records into a calibration and validation set. Indicators such as the root-mean-squared error, the bias, the Nash-Sutcliffe index [34a] are used to qualify the performance of the model. In basins with high runoff variability, the bias component will have a smaller contribution on the Nash-Sutcliffe computation possibly leading to large errors, therefore it is advised to use the multiobjective Kling-Gupta efficiency criterion [35]. Earlier algorithms looked for local minimums [36, 37], but the finding that multiple convergence areas could exist leads to the search for a global optimum [38]. Beven [39] introduced the notion of equifinality, which means that an acceptable model prediction might be achieved in many different ways. However, for Ebel and Loague [40], the equifinality notion is related to ill-posed problems falling in three categories: identifiability (parameterization too complex); uniqueness (only one set of parameter values can be estimated from a given set of observations and this set represents the observed behavior at different time periods); and stability (small errors in the observed data do not produce large changes in the calibration). The GLUE method (Generalized likelihood Uncertainty Estimation) and the PEST software (Model-Independent Parameter Estimation) make a nonlinear parameter optimization using a Monte Carlo approach. However, those techniques are very difficult or even impossible to apply on physically based and distributed models.

7.1.2.6 Scenario

In their review, March et al. [41] defined scenario analysis as internally consistent stories about ways that a specific system might evolve in the future. Scenarios are plausible accounts of the future rather than forecasts. This is consistent with [42], that said that narrative scenarios are mostly a logical plot about how events will unfold in the future. Narrative scenarios are helping to identify the drivers of change and the implications of current trajectories as well as the options for action [43], and are halfway between facts and speculations in terms of complexity and uncertainty [44]. Swart et al. [45] pinpoint that the legitimacy of quantitative scenarios is linked to a good understanding of the system, the knowledge of its initial state, the persistent dynamics governing changes, and simulation algorithms that are sufficiently accurate. Those conditions are generally not met when working on long-range scenarios of a socio-ecological system. Mahmoud et al. [46] identified two types of scenario. The Exploratory scenarios describe the future according to known processes of change and extrapolation from the past. The Anticipatory scenarios are based on different desired or feared visions of the future. Alcamo [47] has noticed two threads of environmental scenario analysis; one is the scenario analysis that is used by the scientific community as a research tool by varying model inputs to mimic future changes in the driving forces of the system. The other thread is the one used with a planning goal where stakeholders and/or policy advisers work together to build a qualitative scenario. Alcamo et al. [47–49] emphasize that the qualitative and quantitative threads should combine so that the model calculations complement the storyline by presenting numerical estimates of the environmental indicators of possible futures. In fact, quantitative scenarios should help to define when, where, and how a situation might occur, and to evaluate the impact of anthropogenic actions. Our experience with qualitative approaches led us to the same conclusion as [50]. Despite its texture, richness, and insight [45], the qualitative approach is vague regarding spatial explicitness and poor regarding the timing and quantification of processes. Methods for translating from a narrative into a quantitative scenario are needed, and the uncertainty of such translations should be assessed. To our knowledge, several studies have omitted a translation phase and reduced the scenarios to incremental coefficients [51, 52]. In ecological studies, scenario analysis over the past 25 years has focused on climate change projections, while the impact on land use and land cover has been neglected so that only about 11% of the 2313 studies analyzed have included both land cover and climate changes [53]. Booth et al. [54] remarked that the combination of narrative and quantitative scenarios has mostly been applied at global to national scales. In their review of water scarcity scenarios, March et al. [41] pointed out the "hegemony" of climate as a driver of change, but they also identified several studies which incorporated other drivers of change. Booth et al. [54] note that there is a growing need for methodologies to translate qualitative scenarios into quantitative drivers. Mallampalli et al. [55] have identified various methods for translating narrative scenarios into quantitative ones, especially concerning land use changes. They include methods such as agent-based modeling, Bayesian networks, and fuzzy set modeling. These may possibly be connected to cellular automata to account for geographical interrelationships [56]. Others, such as [48], have opted for simpler trend models that can be "bent" according to policy scenarios. Such a simple method has the advantage of being parsimonious in terms of data preparation and computing so that it can be easily understood and used by water managers.

7.1.2.7 Tools

In their chapter, Stärz et al. [57] separated four classes of decision support systems available for IWRM: knowledge base; geographical information systems; mathematical models; and complex DSS. As the decision process relies on data, information, and spatial and temporal inferences, all of these tools help the process, but the one actually used for IWRM tends to be a more complex system composed of several components, integrating a database, and being accessible through a graphical user interface so that different levels of user can interact with the tool. In earlier works, various types of IWRM have been developed with the iThink/STELLA, a tool dedicated to general dynamic modeling, similar to, for example, the works of Simonovic et al. in Egypt [58] and Huerta in Mexico [59]. Dedicated tools have emerged [60–62]. Borden et al. [63] have made an excellent review of the tools available for water management and separated them into five main classes: water allocation and planning; flood and river water quality management; real-time flood warning system; water quality and sediment transport; and groundwater management. All those aspects are part of IWRM, but as we are focusing on the planning aspect, we only retained the first class, which is separated into two sections: the node and branches models (also identified as conceptual river basin models or RBMs); and the physically based models (also known as distributed hydrological models, or DHMs). As the authors point out, "while simpler, RBM structures can be very effective at planning water resources as they tend to be easy to operate, have short simulation times, and are conceptually easy to understand." DHMs are deterministic, fully distributed, physically based modeling systems where the data required and the computational power increase. However, they also offer the possibility to combine simple and advanced approaches enabling a compromise between result resolution and data availability constraints. The main representatives of RBMs are MIKE HYDRO Basin (DHI, Denmark), RIBASIM (Deltares, Netherlands), Source (eWater, Australia), WEAP (SEI, Sweden), and AQUATOOL (UPV, Spain). The different tools provide a GUI and have more or less the same capacities regarding allocation, rainfall-runoff models, demand calculator, irrigation management, groundwater linking, and water quality assessment. They are also well maintained by their provider and have large communities of users. WEAP is linked to a GIS interface to build up the topology of the entire basin and the links between demand and supply nodes. WEAP operates on the basic principles of a water balance. It is freely available to academic, governmental, and nonprofit organizations in developing countries. There are many different DHM tools, but the most widely used is SWAT (Soil and Water Assessment Tool) [64]. This powerful tool simulates most of the hydrological processes inside a river basin at the level of hydrologic response units (HRUs). HRUs are areas with homogeneous physical characteristics, land use, management practices, and soil types. In some aspects, HRUs can be compared to the lumped basins of the RBMs. The lumped nature of HRUs can, however, be perceived as a limitation to the design of a scenario. The model also includes additional routines that enable quantifying of the outflow and soil moisture content, and the fate of the sediment, nutrients, and pesticides, channeled toward the main gages in each sub-basin.

7.2 An analysis of the research situation in the Mediterranean context

In this section, a brief overview of past research works over the last decade on Integrated Modeling of Water Resource (IMWR) for decision support systems (DSSs) in the Mediterranean region is provided. The actual permeation of IWRM in the legislative and institutional levels is very heterogeneous between nations. In Europe, the Water Framework Directive (WFD) from 2000 established a general framework for the protection of water resources, in particular to improve aquatic ecosystems and mitigate the effects of floods and droughts. In the southern countries, IWRM implantation is more heterogeneous. For example, Morocco included the principles in the Water Law of 1995, while in Tunisia, water management is still very concentrated into the central state. IMWR-DSSs, by using models and including information management capabilities and involving stakeholders, are generally implemented to simulate the functioning and possible evolution of hydrosystems (surface water, groundwater, reservoirs) under natural and socio-economic drivers, to help decision-makers and end-users in solving complex water resource management problems and selecting the best decisions. The DSSs are targeted according to the role they might play in IWRM processes.

7.2.1 Main goals of integrated modeling for DSS

Implementations of IMWR-DSS in the Mediterranean context have reflected the growing interest in developing interdisciplinary researches, linking science with policy, and bringing insights and concerted solutions to address more effectively the complexity of water resources systems along with the issues faced by society and ecosystems. IMWR-DSS was designed and applied in watersheds to analyze and forecast the effects on surface water and groundwater availability, on water quality and water economics, of various constraints such as droughts, floods, agricultural practices, water management policies, and climate change.

Drought management plans has been implemented within the WFD framework. Among others, Spain which the most arid country of Europe, and France have elaborated drought management strategies which are integrated into a legal and institutional arsenal. In France, the main principles of water management were established in 1992 with a national water law. Master management plans are elaborated for each hydrological region (SDAGE) and local scales (SAGE) where quantitative objectives of river flows and groundwater levels are fixed. Low-Flow Management Plans (PGE, Plan de Gestion des Etiages) aim at enforcing those objectives. The PGE is a contractual document for participation between different water stakeholders. The PGE of the Garonne-Ariège region and the Charente basin relies on hydrological modeling to forecast river flows. The modeling has been assigned to a private company; the management partners feed the model on a day-to-day basis. The stakeholders can use the online e-tiage tool (http://www-e-tiage.com) to evaluate the actual river flows, critical levels, weather maps and forecast, and river flow forecast. During the summer, a weekly meeting involving water managers and elected representatives takes place and it is decided if the release of dam water is necessary for flow support. In Spain, the WFD was transposed in the national legal framework, the Spanish Water Act, in 2003. The management of drought includes the diagnostic of drought through adequate indicators, a national program of river and groundwater measurements, and management organisms. The national drought policy is defined by the central government, the River Basin District Administrations are responsible for drought planning and operational management [65], and drought events are managed by a Permanent Drought Commission (PDC), which has evolved [66], and now involves a multisectorial partnership which is closer to an integrated participative approach. In the Jucar basin, the AQUATOOL DSS [67] is used to assess the vulnerability of the system by feeding the system with streamflow scenarios and water demand scenarios [68]. In the management phase of drought, two approaches are used. The early warning system is based on Standardized Operative Drought Monitoring Indicators (SODMIs), which are computed monthly based on real-time in situ monitoring of reservoirs, aquifers, rivers, and precipitation. The AQUATOOL DSS is further used to simulate the impact of different options of water allocation that the Water Allocation Committee wants to examine. AQUATOOL also offers the possibility to carry out a risk evaluation [69].

Groundwater depletion as a main and persistent problem in the Mediterranean area has been targeted by various IMWR-DSSs. The effects of droughts on groundwater recharge and heads were analyzed in a Syrian watershed by DSS-Zabadani [70]. In Morocco, DSS-Tensift [71] using remote sensing was developed to predict the effects of various agricultural schemas on groundwater decrease in a semiarid basin. In Tunisia, the objective of DSS-Zeuss [72] was to assess the impact of a desalination project on the groundwater sustainability of a coastal aquifer.

Reservoirs management is an issue in terms of ensuring efficient and equitable water supplies, especially in wide river basins. In the Nile basin, integrated modeling [73] was used to determine release sequences from system reservoirs so that basins objectives (meeting water supply targets, avoiding water shortages, minimizing losses, maintaining land use patterns, regulating river flows, avoiding spillage, and generating energy) are suitably met. In the south of Portugal, multiobjective water allocation exercises showed the effects of combining IWRM and optimization for dam management [74]. Simonovic et al. [58] selected the management of the High Aswan Dam as an illustration of their approach.

Agriculture water management was recently addressed by means of DSSs usually integrating in addition to irrigation, meteorology, hydrology, crops functioning, and economy. DSS-Aquater [75] has integrated the information deriving from soil, climatic database, and remote sensing, with a crop simulation model to schedule irrigation at the district level in Italy. In Spain, DSS-WISCHE [76] was developed with an agriculture community to provide dynamic scheduling of the daily irrigation for a given land area by taking into account the irrigation network topology, the water volume technical conditions, and the logistical operations. In another Spanish basin, economic and hydrology models [77] were integrated to implement different policy scenarios with the purpose of assessing the impacts on groundwater storage and analyzing the responses of different farmers to diverse agricultural water policies. In semiarid areas, since irrigation relies increasingly upon groundwater abstraction, questions are inevitably raised concerning the future of local agriculture and the groundwater resource [78].

Climate change effects, as a common issue around the Mediterranean, were also addressed by several other DSSs to assess the possible long-term impacts on water resources. An Integrated Hydrological Modeling System (IHMS) [79] has been developed to study the impact of changes in climate, land use, and water management on groundwater and seawater intrusion into coastal areas in Cyprus and Italy. Climate change effects were also analyzed by coupled human-natural system modeling [80], the Adda river model, which was applied to an irrigation basin from a regulated lake in Italy. The objective is to study both systems (natural and human) and their coevolution under future changing climate conditions as well as to analyze different policy adaptation options.

Finally, it is also interesting to note that while the urban topic is inherent to all models, few IWRM studies focus on it specifically.

7.2.2 Modeling approaches and decision process

Various approaches have been used for developing integrated modeling systems for DSSs dedicated to water resources management and planning in Mediterranean countries. These approaches were featured by the complexity of the concerned hydrosystems, the targeted goals, the available data, the level of knowledge and technology, and the degree of users' and stakeholders' involvement.

IMWR-DSSs are mostly based on two master parts: integrated models and the decision-making processes. Therefore, the quality of IMWR-DSSs depends on the quality of the integrated modeling (models' suitability, calibration and validation, structure and organization of the workflow) [81] and on the robustness of the decision process (selection of stakeholders and users, building alternative policies, making sound decisions).

7.2.2.1 Integrated modeling

In IMWR-DSS, modeling platforms may encompass atmospheric, hydrologic, biologic, environmental, social, and economic modeling capabilities. The number and types of models in a DSS are generally chosen depending on the addressed processes and available technologies. The research works published during the last decade show that IMWR-DSSs in the Mediterranean were composed either of a single multimodules model or by several individual models embodied within a simulation system.

SWAT and WEAP are among the most used multimodules models; both are river basin simulators. Fig. 7.2 shows a bibliometry of the use of the two tools in the Mediterranean area.

The WEAP website lists more than 500 references (www.weap21.org), among which 16% correspond to the Mediterranean area for the period 2002–2018. Almost 60% of those references were published in peer-reviewed journals. Two-thirds of the references come from Arabic-speaking countries, which are also related to international collaborations like the ACSAD-BGR project (2008–2012), the GLOWA Jordan river project (2000–2010), and the LMI TREMA project (2010–2020). European countries have not chosen this solution, except for the Guadiana basin in Spain and a dozen more studies in Greece. The most common topics are the assessment of water resources (40%) and the impact of climate change (31%).

The SWAT website lists more than 3500 references during the same period (https://www.card.iastate.edu/swat_articles). Only 6% (225) of the articles concern the Mediterranean area. SWAT is used much more in northern countries (61%). The use of SWAT is quite different from WEAP. One-quarter of the studies were carried out over small watersheds ($<250\,km^2$). Apart from the main hydrologic assessment topic, SWAT is mostly used for pollutant assessment (cycling, loss, and transport), in particular for nutrients (nitrogen, nitrate, phosphorus), pathogen, and pesticides. SWAT has also been used for climate change and land use change. Only 10% of the studies concern groundwater.

Integrating individual models is also a widely used approach. The IHMS system [79] comprised three models: DiCaSM, or the Distributed Catchment Scale Model; the groundwater flow model MODFLOW; and SWI, or the Sea Water Intrusion model. In addition to estimating water balance components, DiCaSM produces the recharge data that are used as input to MODFLOW. Subsequently, MODFLOW generates the head distribution and groundwater flows that are used as input to the SWI model. The three models operate at different spatial and temporal scales. WISCHE for irrigation scheduling consists also of three mathematical models: M_ETC for minimizing the irrigation earliness-tardiness cost; MM_S for minimizing the maximum water speed; and

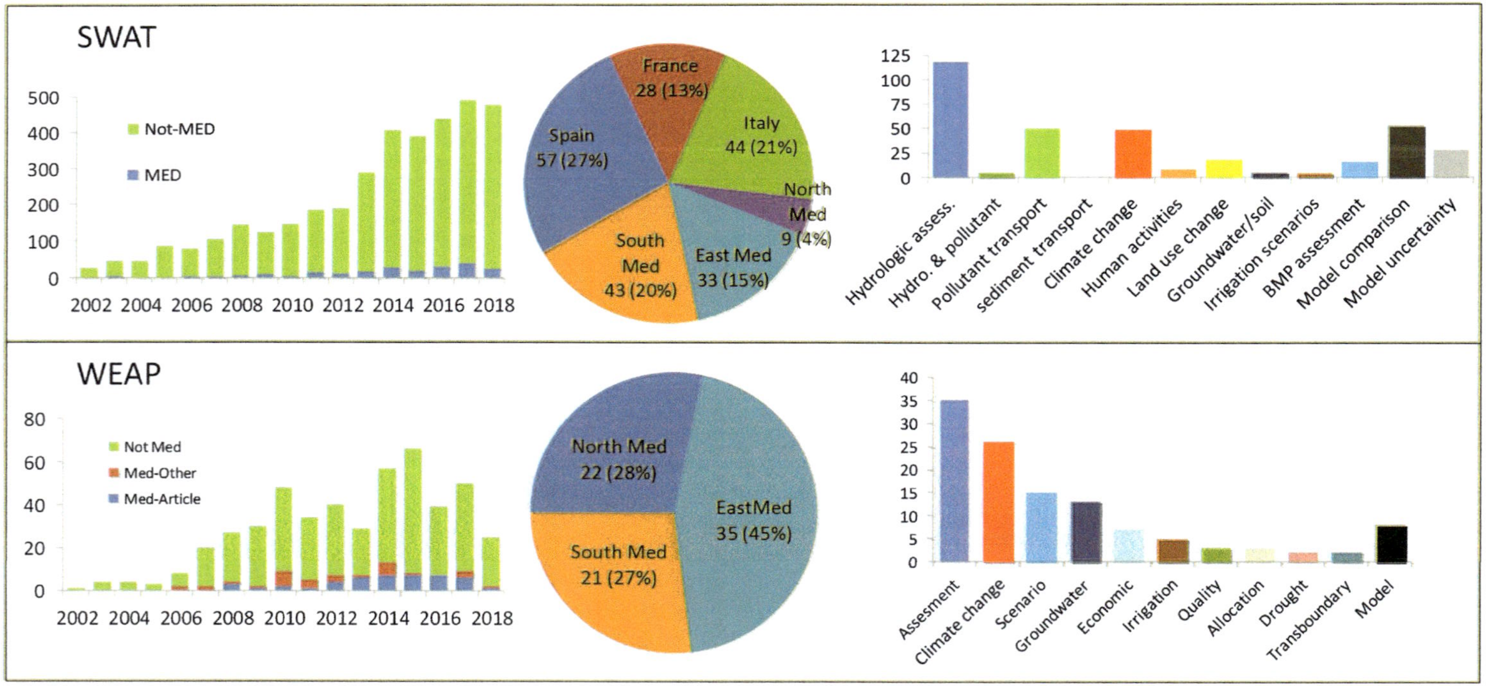

Fig. 7.2 A bibliometry of WEAP and SWAT studies in the Mediterranean area between 2002 and 2018.

MM_P for minimizing the maximum water pressure and an optimization module [76]. The Adda river modeling system includes a lumped, conceptual rainfall-runoff model, a mass balance model of regulated lake dynamics subject to human regulation, a routing model of the water released from the lake outlets to the intake of the irrigation canals, and a spatially distributed agricultural model simulating soil-crop water balance, crop growth stages, and final yield in each irrigation unit [80]. In the MedAction tool [82], up to 12 individual models were linked into a single system model simulating the linked bio-physical and socio-economic processes. It was applied to a watershed in Spain to assess the impact of a changing climate on the water resources and agriculture, and to determine the measures to mitigate the negative effects.

Individual models of a DSS might be coded in different programming languages, with different modeling techniques, and could operate on different spatial and temporal scales. The integration of the models is generally based on the dynamic coupling between models, so that the results of one model are transferred as input data to the other. Generally, the elaborated systems are sealed by a customized user interface whose role is to enable easy access to data and models, to allow pre-processing and post-processing via numerical data, diagrams, and thematic maps.

7.2.2.2 Involvement of stakeholders and decision process

The purpose of DSSs is to be used by end-users and stakeholders to support the design of present and future actions and policies in the field of IWRM. Therefore, the concerned water actors should be involved in the whole process of the design, implementation, and exploitation of DSSs. End-users are composed of representatives of local independent users (inhabitants, farmers, etc.), the community of users from the water sector (e.g., irrigation, tourism), water supply authorities, water management agencies, etc. In most cases, involvement is achieved via interviews, questionnaires, or a series of participatory workshops. Technically skilled stakeholders could provide organized data, documents, and technical studies to inform models, policies, and scenarios. Ultimately, the stakeholders are also supposed to take over the elaborated tools to keep monitoring their water systems and assessing their management alternatives.

Building scenarios illustrating management alternatives and assessing their impact is a milestone in the process of decision-making. The scenarios could be either narrative or quantitative; generally, qualitative ones are more targeted (e.g., Refs. [77, 83, 84]). For supporting decisions, some DSSs have integrated optimization procedures to select the best management alternatives. Two recent studies are given as examples for illustrating the use of optimization. In a water-scarce basin in Turkey where irrigation is the major water-consuming activity, sets of alternative water management scenarios were generated in a DSS. Their outcomes were evaluated by a set of social, economic, and environmental indicators on which a multicriteria analysis was applied to select the best option [85]. In another case study in Spain, economic indicators were used in addition to hydrological and land use indicators to evaluate the impacts on water resources of climate change and policy scenarios; a selection of the best alternative was performed using objective functions [86].

7.3 Two examples in the Mediterranean context

The following two examples illustrate two important concerns of IWRM: quantitative and qualitative management of water resources. In the Moroccan example, there has been a rapid depletion of the groundwater table since the end of the 20th century. The WEAP tool, which provides a graphical approach to build a model of the system, facilitates the participation of many stakeholders into the building of the model, even if they have little scientific or technical knowledge. The model building can be resumed in three phases. At first, institutional representatives of the main sectors (agriculture, city, high-education, forest, etc.) were gathered in a workshop. The modeler acted as a guide for the stakeholders, focusing on questions such as what is important, what complexity is needed, what should be the time step regarding the goals, etc. After a sketch of the watershed functioning was accepted by the stakeholders, the second phase began. This typical technical process involved several bilateral interactions with the different stakeholders: data gathering, calibration and validation, etc. The running version of the model was discussed in a new workshop with stakeholders: conception issues were pinpointed, data and calibration issues were discussed, giving way to new technical rounds. The scenario phase was the last part of the process. In this case, the development of the scenarios took place separately. It appeared that the translation of those narrative scenarios to quantitative and spatialized scenarios would be difficult. An approach residing on the use of remote sensing imagery to understand the local trends is presented here.

The Tunisian example focuses on the pollution mitigation of a lake classified at the UNESCO World Heritage. The emission of pollutants and sediments has been quantified from in situ analysis. A hydrological model (SWAT) was used to simulate the transportation of the pollutants and sediments throughout the whole watershed into the lake. An analysis of sensibility was carried out and the model was calibrated. The impact of Best Management Practices (BMPs) was evaluated. The results are expected to raise awareness (with a very extreme scenario of fertilizer reduction) and to suggest possible options to improve the water quality of the lake.

7.3.1 Using remote sensing imagery for IWRM: The example of the Haouz-Mejjate aquifer in Morocco

The development of cropped areas and irrigation is putting high pressure on water resources of the semiarid south-Mediterranean area. In the following example in Morocco, time series of remote sensing images are used to estimate evapotranspiration and irrigation requirements. A statistical model of crop coefficients is used to evaluate future scenarios which incorporate anthropogenic and climate changes.

7.3.1.1 The Haouz-Mejjate context

At the foot of the Atlas ranges in Morocco, the Tensift basin is a 6000-km^2 alluvial plain containing the Haouz aquifer and its western spread, the Mejjate aquifer. The climate is semiarid with rainfall of 240 mm/year and potential evapotranspiration of 1600 mm/year; it has mild, wet winters and very hot summers. Irrigated agriculture represents about 85% of water demand (mainly olive and orange orchards, and winter wheat). While urban water demand is increasing, mainly driven by Marrakech demographic growth, agriculture is also densifying and thus relying more than ever on groundwater. The piezometric network is witnessing groundwater table falls from 1 to 3 m/year, and the mean annual groundwater deficit (about 100 hm^3 since 2000) is equivalent to 50% of the reserves lost during the previous 40 years. Since irrigation relies increasingly upon groundwater abstraction, questions are inevitably raised concerning the future of local agriculture and the groundwater resource. The irrigated area is commonly divided into three parts: the "traditional" irrigated areas originally received their water from the Atlassian wadis, but are more and more dependent on groundwater. The "modern" irrigated areas receive water from dams either inside the watershed, either by a watershed transfer through a 140-km long canal. The rest of the irrigated area, which relies only on groundwater, is called "private irrigation."

In Morocco, the Water Law promulgated in 1995 (Law 10-95) takes up the main principles of IWRM. The law aims to establish a national water policy based on a forward-looking vision through Integrated Water Resources Management Plans (PDAIRE) that take into account both the evolution of resources and national water needs. The law also set up decentralized water management through basin agencies which have the mission to assess, plan, and manage water resources. However, the exercise of the law has proven to be delicate, especially concerning the control of groundwater abstraction [87]. In 2008, the Green Morocco Plan (Plan Maroc Vert, or PMV) was launched, with the ambition to double the agricultural GDP in 10–15 years, create 1.5 million jobs in the sectors, and increase agricultural income by a factor of two to three at the national level. Those two initiatives have been presented as a virtuous circle. In fact, the PMV would help modernize the irrigated sector, in particular by subsidizing the reconversion of surface irrigation to drip irrigation. The water savings could then be used to shift to more productive crops. Actually, capitalistic farmers were fast enough to obtain the subsidies in order to accelerate the growth of the irrigated area even as intensification of irrigated crops was occurring [88, 89].

7.3.1.2 Using remote sensing imagery to estimate evapotranspiration

Regarding the Haouz plain, several works [90] have intended to use Earth Observation for crop monitoring [91], irrigation water budget [92], soil moisture [93–95], irrigation assessment, [96, 97], and irrigation scheduling [98–100]. Based on these different works, Le Page et al. [71] have proposed to use the MODIS data set to address the consumption of water by irrigation in Integrated Modeling of Water Resources. The MOD13Q1 product (https://earthexplorer.usgs.gov) is a global data set provided every 16 days at 250 m spatial resolution which relies on level 2 daily surface reflectance products, which are corrected for molecular scattering, ozone absorption, and aerosols [101]. The Normalized Difference Vegetation Index (NDVI) was used with three objectives:

1. Apply a decision tree algorithm to separate the main crop classes for each studied year.
2. Use this multiannual classification map to determine the areas that are probably irrigated (an area is considered irrigated if the "tree" class is present 6 out of the 9 years or if 6 out of 9 "annual" years occurs on modern areas).
3. Combine the land cover map to NDVI time series by using empirical relations previously identified from field experiments [102–105] to obtain the monthly evolution of crop coefficients from irrigated subareas of the plain.

Different intermediate validations were done. The land cover map was compared to field trip observations, obtaining 90% of correspondence. The monthly evapotranspiration of irrigated crops was compared to field measurements obtained from previous flux measurements gathered over the studied area. Despite the scale difference of the measurement and the estimate, the results were generally very coherent with the observations (Fig. 7.3).

Those different estimations finally led to a monthly evapotranspiration data set of each irrigated area for the 8 years of study. The estimated performance of the irrigated areas was combined with evapotranspiration to obtain the water demand. This net water demand was finally faced off to the actual distributed surface water to derive the groundwater abstraction. The WEAP-21 tool was used to allocate the available offer to the water demand and a MODFLOW model was used for the groundwater transfers.

The modeling showed the capacity to reproduce the river flow at the outlet of the whole basin and the water table depth at different points of the aquifer. Interestingly, the authors estimated the sensibility of the model to different parameters and variables. As expected, the model is highly sensitive to evapotranspiration (Fig. 7.4), which exerts a major influence on the groundwater pumping.

7.3.1.3 Using remote sensing imagery to simulate cropping scenarios

Long-term scenarios of water management take into account many different aspects. Among them, the effects of anthropogenic activities on the water demand may be one of the most important components of a hydrologic disequilibrium. In the Mediterranean area, where irrigation may account for 80%–90% of the consumptive water demand, the dynamic of irrigated areas is utterly important. Long-term time series of

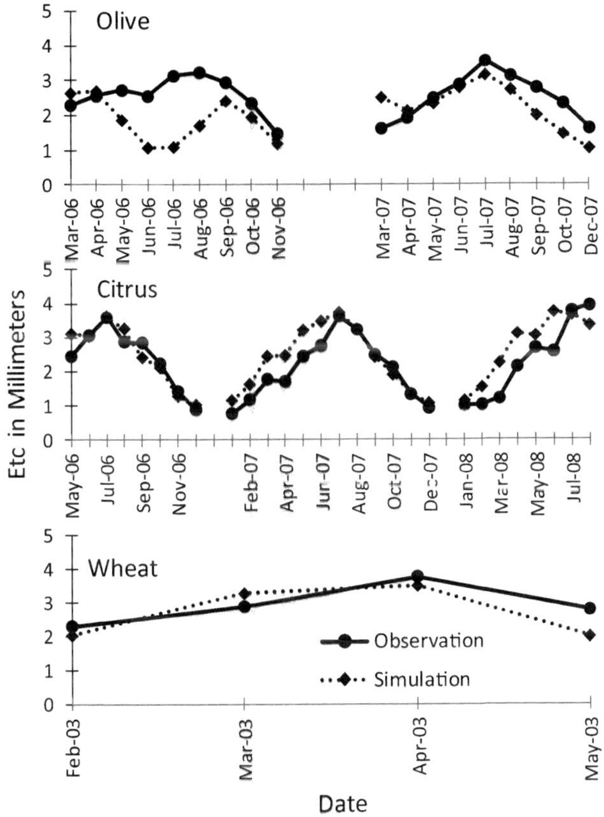

Fig. 7.3 Comparison of simulated and measured average daily evapotranspiration for olive, citrus, and wheat crops in Morocco. *Adapted from M. Le Page, B. Berjamy, Y. Fakir, F. Bourgin, L. Jarlan, A. Abourida, et al., An integrated DSS for groundwater management based on remote sensing. The case of a semi-arid aquifer in Morocco, Water Resour. Manag. 26 (2012) 3209–3230, https://doi.org/10.1007/s11269-012-0068-3.*

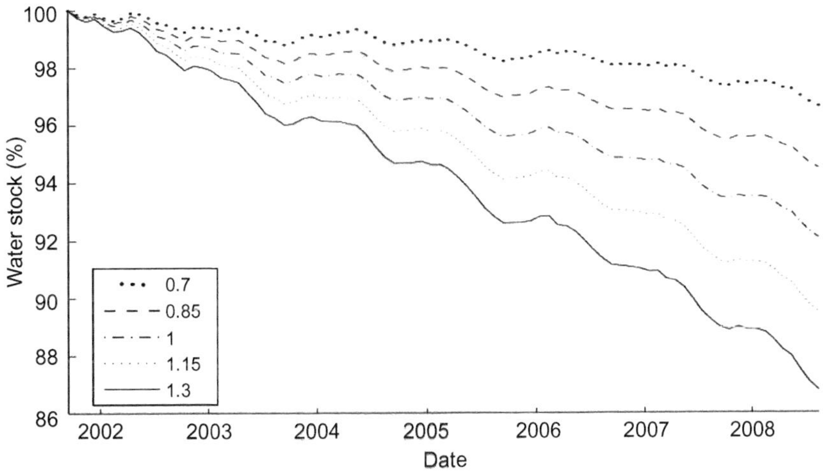

Fig. 7.4 Impact of 10% to 30% variation of evapotranspiration on the groundwater storage. *Adapted from M. Le Page, B. Berjamy, Y. Fakir, F. Bourgin, L. Jarlan, A. Abourida, et al., An integrated DSS for groundwater management based on remote sensing. The case of a semi-arid aquifer in Morocco, Water Resour. Manag. 26 (2012) 3209–3230, https://doi.org/10.1007/s11269-012-0068-3.*

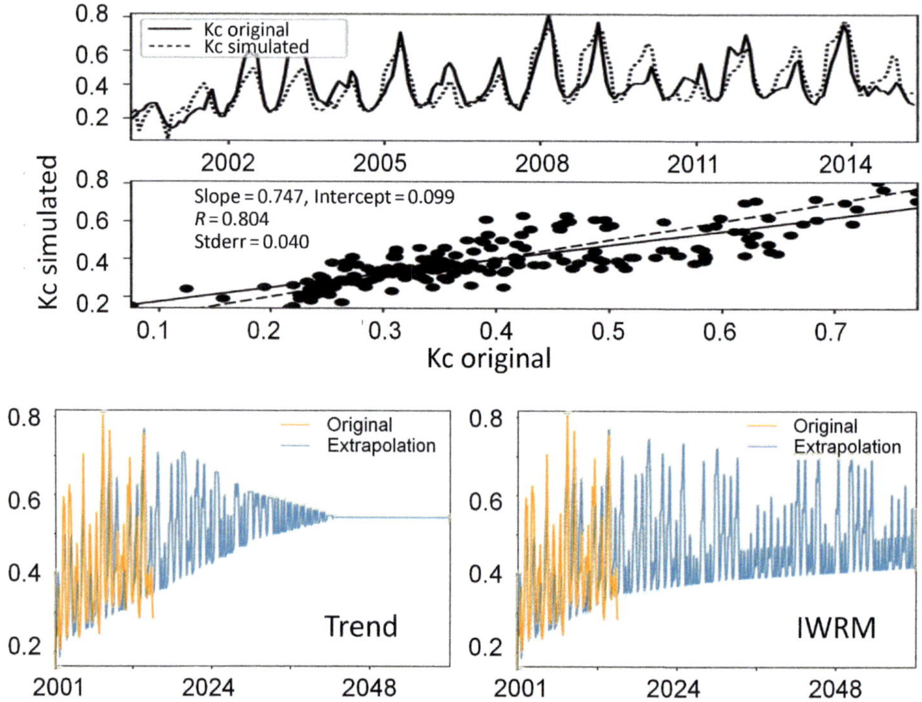

Fig. 7.5 *Top*: original crop coefficients obtained from Earth Observation *(full line)* and the simulated ones *(dotted line)* for an irrigated sector of the Haouz plain. *Bottom*: long-term simulation of crop coefficient following the trend *(left)* or according to a reduction coefficient translating IWRM narratives *(right)*.

remote sensing observations allow to study trends. In the example below, the 20 years long-time series of NDVI is used to compute the cropping trend of different irrigated areas in the Haouz plain. This is combined with yearly rainfalls and percentage of tree crops to extrapolate the crop coefficient into the future. An equation of irrigation efficiency can also be used to simulate alternatives of water demands. The statistical model is able to reproduce fairly well the crop coefficient evolution as shown in Fig. 7.5 (top). Fig. 7.5 (bottom left) shows how the crop coefficient is extrapolated and eventually saturates according to full coverage of tree crops. On the right, measures are taken to constrain the expansion of irrigated area and the extension of tree crops. Those measures have been translated into a simple reduction coefficient into the statistical model (Fig. 7.5, bottom right).

7.3.1.4 Combined impact of climate change and land cover change

Eight data sets (four models, two scenarios) were extracted from the MedCordex database and corrected by two statistical methods. The RCP4.5 scenario assumes that the increase in the concentration of greenhouse gases in the atmosphere gradually stabilizes at around 570 ppm by 2070. In the RCP8.5 scenario, greenhouse gas emissions continue to increase.

In Marrakech, the temperature is expected to increase by 1.5°C (RCP4.5) to 2.5°C (RCP8.5) by 2050. Temperature evolution is consistent between climate models and regionalization methods, and is therefore assumed to be robust. The variations of precipitation are relatively small for the RCP4.5 scenario, while the RCP8.5 scenario forecasts a decrease in precipitation between −22.4% and −27.8% by 2050. However, the inter-model variability is significant and much higher than the expected average change in precipitation, meaning that it is difficult to give a robust evolution of precipitation.

Those climatic trends have been used by Ref. [106] to assess the impact on river discharge, which is expected to reach a reduction of up to 30% by the year 2100. This will reduce the water available for traditional irrigation and aquifer recharge, and will also reduce the water available for reservoir storage. At the same time, urban water demand increases. The effect of temperature increase and precipitation decrease can then be analyzed concurrently to land cover scenarios (Fig. 7.6). In Ref. [107], the impact of the reconversion from surface irrigation to drip irrigation is also analyzed.

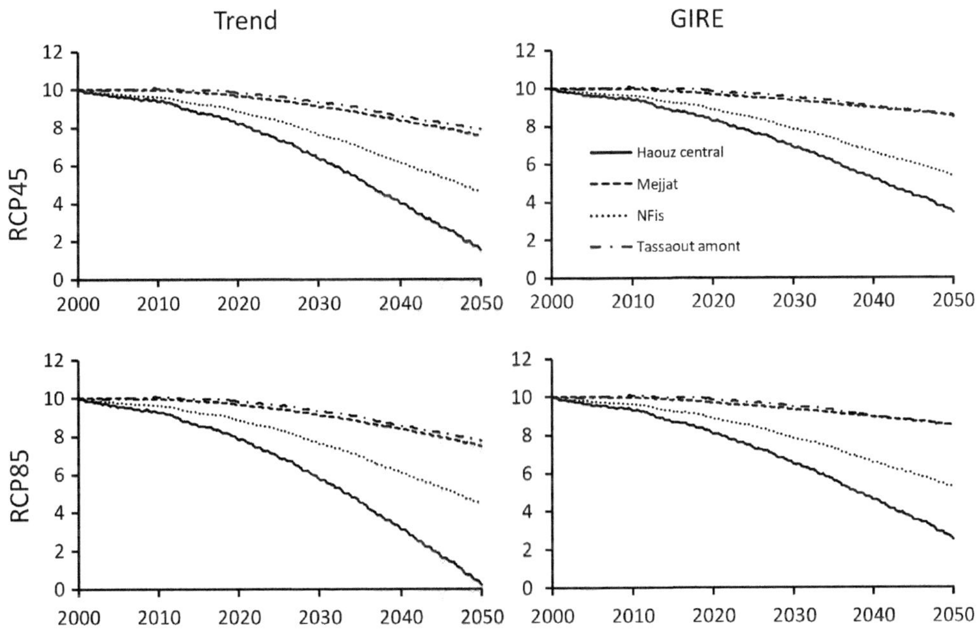

Fig. 7.6 Impact of climate change scenarios versus land cover scenarios on the water content of four subareas of the Haouz-Mejjate aquifer. The discharge is simulated from a hypothetical 10 billion m^3 at the beginning of the period.

7.3.1.5 Conclusion

Remote sensing estimation of monthly evapotranspiration over the study area was of great interest as a scientific approach and also for water managers, as it led to obtaining of the irrigation water demand and derivation of the groundwater abstraction. Groundwater variation in both time and space was decisive for past and predictive assessment of water budgets as well as for setting water management actions. However, applying management actions for groundwater on the terrain remains very difficult due to the uncontrolled withdrawals and to the continuous intensification and expansion of groundwater-irrigated agriculture.

7.3.2 Managing pollutants and erosion in Tunisia with the SWAT model

Land use change and increased use of fertilizers and pesticides impact the water quality and quantity in the Mediterranean basins. In this example study, the diffuse pollution and erosion risks are evaluated using experimental monitoring and hydrological modeling. After a sensibility assessment and a calibration and validation process, the model is used to assess Best Management Practices (BMPs) scenarios of the decrease of diffuse pollution by nitrate and sediment loads.

7.3.2.1 The Ichkeul Lake and Joumine watershed

Agriculture development in the north of Tunisia ensures crop productivity but deteriorates water quality with the discharge nutrients in surface water that will eventually reach lakes and reservoirs. Diffuse pollution is strongly affected by watershed land use change [108].

The Ichkeul Lake (Fig. 7.7A) has been a UNESCO World Heritage site since 1980. The salinity of the lake varies between marine salinity during the hot season and freshwater during the wet season [109]. The lake is therefore recognized for the value of its biological heritage [110]. The watershed of Ichkeul Lake drains an area of 2080 km^2, through seven wadis, with the highest pollution of nitrates coming from the Joumine wadi [111]. The Joumine watershed (418 km^2), with elevation ranging between 35 and 716 m, is an agricultural catchment with predominant cultures of oats, winter wheat, and sunflowers. The climate is subhumid with an average precipitation of 750 mm/year, average temperature ranging from 25°C in August to 8°C in January, and average evaporation of 1600 mm/year. The major soil types are vertisol with silty-clay texture. The hydrological regime for the main channel is characterized by permanent streamflow in the wet periods and a low flow in dry periods. The Joumine reservoir was built in 1983 with an initial storage capacity of 130 Mm^3. Its purpose was to control floods, supply water for drinking and irrigation, and preserve the Ichkeul Lake ecosystems. The annual flow is about 116 Mm^3. Soil erosion and sediment production on the Joumine watershed, deposited in the reservoir, are the consequences of a loss in the reservoir storage capacity by about 500,000 m^3 per year [112]. In addition to

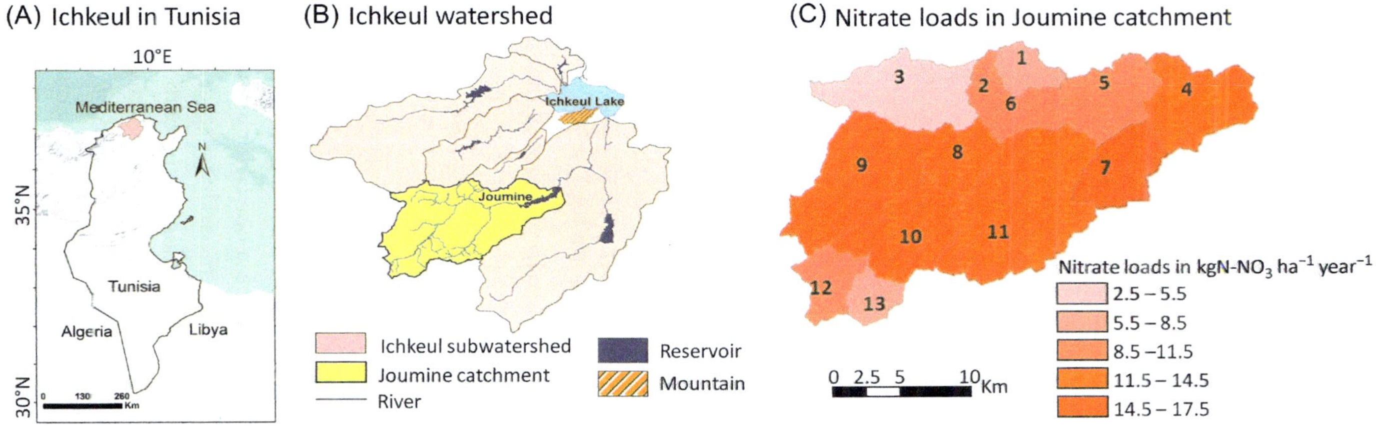

Fig. 7.7 (A) Location of the Ichkeul watershed in Tunisia. (B) The Joumine watershed inside the Ichkeul watershed. (C) Hydrology and nitrate loads in the Joumine catchment. *Adapted from J. Aouissi, Z.L. Chabaane, S. Benabdallah, C. Cudennec, Assessing the hydrological impacts of agricultural changes upstream of the Tunisian World Heritage sea-connected Ichkeul Lake, Proc. IAHS 365 (2015) 61–5, doi:10.5194/piahs-365-61-2015 and S. Mtibaa, N. Hotta, M. Irie, Analysis of the efficacy and cost-effectiveness of best management practices for controlling sediment yield: a case study of the Joumine watershed, Tunisia. Sci. Total Environ. 616–617 (2018) 1–16, https://doi.org/10.1016/j.scitotenv.2017.10.290.*

erosion risk, the water in the Joumine reservoir has been affected by toxic pollutants as a consequence of intensive agriculture in the watershed [113]. These problems deteriorate the water quality in the Joumine reservoir and subsequent water uses.

7.3.2.2 Simulation of sediment production and diffuse pollution

The SWAT model mentioned earlier is believed to be an effective tool to evaluate the impact of land management practices on water, sediment and load of chemical yields, pesticides on different, complex watershed scales, and environmental situations [114, 115]. It is used to simulate the quantity and quality of surface water and predicts the impact of agriculture practices, land use change, and climate change.

The impact of land use change and the nitrates load of the study watershed were simulated by SWAT [116, 117]. The main input data of the SWAT model are the Digital Elevation Model (DEM), land use, agricultural management, soil properties, and climatic data. The soil property was determined from the national soil authority. Land use data for 2010 were interpreted with Landsat series images. Daily weather data were obtained from local weather stations. Agriculture management data were obtained after many field surveys between 2006 and 2010 to provide the agriculture practices at the Joumine watershed. An experimental approach was used for monitoring the surface water quality at five locations for water samples. Nitrate loads ranged between 2.5 and 17.5 $kgN\text{-}NO_3\ ha^{-1}\ year^{-1}$ (Fig. 7.7C). Sediment production was also evaluated by Mtibaa et al. [118], and showed that about 34% of the catchment areas were affected by high to very high soil erosion risk. SWAT allowed identification of critical source areas and simulation of sediment yield according to a soil loss equation [119]. The only data available are the average sediment volume deposited in the Joumine reservoir, based on bathymetry surveys. An intermediary empirical relation was fitted to obtain a monthly suspended sediment flux to be used as the reference.

7.3.2.3 Sensitivity analysis and calibration of the hydrological model

A sensitivity analysis is required to determine the most influential parameters of the simulations. The method of the Latin hypercube at one factor was used. This method is based on a Monte Carlo random sampling, but the model hyperspace is subdivided into N strata with a probability of occurrence of $1/N$. Random values of the parameters are generated such that each range is sampled only once. The most important parameters are then separated, showing the six parameters that were affecting streamflow most: base flow factor; runoff curve number for moisture condition II; channel conductivity; Manning's n value for main channel; soil evaporation compensation coefficient; and soil depth and available water capacity of soil layer.

Following the sensitivity analysis, two steps are needed. Calibration consists of reducing the difference between the observed and simulated streamflow by adjusting model parameters. Validation consists of using the obtained parameters over another time period. In our case, we used the Parasol method to calibrate the most sensitive parameters (here 1 to 7) over a 17-year period (1990–2006). The model was validated for the period 2007–2009. The model performance was studied using a statistical index: Nash-Sutcliffe efficiency (NSE), r^2, and percent bias (PBIAS) [120].

$$NSE = 1 - \frac{\sum_i^N (obs_i - sim_i)^2}{\sum_i^N (obs_i - obs_{my})^2}$$

$$r^2 = \frac{\left[\sum_i^N (obs_i - obs_{my})(sim_i - sim_{my})\right]^2}{\sum_i^N (sim_i - sim_{my})^2 \sum_i^N (obs_i - obs_{my})^2}$$

$$PBIAS = \left[\frac{obs_{my} - sim_{my}}{obs_{my}} * 100\right]$$

where obs_i is the observed value, sim_i is the simulated value, obs_{my} and sim_{my} are the mean of observed and the mean of simulated values, respectively, and N is the number of observations.

NSE ranges between $-\infty$ and 1 (optimal value). Values between 0 and 1 are generally viewed as acceptable levels of performance. Negative values indicate unacceptable performance. The correlation coefficient, which ranges from -1 to 1, is an index of the degree of linear relationship between observed and simulated data. Percent bias (PBIAS) measures the average tendency of the simulated data to be larger or smaller than their observed counterparts [121]. The PBIAS is expressed as a percentage with an optimal value to 0.

The SWAT model simulation shows a good agreement between monthly measured and simulated streamflow before calibration with an NSE and r^2 values for 0.76 and 0.86, respectively. However, model runs reproduced a peak flow with an over-estimation of low flow. After calibration, the statistical indices were improved with an NSE value of 0.9 and r^2 value of 0.92 (Fig. 7.8). The calibration model shows a reduction of bias values from 58% to 31%, before and after calibration, respectively. Regarding the performance of the model, NSE, r^2 and bias for the validation period from 2007 to 2009 are 0.89%, 0.92%, and 19% respectively (Fig. 7.9). Generally, the model overestimated the low streamflow. The error in SWAT simulation can be related to instrument measure and the input uncertainty.

After calibration and validation of hydrological process, the model was used to simulate nitrate and sediment loads. The comparison between observed and simulated values was realized at the water sampling locations, showing that they vary in the same range

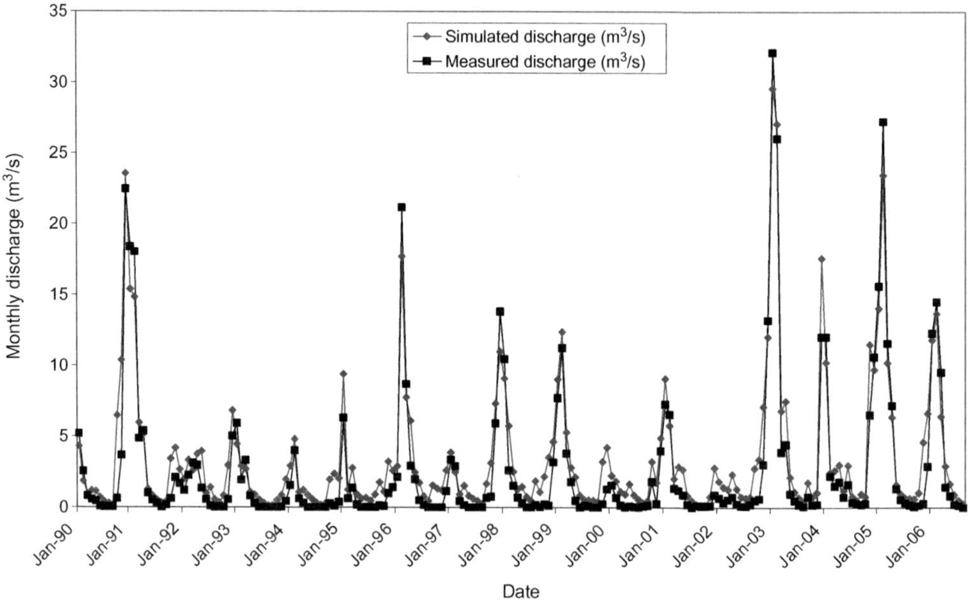

Fig. 7.8 Simulated and observed monthly discharge at the outlet of the Joumine basin after calibration [117].

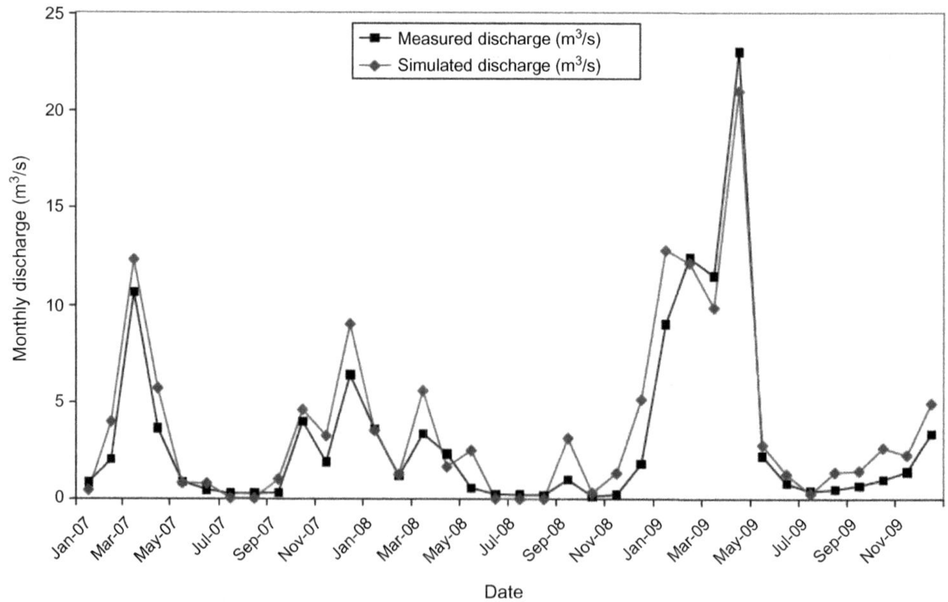

Fig. 7.9 Simulated and measured monthly discharge at the outlet of the Joumine watershed during the validation periods (2007–2009) [117].

from 0 to 5 mg/L. The higher nitrates concentration comes from the part of the basin devoted to agriculture activities. Steep areas have a higher erosion risk. They require the implementation of mitigation practices for reducing sediment yield.

7.3.2.4 Best management practices scenarios

Four scenarios of agricultural BMPs were tested to determine their impact in nitrates reduction: **SC1**: 20% of *N* fertilizer reduction; **SC2**: 100% of *N* fertilizer reduction; **SC3**: implementation of filter strips (width = 50 m) applied only to sub-basins draining directly to the main stream; and **SC4**: combination between SC1 and SC3. The 100% reduction of fertilizer has a huge impact on wheat yield; it was only intended for awareness purposes, showing 72% reduction of NO_3-N loads [117]. The optimal scenario (BMP) should reduce nitrate loads in the rivers and have a low impact on crop yields, which can be obtained with SC4.

For reducing the erosion risk at the Joumine watershed, Mtibaa et al. [118] constructed six BMP scenarios: (i) detention/sediment ponds; (ii) contour ridges; (ii) grass strip cropping; (iv) no-till farming with residue management; (v) buffer strips; and (vi) land use conversion. The simulations show that the sediment yield can be reduced by implementing BMPs. Ponds and contour ridges reduce the sediment production by 10% and 60%, respectively. The agriculture activities change from cropland to olive orchards with slopes >20% reduced sediment yield by 71.1%. In fact, the olive orchards generate less sediment than cropland, because the interception of rainfall protects the soil surface from detachment caused by raindrop splashes. The implementation of buffer strips along the main channel across croplands reduces the sediment yield, for 5 and 20 m buffer strip width to 20% and 30%, respectively, at the watershed level.

7.3.2.5 Conclusion

In this basin, the SWAT model has been used to simulate the production, yield, and transportation of pollutants and erosion. The sensibility of the hydrological model has been studied, so that the most important parameters were calibrated against the actual discharges. The model was then used to simulate nitrate and sediment loads, which are finally compared to observed values at the water sampling locations. Finally, different scenarios have been considered to reduce diffuse pollution by nitrate and sediment loads.

7.4 Conclusion

Integrated Water Resource Management is a holistic approach involving technical, economic, financial, and institutional aspects, and where a participatory approach facilitates the best decision-making. Born in the 1990s, the concept has evolved toward a more practical framework during the last 30 years. In the Mediterranean area, different integrated modeling tools have been developed in attempts to solve simultaneously the need for a shared vision of the system between stakeholders and a faithful simulation of the system. The modeling processes are currently confronting several technical difficulties. The main ones are the complexity of the hydrosystems, the multiobjective approach, availability and collection of data, error propagation, and uncertainties. Many studies have focused on the impact of climate change. Fewer have focused on the impact of anthropogenic-induced changes like land use change. In addition, as water resources multimodeling-based DSSs are usually performed by distinct teams (users, scientists, and programmers) for very specific case studies, the technologies are diverse and difficult to replicate.

The management of water resources has been the source of the first human rules and laws. It is therefore no surprise that the IWRM framework overlaps and often conflicts with past water management policies. In various aspects, the new paradigm implies a redistribution of cards between players. The water basin, which has seemed to be the best level of administration, has often been incompatible with previous administrative divisions, or traditional management areas, but this level of management might be questioned in a future made of more water transfer or nonconventional water resources. The decision process not only is reversed from top-down to bottom-up, but also becomes very horizontal and involves many different objectives. With the growth of public concern, and even the summoning to empower each individual regarding the preservation of natural resources and the impacts of global planetary change, the definition of the stakeholder is becoming blurry. In this increasingly complex arena, modeling for IWRM must respond with adequate timing, data, complexity, and uncertainty. Currently, one major challenge in the Mediterranean is the gap existing between the theory of IWRM (research, technical tools, and cooperation) and the reality of implementation of IWRM on the terrain. This is particularly observed in southern Mediterranean countries facing water scarcity, socio-economic problems, and political issues.

References

[1] Global Water Partnership, Integrated Water Resources Management, Global Water Partnership, Stockholm, 2000.

[2] FAO (Ed.), Coping With Water Scarcity: An Action Framework for Agriculture and Food Security, Food and Agriculture Organization of the United Nations, Rome, 2012.

[3] D. Seckler, U. Amarasinghe, D. Molden, R. de Silva, R. Barker, World Water Demand and Supply, 1990 to 2025: Scenarios and Issues, International Water Management Institute, Colombo, Sri Lanka, 1998.

[4] J.S. Wallace, Increasing agricultural water use efficiency to meet future food production, Agric. Ecosyst. Environ. 82 (2000) 105–119, https://doi.org/10.1016/S0167-8809(00)00220-6.

[5] F.R.R. Rijsberman, Water scarcity: fact or fiction? Agric. Water Manag. 80 (2006) 5–22, https://doi.org/10.1016/j.agwat.2005.07.001.

[6] M. Bazza, M. Kay, C. Knutson, FAO, Drought Characteristics and Management in North Africa and the Near East, 2018.

[7] G. Wang, S. Mang, H. Cai, S. Liu, Z. Zhang, L. Wang, et al., Integrated watershed management: evolution, development and emerging trends, J. For. Res. 27 (2016) 967–994, https://doi.org/10.1007/s11676-016-0293-3.

[8] D.P. Loucks, E. van Beek, J.R. Stedinger, J.P.M. Dijkman, M.T. Villars, E.V. Beek, et al., Water Resources Systems Planning and Management and Applications: An Introduction to Methods, Models and Applications, vol. 51, UNESCO, 2005.
[9] S. Darghouth, C. Ward, G. Gambarelli, E. Styger, J. Roux, Watershed Management Approaches, Policies, and Operations: Lessons for Scaling Up, World Bank, 2008.
[10] D. Borchardt, J. Bogárdi, R.B. Ibisch (Eds.), Integrated Water Resources Management: Concept, Research and Implementation, Springer, Cham, Heidelberg, New York, 2016.
[11] UN-Division of Sustainable Development. Agenda 21 1992.
[12] UN-Water. Status Report on IWRM and Water Efficiency Plans for CDS16, 2008.
[13] A.K. Biswas, Integrated water resources management: a reassessment: a water forum contribution, Water Int. 29 (2004) 248–256, https://doi.org/10.1080/02508060408691775.
[14] F. Molle, Nirvana concepts, narratives and policy models: insights from the water sector, Water Altern. 1 (2008) 26.
[15] B.T. Hart, J. Doolan, Decision Making in Water Resources Policy and Management: An Australian Perspective, Academic Press, an imprint of Elsevier, London, United Kingdom, 2017.
[16] D. Borchardt, R. Ibisch, Integrated Water Resources Management in a Changing World: Lessons Learnt and Innovative Perspectives, vol. 12, IWA Publishing, 2013.
[17] W.J. Werick, W. Whipple Jr., Managing Water for Drought. National Study of Water Management During Drought, U.S. Army Corps of Engineers, 1994.
[18] A. Serrat-Capdevila, J.B. Valdes, H.V. Gupta, Decision support systems in water resources planning and management: stakeholder participation and the sustainable path to science-based decision making, in: C. Jao (Ed.), Efficient Decision Support Systems—Practice and Challenges From Current to Future, InTech, 2011, https://doi.org/10.5772/16897.
[19] J. Bielsa, I. Cazcarro, Implementing integrated water resources management in the Ebro River Basin: from theory to facts, Sustainability 7 (2014) 441–464, https://doi.org/10.3390/su7010441.
[20] D.J.H. Philips, J.A. Allan, M. Claasen, J. Granit, A. Jägerskog, E. Kistin, et al., The TWO Analysis Introducing a Methodology for the Transboundary Waters Opportunity Analysis, Stockholm International Water Institute (SIWI), Stockholm, 2008.
[21] G. Blöschl, M. Sivapalan, Scale issues in hydrological modelling: a review, Hydrol. Process. 9 (1995) 251–290, https://doi.org/10.1002/hyp.3360090305.
[22] S. Trzaska, E. Schnarr, A Review of Downscaling Methods for Climate Change Projections, Center for International Earth Science Information Network (CIESIN), 2014.
[23] Y. Liu, H. Gupta, E. Springer, T. Wagener, Linking science with environmental decision making: experiences from an integrated modeling approach to supporting sustainable water resources management, Environ. Model Softw. 23 (2008) 846–858, https://doi.org/10.1016/j.envsoft.2007.10.007.
[24] S. Buis, A. Piacentini, D. Déclat, The PALM Group, PALM: a computational framework for assembling high-performance computing applications, Concurr. Comput. Pract. Exp. 18 (2006) 231–245, https://doi.org/10.1002/cpe.914.
[25] E. Beheim, G.S. Rajwar, M.J. Haigh, J. Krečeк, Integrated Watershed Management, Springer, Dordrecht, 2010, https://doi.org/10.1007/978-90-481-3769-5.
[26] J. Vaze, P. Jordan, R. Beecham, A. Frost, G. Summerell, Guidelines for Rainfall-Runoff Modelling: Towards Best Practice Model Application, eWater Cooperative Research Centre, 2012.
[27] J. Estévez, P. Gavilán, J.V. Giráldez, Guidelines on validation procedures for meteorological data from automatic weather stations. J. Hydrol. 402 (2011) 144–154, https://doi.org/10.1016/j.jhydrol.2011.02.031.
[28] N. Vuichard, D. Papale, Filling the gaps in meteorological continuous data measured at FLUXNET sites with ERA-Interim reanalysis, Earth Syst. Sci. Data 7 (2015) 157–171, https://doi.org/10.5194/essd-7-157-2015.
[29] D. Loucks, Model Complexity—What Is the Right Amount? https://i2insights.org/2016/11/03/model-complexity/, 2016. (Accessed 31 August 2018).
[30] W.E. Walker, P. Harremoës, J. Rotmans, J.P. van der Sluijs, M.B.A. van Asselt, P. Janssen, et al., Defining uncertainty: a conceptual basis for uncertainty management in model-based decision support, Integr. Assess. 4 (2003) 5–17, https://doi.org/10.1076/iaij.4.1.5.16466.

[31] F. Pianosi, K. Beven, J. Freer, J.W. Hall, J. Rougier, D.B. Stephenson, et al., Sensitivity analysis of environmental models: a systematic review with practical workflow, Environ. Model Softw. 79 (2016) 214–232, https://doi.org/10.1016/j.envsoft.2016.02.008.

[32] J.C. Refsgaard, J.P. van der Sluijs, A.L. Højberg, P.A. Vanrolleghem, Uncertainty in the environmental modelling process—a framework and guidance, Environ. Model. Softw. 22 (2007) 1543–1556, https://doi.org/10.1016/j.envsoft.2007.02.004.

[33] European Food Safety Authority, A. Hart, L. Maxim, M. Siegrist, N. Von Goetz, C. da Cruz, et al., Guidance on communication of uncertainty in scientific assessments, EFSA J. 17 (2019), https://doi.org/10.2903/j.efsa.2019.5520.

[34] A. van Griensven, T. Meixner, S. Grunwald, T. Bishop, M. Diluzio, R. Srinivasan, A global sensitivity analysis tool for the parameters of multi-variable catchment models, J. Hydrol. 324 (2006) 10–23, https://doi.org/10.1016/j.jhydrol.2005.09.008.

[34a] J.E. Nash, J.V. Sutcliffe, River flow forecasting through conceptual model. Part 1—a discussion of principles, J. Hydrol. 10 (1970) 282–290, https://doi.org/10.1016/0022-1694(70)90255-6.

[35] H.V. Gupta, H. Kling, K.K. Yilmaz, G.F. Martinez, Decomposition of the mean squared error and NSE performance criteria: implications for improving hydrological modelling, J. Hydrol. 377 (2009) 80–91, https://doi.org/10.1016/j.jhydrol.2009.08.003.

[36] J.A. Nelder, R. Mead, A simplex method for function minimization, Comput. J. 7 (1965) 308–313, https://doi.org/10.1093/comjnl/7.4.308.

[37] H.H. Rosenbrock, An automatic method for finding the greatest or least value of a function, Comput. J. 3 (1960) 175–184, https://doi.org/10.1093/comjnl/3.3.175.

[38] K. Beven, A. Binley, The future of distributed models: model calibration and uncertainty prediction, Hydrol. Process. 6 (1992) 279–298, https://doi.org/10.1002/hyp.3360060305.

[39] K. Beven, Prophecy, reality and uncertainty in distributed hydrological modelling, Adv. Water Resour. 16 (1993) 41–51, https://doi.org/10.1016/0309-1708(93)90028-E.

[40] B.A. Ebel, K. Loague, Physics-based hydrologic-response simulation: seeing through the fog of equifinality. Hydrol. Process, 20 (2006) 2887–2900, https://doi.org/10.1002/hyp.6388.

[41] H. March, O. Therond, D. Leenhardt, Water futures : reviewing water-scenario analyses through an original interpretative framework, Ecol. Econ. 82 (2012) 126–137, https://doi.org/10.1016/j.ecolecon.2012.07.006.

[42] G. Gallopin, A. Hammond, P. Raskin, R. Swart, Global scenarios and human choice, Stockholm Environment Institute, 1997. ISBN 91-88714.

[43] P. Raskin, G. Gallopin, P. Gutman, A. Hammond, R. Swart, Bending the Curve : Toward Global Sustainability, Stockholm Environment Institute. 1998.

[44] M. van Dijk, A Review of Global Scenario Exercises for Food Security Analysis: Assumptions and Results, FOODSECURE Working Paper 2, 2012.

[45] R.J. Swart, P. Raskin, J. Robinson, The problem of the future: sustainability science and scenario analysis, Glob. Environ. Chang. 14 (2004) 137–146, https://doi.org/10.1016/j.gloenvcha.2003.10.002.

[46] M. Mahmoud, Y. Liu, H. Hartmann, S. Stewart, T. Wagener, D. Semmens, et al., A formal framework for scenario development in support of environmental decision-making, Environ, Model Softw. 24 (2009) 798–808, https://doi.org/10.1016/j.envsoft.2008.11.010.

[47] J. Alcamo, Environmental Futures: The Practice of Environmental Scenario Analysis, vol. 2, (2008) https://doi.org/10.1016/S1574-101X(08)00414-6.

[48] J. Alcamo, T. Henrichs, T. Rösch, World Water in 2025: Global Modeling and Scenario Analysis for the World Commission on Water for the 21st Century, University of Kassel: Center for Environmental Systems Research, 2000.

[49] J. Alcamo, Scenarios as Tools for International Environmental Assessments, Office for Official Publications of the European Communities, Luxembourg, 2001.

[50] J. Reinhardt, S. Liersch, M.A. Abdeladhim, M. Diallo, C. Dickens, S. Fournet, et al., Systematic evaluation of scenario assessments supporting sustainable integrated natural resources management: evidence from four case studies in Africa, Ecol. Soc. 23 (2018), https://doi.org/10.5751/ES-09728-230105.

[51] M. Milano, D. Ruelland, A. Dezetter, J. Fabre, S. Ardoin-Bardin, E. Servat, Assessing the capacity of water resources to meet current and future water demands over the Ebro catchment (Spain), in: Proceeding of H09, IAHS-IAPSO-IASPEI Assembly, Gothenburg, Sweden, 2013, pp. 199–206.

[52] I. Alameddine, A. Fayyad, M. Abou Najm, M. El-Fadel, Sustainability of basin level development under a changing climate, Int. J. Sustain. Dev. Plan. 13 (2018) 394–405, https://doi.org/10.2495/SDP-V13-N3-394-405.

[53] N. Titeux, K. Henle, J.B. Mihoub, A. Regos, I.R. Geijzendorffer, W. Cramer, et al., Biodiversity scenarios neglect future land-use changes, Glob. Chang. Biol. 22 (2016) 2505–2515, https://doi.org/10.1111/gcb.13272.

[54] E.G. Booth, J. Qiu, S.R. Carpenter, J. Schatz, X. Chen, C.J. Kucharik, et al., From qualitative to quantitative environmental scenarios: translating storylines into biophysical modeling inputs at the watershed scale, Environ. Model. Softw. 85 (2016) 80–97, https://doi.org/10.1016/j.envsoft.2016.08.008.

[55] V.R. Mallampalli, G. Mavrommati, J. Thompson, M. Duveneck, S. Meyer, A. Ligmann-Zielinska, et al., Methods for translating narrative scenarios into quantitative assessments of land use change, Environ. Model. Softw. 82 (2016) 7–20, https://doi.org/10.1016/j.envsoft.2016.04.011.

[56] T. Houet, C. Marchadier, G. Bretagne, M.P. Moine, R. Aguejdad, V. Viguié, et al., Combining narratives and modelling approaches to simulate fine scale and long-term urban growth scenarios for climate adaptation, Environ. Model. Softw. 86 (2016) 1–13, https://doi.org/10.1016/j.envsoft.2016.09.010.

[57] C. Stärz, S. Kaden, B. Klauer, L. Leben (Eds.), Decision Support in the Context of IWRM: Lessons Learnt from Research Projects in Developing and Emerging Countries, Springer, Cham, Heidelberg, New York, 2016.

[58] S.P. Simonovic, H. Fahmy, A. El-Shorbagy, The use of object-oriented modeling for water resources planning in Egypt, Water Resour. Manag. (1997) 243–261.

[59] J.M. Huerta, A System dynamics approach to conflict resolution in water resources: the model of the Lerma-Chapala watershed, in: Systems Dynamics Society International Conference, Keble College, University of Oxford, Oxford, UK, 2004, p. 16.

[60] H. Assaf, E. Van Beek, C. Borden, P. Gisbers, A. Jolma, S. Kaden, et al., Generic simulation models for facilitating stakeholder involvement in water resources planning and management: a comparison, evaluation, and identification of future needs, in: A.J. Jakeman, A.A. Voinov, A.E. Rizzoli, S.H. Chen (Eds.), Environmental Modelling, Software and Decision Support: State of the art and New Perspectives, 2008.

[61] C. Giupponi, A. Sgobbi, Decision support systems for water resources management in developing countries: learning from experiences in Africa, Water 5 (2013) 798–818, https://doi.org/10.3390/w5020798.

[62] C. Teodosiu, C. Ardeleanu, L. Lupu, An overview of decision support systems for integrated resources management. Environ. Eng. Manag. J. 8 (2009) 153–162, https://doi.org/10.30638/eemj.2009.023.

[63] C. Borden, A. Gaur, R. Singh, Water Resource Software, The World Bank, 2016.

[64] J.G. Arnold, R. Srinivasan, R.S. Muttiah, J.R. Williams, Large area hydrologic modeling and assessment part I: model development, J. Am. Water Resour. Assoc. 34 (1998) 73–89, https://doi.org/10.1111/j.1752-1688.1998.tb05961.x.

[65] R. Maia, S. Vicente-Serrano, Drought planning and management in the Iberian Peninsula, in: Drought and Water Crises, CRC Press, 2017, pp. 481–506, https://doi.org/10.1201/9781315265551-28.

[66] M. Carmona, M. Máñez Costa, J. Andreu, M. Pulido-Velazquez, D. Haro-Monteagudo, A. Lopez-Nicolas, et al., Assessing the effectiveness of multi-sector partnerships to manage droughts: the case of the Jucar river basin: assessing governance performance, Earth's Future 5 (2017) 750–770, https://doi.org/10.1002/2017EF000545.

[67] J. Andreu, J. Capilla, E. Sanchís, AQUATOOL, a generalized decision-support system for water-resources planning and operational management, J. Hydrol. 177 (1996) 269–291, https://doi.org/10.1016/0022-1694(95)02963-X.

[68] J. Andreu, J. Ferrer-Polo, M.A. Pérez, A. Solera, J. Paredes-Arquiola, Drought planning and management in the Júcar River Basin, Spain, in: K. Schwabe, J. Albiac, J.D. Connor, R.M. Hassan, L. Meza González (Eds.), Drought in Arid and Semi-Arid Regions, Springer, Dordrecht, 2013, pp. 237–249, https://doi.org/10.1007/978-94-007-6636-5_13.

[69] J. Andreu, D. Haro, A. Solera, J. Paredes, A. Iglesias, D. Assimacopoulos, V.L. HAJ (Eds.), Experiences in Proactive and Participatory Drought Planning and Management in the Jucar River Basin, Spain, John Wiley & Sons, Ltd, Drought, Chichester, UK, 2018, pp. 217–237, https://doi.org/10.1002/9781119017073.ch12.

[70] A. Droubi, M. Al-Sibai, A. Abdallah, S. Zahra, M. Obeissi, J. Wolfer, et al., A decision support system (DSS) for water resources management—design and results from a pilot study in Syria, in: F. Zereini, H. Hötzl (Eds.), Climatic Changes and Water Resources in the Middle East and North Africa, Springer, Berlin, Heidelberg, 2008, pp. 199–225, https://doi.org/10.1007/978-3-540-85047-2_16.

[71] M. Le Page, B. Berjamy, Y. Fakir, F. Bourgin, L. Jarlan, A. Abourida, et al., An integrated DSS for groundwater management based on remote sensing. The case of a semi-arid aquifer in Morocco, Water Resour. Manag. 26 (2012) 3209–3230, https://doi.org/10.1007/s11269-012-0068-3.

[72] R. Hadded, I. Nouiri, O. Alshihabi, J. Maßmann, M. Huber, A. Laghouane, et al., A decision support system to manage the groundwater of the Zeuss Koutine aquifer using the WEAP-MODFLOW framework, Water Resour. Manag. 27 (2013) 1981–2000, https://doi.org/10.1007/s11269-013-0266-7.

[73] A.P. Georgakakos, Decision support systems for integrated water resources management with an application to the nile basin, in: Topics on System Analysis and Integrated Water Resources Management, Elsevier, 2007, pp. 99–116, https://doi.org/10.1016/B978-008044967-8/50005-1.

[74] R. Fragoso, V. Bushenkov, C. Marques, Multi-objective water allocation in the Alqueva region, New Medit 8 (2010).

[75] M. Acutis, A. Perego, E. Bernardoni, M. Rinaldi, AQUATER software as a DSS for irrigation management in semi-arid Mediterranean areas, Ital. J. Agron. 5 (2010) 205, https://doi.org/10.4081/ija.2010.205.

[76] M. Almiñana, L.F. Escudero, M. Landete, J.F. Monge, A. Rabasa, J. Sánchez-Soriano, WISCHE: a DSS for water irrigation scheduling, Omega 38 (2010) 492–500, https://doi.org/10.1016/j.omega.2009.12.006.

[77] C. Varela-Ortega, I. Blanco-Gutiérrez, C.H. Swartz, T.E. Downing, Balancing groundwater conservation and rural livelihoods under water and climate uncertainties: an integrated hydro-economic modeling framework, Glob. Environ. Chang. 21 (2011) 604–619, https://doi.org/10.1016/j.gloenvcha.2010.12.001.

[78] K. Boukhari, Y. Fakir, T.Y. Stigter, Y. Hajhouji, G. Boulet, Origin of recharge and salinity and their role on management issues of a large alluvial aquifer system in the semi-arid Haouz plain, Morocco, Environ. Earth Sci. 73 (2015) 6195–6212, https://doi.org/10.1007/s12665-014-3844-y.

[79] R. Ragab, J. Bromley, D.R. D'Agostino, N. Lamaddalena, G.T. Luizzi, G. Dörflinger, et al., Water Resources Management Under Possible Future Climate and Land Use Changes: The Application of the Integrated Hydrological Modelling System, IHMS, Springer, Netherlands, 2012, pp. 69–90, https://doi.org/10.1007/978-94-007-4756-2_5.

[80] M. Giuliani, Y. Li, A. Castelletti, C. Gandolfi, A coupled human-natural systems analysis of irrigated agriculture under changing climate, Water Resour. Res. 52 (2016) 6928–6947, https://doi.org/10.1002/2016WR019363.

[81] M. Volk, S. Lautenbach, H. van Delden, L.T.H. Newham, R. Seppelt, How can we make progress with decision support systems in landscape and river basin management? Lessons learned from a comparative analysis of four different decision support systems, Environ. Manag. 46 (2010) 834–849, https://doi.org/10.1007/s00267-009-9417-2.

[82] H. van Delden, P. Luja, G. Engelen, Integration of multi-scale dynamic spatial models of socio-economic and physical processes for river basin management, Environ. Model Softw. 22 (2007) 223–238, https://doi.org/10.1016/j.envsoft.2005.07.019.

[83] K. Kok, H. van Delden, Combining two approaches of integrated scenario development to combat desertification in the Guadalentín watershed, Spain, Environ. Plan. B Plann. Des. 36 (2009) 49–66, https://doi.org/10.1068/b32137.

[84] V. Pisinaras, C. Petalas, V.A. Tsihrintzis, G.P. Karatzas, Integrated modeling as a decision-aiding tool for groundwater management in a Mediterranean agricultural watershed: integrated modeling for groundwater management, Hydrol. Process. 27 (2013) 1973–1987, https://doi.org/10.1002/hyp.9331.

[85] C.P. Cetinkaya, M.C. Gunacti, Multi-criteria analysis of water allocation scenarios in a water scarce basin, Water Resour. Manag. 32 (2018) 2867–2884, https://doi.org/10.1007/s11269-018-1963-z.

[86] M.T. Kahil, F.A. Ward, J. Albiac, J. Eggleston, D. Sanz, Hydro-economic modeling with aquifer–river interactions to guide sustainable basin management, J. Hydrol. 539 (2016) 510–524, https://doi.org/10.1016/j.jhydrol.2016.05.057.

[87] F. Molle, O. Tanouti, La micro-irrigation et les ressources en eau au Maroc: un coûteux malentendu, Alternatives Rurales 18 (2017).
[88] R.Q. Grafton, J. Williams, C.J. Perry, F. Molle, C. Ringler, P. Steduto, et al., The paradox of irrigation efficiency, Science 361 (2018) 748–750, https://doi.org/10.1126/science.aat9314.
[89] F. Molle, Conflicting Policies: Agricultural Intensification, vs. Water Conservation in Morocco, 2017.
[90] S. Khabba, L. Jarlan, S. Er-Raki, M. Le Page, J. Ezzahar, G. Boulet, et al., The SudMed program and the joint international laboratory TREMA: a decade of water transfer study in the soil-plant-atmosphere system over irrigated crops in semi-arid area, Procedia Environ. Sci. 19 (2013) 524–533, https://doi.org/10.1016/j.proenv.2013.06.059.
[91] V. Simonneaux, B. Duchemin, D. Helson, S. Er-Raki, A. Olioso, A.G. Chehbouni, The use of high-resolution image time series for crop classification and evapotranspiration estimate over an irrigated area in central Morocco. Int. J. Remote Sens. 29 (2008) 95–116, https://doi.org/10.1080/01431160701250390.
[92] L. Olivera-Guerra, O. Merlin, S. Er-Raki, S. Khabba, M.J. Escorihuela, Estimating the water budget components of irrigated crops: combining the FAO-56 dual crop coefficient with surface temperature and vegetation index data, Agric. Water Manag. 208 (2018) 120–131, https://doi.org/10.1016/j.agwat.2018.06.014.
[93] A. Amazirh, O. Merlin, S. Er-Raki, Q. Gao, V. Rivalland, Y. Malbeteau, et al., Retrieving surface soil moisture at high spatio-temporal resolution from a synergy between Sentinel-1 radar and Landsat thermal data: a study case over bare soil, Remote Sens. Environ. 211 (2018) 321–337, https://doi.org/10.1016/j.rse.2018.04.013.
[94] Y. Malbéteau, O. Merlin, B. Molero, C. Rüdiger, S. Bacon, DisPATCh as a tool to evaluate coarse-scale remotely sensed soil moisture using localized in situ measurements: application to SMOS and AMSR-E data in Southeastern Australia, Int. J. Appl. Earth Obs. Geoinf. 45 (2016) 221–234, https://doi.org/10.1016/j.jag.2015.10.002.
[95] O. Merlin, Y. Malbéteau, Y. Notfi, S. Bacon, S. Khabba, L. Jarlan, Performance metrics for soil moisture downscaling methods: application to dispatch data in central Morocco, Remote Sens. 7 (2015) 3783–3807, https://doi.org/10.3390/rs70403783.
[96] M.H. Kharrou, M. Le Page, A.A. Chehbouni, V. Simonneaux, S. Er-Raki, L. Jarlan, et al., Assessment of equity and adequacy of water delivery in irrigation systems using remote sensing-based indicators in semi-arid region, Morocco, Water Resour. Manag. 27 (2013) 4697–4714, https://doi.org/10.1007/s11269-013-0438-5.
[97] M.H. Kharrou, S. Er-Raki, A. Chehbouni, B. Duchemin, V. Simonneaux, M. Le Page, et al., Water use efficiency and yield of winter wheat under different irrigation regimes in a semi-arid region, Agric. Sci. 02 (2011) 273–282, https://doi.org/10.4236/as.2011.23036.
[98] S. Belaqziz, S. Mangiarotti, M. Le Page, S. Khabba, S. Er-Raki, T. Agouti, et al., Irrigation scheduling of a classical gravity network based on the covariance matrix adaptation—evolutionary strategy algorithm, Comput. Electron. Agric. 102 (2014) 64–72, https://doi.org/10.1016/j.compag.2014.01.006.
[99] S. Belaqziz, S. Khabba, S. Er-Raki, L. Jarlan, M. Le Page, M.H. Kharrou, et al., A new irrigation priority index based on remote sensing data for assessing the networks irrigation scheduling, Agric. Water Manag. 119 (2013) 1–9, https://doi.org/10.1016/j.agwat.2012.12.011.
[100] M. Le Page, J. Toumi, S. Khabba, O. Hagolle, A. Tavernier, M. Kharrou, et al., A life-size and near real-time test of irrigation scheduling with a Sentinel-2 like time series (SPOT4-Take5) in Morocco, Remote Sens. 6 (2014) 11182–11203, https://doi.org/10.3390/rs61111182.
[101] C.O. Justice, J.R.G. Townshend, E.F. Vermote, E. Masuoka, R.E. Wolfe, N. Saleous, et al., An overview of MODIS Land data processing and product status, Remote Sens. Environ. 83 (2002) 3–15, https://doi.org/10.1016/S0034-4257(02)00084-6.
[102] B. Duchemin, R. Hadria, S. Er-Raki, G. Boulet, P. Maisongrande, A. Chehbouni, et al., Monitoring wheat phenology and irrigation in Central Morocco: on the use of relationships between evapotranspiration, crops coefficients, leaf area index and remotely-sensed vegetation indices, Agric. Water Manag. 79 (2006) 1–27, https://doi.org/10.1016/j.agwat.2005.02.013.
[103] S. Er-Raki, A. Chehbouni, S. Khabba, V. Simonneaux, L. Jarlan, A. Ouldbba, et al., Assessment of reference evapotranspiration methods in semi-arid regions: can weather forecast data be used as alternate of ground meteorological parameters? J. Arid Environ. 74 (2010) 1587–1596, https://doi.org/10.1016/j.jaridenv.2010.07.002.

[104] S. Er-Raki, A. Chehbouni, J. Hoedjes, J. Ezzahar, B. Duchemin, F. Jacob, Improvement of FAO-56 method for olive orchards through sequential assimilation of thermal infrared-based estimates of ET, Agric. Water Manag. 95 (2008) 309–321, https://doi.org/10.1016/j.agwat.2007.10.013.
[105] S. Er-Raki, A. Chehbouni, N. Guemouria, B. Duchemin, J. Ezzahar, R. Hadria, Combining FAO-56 model and ground-based remote sensing to estimate water consumptions of wheat crops in a semi-arid region, Agric. Water Manag. 87 (2007) 41–54, https://doi.org/10.1016/j.agwat.2006.02.004.
[106] A. Marchane, Y. Tramblay, L. Hanich, D. Ruelland, A. Marchane, Y. Tramblay, et al., Climate change impacts on surface water resources in the Rheraya catchment (High Atlas, Morocco), Hydrol. Sci. J. (2017) 1–17, https://doi.org/10.1080/02626667.2017.1283042.
[107] L. Jarlan, S. Khabba, S. Er-Raki, M. Le Page, L. Hanich, Y. Fakir, et al., Remote sensing of water resources in semi-arid Mediterranean Areas: the joint international laboratory TREMA, Int. J. Remote Sens. (2015) 1–39, https://doi.org/10.1080/01431161.2015.1093198.
[108] W. Ouyang, H. Huang, F. Hao, B. Guo, Synergistic impacts of land-use change and soil property variation on non-point source nitrogen pollution in a freeze–thaw area, J. Hydrol. 495 (2013) 126–134, https://doi.org/10.1016/j.jhydrol.2013.04.037.
[109] C. Casagranda, C.F. Boudouresque, A first quantification of the overall biomass, from phytoplankton to birds, of a Mediterranean brackish lagoon: revisiting the ecosystem of Lake Ichkeul (Tunisia), Hydrobiologia 637 (2010) 73–85, https://doi.org/10.1007/s10750-009-9986-3.
[110] M. Shaiek, M.-S. Romdhane, F. Le Loc'h, Study of the ichthyofauna diet in the Ichkeul Lake (Tunisia), Cybium (2015), https://doi.org/10.26028/cybium/2015-393-003.
[111] A. Yazidi, S. Saidi, N. Ben Mbarek, F. Darragi, Contribution of GIS to evaluate surface water pollution by heavy metals: Case of Ichkeul Lake (Northern Tunisia), J. Afr. Earth Sci. 134 (2017) 166–173, https://doi.org/10.1016/j.jafrearsci.2017.06.009.
[112] M. Irie, K. Kashiwagi, K. Ujiie, I. Nsiri, S. Bouguerra, J. Tarhouni, Feasibility of exploitation of the sediment in the reservoirs for the sustainability of surface water resource in Tunisia, J. Jpn. Soc. Civ. Eng. Ser. G (Environ. Res.) 68 (2012) II_41–II_46, https://doi.org/10.2208/jscejer.68.II_41.
[113] A. Fathalli, A. Ben Rejeb Jenhani, C. Moreira, M. Welker, M. Romdhane, A. Antunes, et al., Molecular and phylogenetic characterization of potentially toxic cyanobacteria in Tunisian freshwaters, Syst. Appl. Microbiol. 34 (2011) 303–310, https://doi.org/10.1016/j.syapm.2010.12.003.
[114] V. Krysanova, M. White, Advances in water resources assessment with SWAT—an overview, Hydrol. Sci. J. (2015) 1–13, https://doi.org/10.1080/02626667.2015.1029482.
[115] W. Ouyang, X. Hao, L. Wang, Y. Xu, M. Tysklind, X. Gao, et al., Watershed diffuse pollution dynamics and response to land development assessment with riverine sediments, Sci. Total Environ. 659 (2019) 283–292, https://doi.org/10.1016/j.scitotenv.2018.12.367.
[116] J. Aouissi, Z.L. Chabaane, S. Benabdallah, C. Cudennec, Assessing the hydrological impacts of agricultural changes upstream of the Tunisian World Heritage sea-connected Ichkeul Lake, Proc. IAHS 365 (2015) 61–65, https://doi.org/10.5194/piahs-365-61-2015.
[117] J. Aouissi, S. Benabdallah, Z.L. Chabaâne, C. Cudennec, Modeling water quality to improve agricultural practices and land management in a Tunisian catchment using the soil and water assessment tool, J. Environ. Qual. 43 (2014) 18, https://doi.org/10.2134/jeq2011.0375.
[118] S. Mtibaa, N. Hotta, M. Irie, Analysis of the efficacy and cost-effectiveness of best management practices for controlling sediment yield: a case study of the Joumine watershed, Tunisia, Sci. Total Environ. 616–617 (2018) 1–16, https://doi.org/10.1016/j.scitotenv.2017.10.290.
[119] J.R. Williams, Sediment routing for agricultural watersheds, J. Am. Water Resour. Assoc. 11 (1975) 965–974, https://doi.org/10.1111/j.1752-1688.1975.tb01817.x.
[120] D.N. Moriasi, J.G. Arnold, M.W. Van Liew, R.L. Bingner, R.D. Harmel, T.L. Veith, et al., Model evaluation guidelines for systematic quantification of accuracy in watershed simulations, Trans. ASABE 50 (2007) 885–900, https://doi.org/10.13031/2013.23153.
[121] H.V. Gupta, S. Sorooshian, P.O. Yapo, Status of automatic calibration for hydrologic models: comparison with multilevel expert calibration, J. Hydrol. Eng. 4 (1999) 135–143, https://doi.org/10.1061/(ASCE)1084-0699(1999)4:2(135).

CHAPTER 8

Floods in the Mediterranean area: The role of soil moisture and precipitation

Christian Massari[a], Stefania Camici[a], Luca Ciabatta[a], Daniele Penna[b], Anna Cinzia Marra[c], Giulia Panegrossi[c]
[a]CNR, Istituto di Ricerca per la Protezione Idrogeologica - Via della Madonna Alta, Perugia, Italy
[b]Department of Agriculture, Food, Environment and Forestry, University of Florence, Florence, Italy
[c]Institute of Atmospheric Sciences and Climate, National Research Council (ISAC/CNR), Rome, Italy

8.1 Introduction

Floods are among the most dangerous and costly natural disasters and cause extensive economic and social damages worldwide [1–3].

In the Mediterranean area (MA), floods are, after droughts, the most dangerous meteorological hazard as they are frequent enough to be considered as a component of the local climate [4]. Among different types of floods, MA is characterized by a relatively high number of flash floods (FFs), with more than half of the events occurring in only three countries (Italy (36%), Spain (15%), and France (10%), see Fig. 1 in Ref. [4]).

FFs are normally produced by deep convective systems [5–7] associated with intense precipitation (with rainfall rates up to 180 mm/h measured in only 5 min) with a limited areal extent (typically less than 100 km^2 [8]). These events are very destructive particularly in headwater catchments and lead to severe damages of manmade infrastructures, produce large loss of lives, and adversely affect the opportunities for socio-economic development of Mediterranean countries [3, 9–11].

In addition to FFs, other types of floods are induced by mesoscale convective systems [12, 13], which can produce stationary rain lasting several hours and leading to rainfall amounts exceeding 200 mm in a few hours that can impact areas from hundred to thousand kilometers. Examples of these types of events are those storms that hit the north-western coast of the Mediterranean catchment like the one that affected Genoa in 2011 and 2014 [14, 15]. Finally, synoptic-scale perturbation systems, such as cut-off lows [16, 17], intense extratropical [18, 19], and tropical-like cyclones [20–23] also impact MA. These events can last for several days and are characterized by extreme rainfall accumulations, like the one that hit the Tiber River in November 2005 [24].

To understand flood trends and their drivers in MA, many studies (e.g., Refs. [25–27]) have analyzed the change in the frequency and the severity of these events in a climate change perspective. For Mediterranean catchments, Blöschl et al. [27]

Water Resources in the Mediterranean Region
https://doi.org/10.1016/B978-0-12-818086-0.00008-X

indicate winter floods occurring later with respect to the past 50 years (1960–2010) owing to delayed winter storms whereas, Mangini et al. [28], even if based only on a limited number of stations, noted a tendency toward increasing flood magnitude and decreasing flood frequency. These findings, obtained over stations with negligible human modifications (e.g., dams or reservoirs), were consistent with trends detected by Mediero et al. [29] in Spain and Llasat et al. [30] in Catalonia, whereas Giuntoli et al. [31] found a decreasing trend for a few hydrometric stations in the South of France. If on the one hand the detection can be difficult, on the other hand the attribution of these trends is rather uncertain, and it is not clear how changes in hydrological variables can modulate climatic trends [32]. For instance, Barrera-Escoda and Llasat et al. [33] demonstrated that changes in precipitation alone cannot explain the variation in flood pattern observed in Catalonia, whereas Tramblay et al. [34] in Southern France showed that despite the observed increase in extreme precipitations, there was no general tendency toward more severe floods.

Despite some studies observed changes in the frequency and the magnitude of floods in MA, the attribution of these changes is not easy (e.g., Refs. [35–38]) as they are further complicated by the effects of the various human activities [39] and the complex interactions between the characteristics of the events generating rainfall and the factors that control the catchment response (soil and bedrock types, slopes, and the initial soil moisture conditions) [9]. All these factors, acting from the hillslope scale to the catchment scale, play a key role in the fundamental processes that generate floods.

In this chapter, we review the most important studies on floods and FFs with particular emphasis on the key hydro-meteorological variables controlling the triggering, the generation, and the evolution of floods in MA. In particular, we provide a description of the methods to monitor, observe, and predict these floods. We underline that although this chapter is far from being exhaustive about floods in MA, it still provides the readers with an introduction to the main research problems about floods in MA, specifically focusing on FFs since, as shown in Fig. 8.1, they represent the most frequent (and most challenging to deal with) type of phenomena impacting MA.

The chapter is organized as follows: Section 8.2 describes the main factors involved in the flood triggering and evolution focusing mainly on precipitation and soil moisture; Section 8.3 describes the methods to monitor these two variables along with the associated tools for flood forecasting in MA.

8.2 Floods in the Mediterranean regions, hydro-meteorological variables, and triggering factors

In this section, we revise the main findings about the impact of soil moisture and precipitation on flood generation. Specifically, we focus on the experimental studies carried out both at the hillslope and at the catchment scale while the methods to monitor rainfall and soil moisture as well as their use in the tools for the prediction of floods are discussed in Section 8.3.

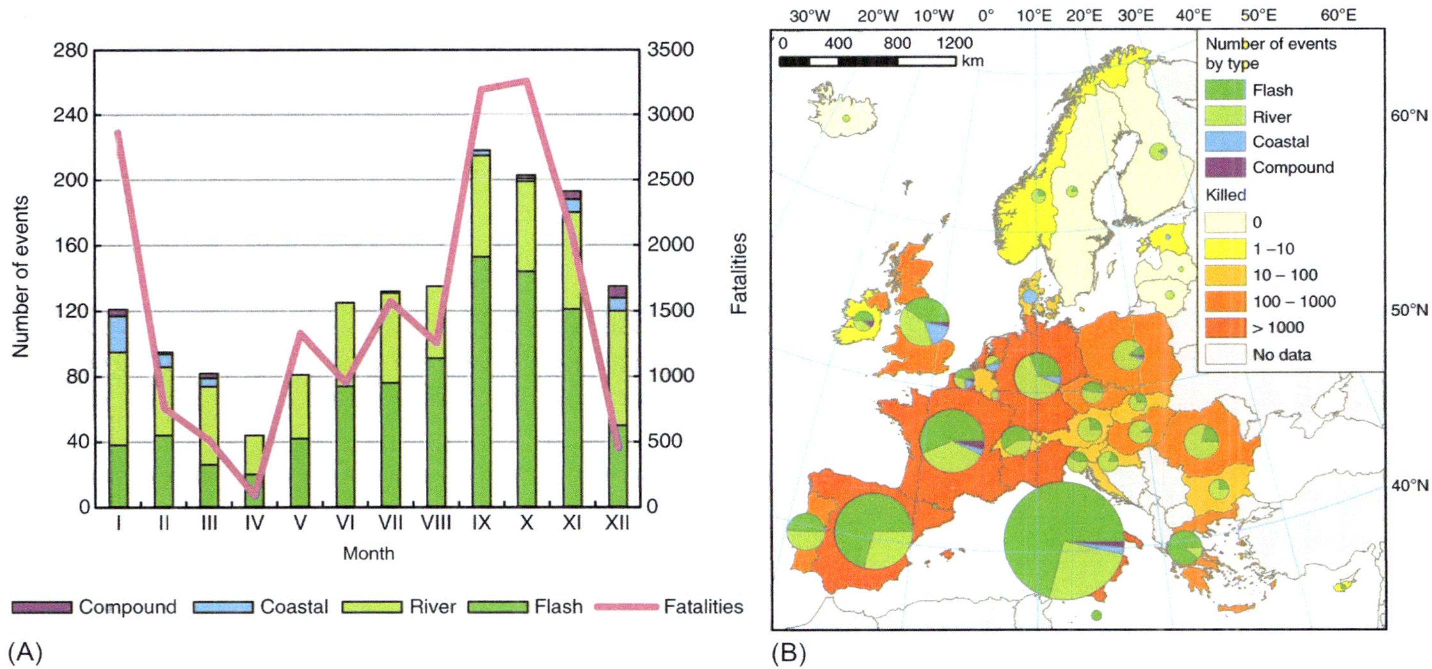

Fig. 8.1 Flood occurrences and fatalities. Total number of flood events and fatalities between 1870 and 2016, (A) by month and (B) by country. *(Based on HANZE database with country borders from PBL82. From D. Paprotny, A. Sebastian, O. Morales-Nápoles, S.N. Jonkman, Trends in flood losses in Europe over the past 150 years, Nat. Commun. 9 (1) (2018) 1985.)*

The small spatial and temporal scales of floods affecting MA, relative to the sampling characteristics of conventional networks for rain and discharge measurement, have considerable impact on the observability of these events. For instance, Marchi et al. [9], in analyzing 25 major flash floods that occurred in Europe in the last 20 years, observed that only about one half of the cases were properly documented by conventional stage measurements. In many cases, the rivers were either ungauged or the stream gauge structures were damaged by the event. This makes particularly difficult to understand the causes and the factors that have triggered the flood and FFs events.

In general, runoff coefficients (i.e., the ratio between the rainfall that has produced the flood and the event runoff) of floods occurring in small catchments (specifically FFs) in MA are characterized by a large range of variability with respect to those of other flood types because of the rapid catchment response to intense rainfalls, modulated by soil moisture, soil hydraulic properties, and high gradients [40]. In other words, runoff generation is highly case-specific as complex interactions among these factors are involved in the generation and evolution of floods. For instance, runoff can be affected by different mechanisms such as overland flow (due to limited infiltration capacity [41]), saturation excess [42], water moving through macropores [43], and deep percolation, or a combination of them [44]. Whether one of these prevails over the others is a function not only of the catchment characteristics but also of rainfall intensity [40], as "even the highest infiltration capacities of forest soils will not accommodate the highest recorded rainfall intensities" [45].

In general, the hillslope response to intense rainfall plays a key role in the activation of these multiple runoff-generation mechanisms [46]. As a result, their characterization is central for the understanding of the FFs response as well as for figuring out the role of other factors like the antecedent wetness conditions of the catchment. Western and Grayson [47] showed the existence of a clear threshold relation between runoff and soil moisture, with sudden increase of runoff coefficients above a certain moisture threshold. Similar results were found by Tromp-van Meerveld McDonnell, Latron and Gallart, and Zehe et al. [48–50] Particularly, Penna et al. [46] related the small runoff ratios of rainfall-runoff events occurred during dry conditions to the response of the nearly saturated riparian zone, whereas during wetter conditions, when a threshold of soil moisture was exceeded, runoff was mainly due to the hillslope subsurface contribution. However, not all studies found an important effect of the antecedent soil moisture on runoff response [51].

Extending the results found at the plot and hillslope scale to the catchment scale is not always straightforward as difficulties exist in adequately representing small-scale process complexity at a large scale in a meaningful way [52]. For instance, the relative importance of factors such as soil type and land use when medium-sized catchments are considered [53] whereas antecedent soil-moisture conditions and the type of climate (i.e., wet, semiarid, etc.) seem to be the main driving factors at these scales. In this respect, Berghuijs et al. [54] developed a method for inferring the relative importance of different flood-generating mechanisms (extreme precipitation, soil moisture excess, and snowmelt) of maximum annual flows across Europe and demonstrated that from 1960 to 2010 relatively few annual floods were caused by annual rainfall peaks. Instead, soil moisture excess almost exclusively caused floods in a broad arc stretching across much of Europe, including large parts of Portugal, Spain, France, Italy, and majority of countries in MA. These results agreed with the ones of Tramblay et al. [34] that, for Mediterranean catchments, indicate that soil moisture is the main driver of trend on annual flood occurrences.

Similarly, Penna et al. [40] examined the runoff coefficients for 25 relevant FFs, that occurred in several regions of Europe (mainly in MA) since 1994, to understand how different landscapes and climate properties filter rainfall into event-based runoff, and investigated the relationship between runoff coefficients and antecedent moisture conditions. They found that almost all the FFs in MA occurred in water-limited conditions and that runoff coefficients, despite being generally small for FFs (most of the coefficients provided values below 0.2), were higher for MA compared to other areas. In addition, they found that antecedent moisture conditions played a significant role in determining land-surface response to extreme rainfall events, confirming the studies of Gaume et al., Sturdevant-Rees et al., and Borga et al. [8, 55, 56]. In particular, FFs events in MA generally occur with dry and moderate antecedent conditions due to the significant rainfall accumulations that characterize MA during convective storms. In contrast, in continental region, FFs generally occur under "normal" and wet conditions and for them to occur saturation must be relatively high because the storms in the continental area are generally short and characterized by smaller accumulation amounts.

The role of the antecedent wetness conditions is so important that many studies investigated the use of in situ and satellite soil moisture observations to characterize the status of the catchments before flood events [50, 57–63]. Notable examples in the MA are the studies of Brocca et al. and Tramblay et al. [59, 62], who investigated the relationship between soil moisture measurements (in situ and from satellite sensors) and the soil potential maximum retention parameter of the Soil Conservation Service—Curve Number method (SCS-CN [64]). They found that soil moisture is a good indicator of the wetness conditions at the catchment scale in MA and more informative with respect to simple Antecedent Precipitation Indices (APIs). The information gathered in these studies was then used to build [65, 66] and improve models via data assimilation [67, 68] (see Section 8.3 for further details).

8.3 Monitoring and modeling floods in the Mediterranean area: The usefulness of rainfall and soil moisture observations from different sources

In this section, we provide an overview of the existing techniques to monitor rainfall and soil moisture, which, based on Section 8.2, are the two most important hydrometeorological variables to consider for monitoring and modeling floods and FFs in MA. In particular, for rainfall we review the existing conventional methods to monitor this variable and describe the associated challenges and the opportunities of these approaches in flood forecasting. For soil moisture, we also describe the different existing monitoring techniques but unlike rainfall we provide a unique section where we discuss the use of soil moisture observations in flood forecasting. Indeed, soil moisture is generally not an input to hydrological models and thus deserves a different treatment. Note that, since the topic is relatively broad, other examples not covering only MA will be mentioned as they are particularly relevant for flood prediction in MA.

8.3.1 Rainfall

8.3.1.1 Rain gauges and radars

Ground networks of rain gauges are commonly considered the most accurate source of information about rainfall [69, 70]. However, point rain gauge measurements, representative only of a limited area [71–74], require interpolation over tens or hundreds of square kilometers [75] for an aerial analysis of precipitation. This potentially brings significant errors in rainfall estimation caused by the very high spatial variability of precipitation in MA and its intermittent nature.

Weather radars are viable tools to track the extreme spatial and temporal characteristics of precipitation thanks to their subhourly and subkilometer temporal and spatial scales [76]. However, radars are also subject to limitations particularly in mountainous and complex terrains, due to various sources of error (such as ground clutter, beam blockage, and spurious effects related to the radar viewing geometry, besides hardware instabilities), which can cause significant biases [77–80] as well as inaccuracies for high rainfall intensities [81].

Use of rain gauges and radars for FFs and flood forecasting

Being very accurate, rain gauges are the most common source of rainfall for flood modeling in MA catchments since for optimal coverage they are able to provide relatively high skills in flood forecasting (see Table 1 [82] for a summary of the studies carried out in this topic). Several studies have discussed the influence of rain gauge density and gauge distribution on the accuracy of precipitation estimation and hydrological modeling [83–85].

For instance, Moulin et al. [86], in a study in France, observed that uncertainty in the mean areal rainfall estimation is a key factor that leads to rainfall-runoff (RR) modeling error. In particular, the results show that a large part of the (RR) modeling errors can be explained by the uncertainties on rainfall estimates, especially in the case of smaller catchments. These errors are a major factor limiting accuracy and sharpness of rainfall runoff simulations, and thus their operational use for flood and especially for FFs forecasting. Other studies confirmed similar findings (also outside the MA), that is, when rain gauge data are used to estimate the mean areal precipitation, the major source of input uncertainty comes from the lack of representativeness of a discrete set of gauges of a network [69].

The use of radars is particularly critical for FFs, which often occur in headwater mountainous catchments with very intense rainfall regimes, and thus can determine significant errors in flood forecasting [87, 88]. Additional difficulties in the use of weather radar are (1) the need for recalibration of parameters and equations used to convert radar measurements to rain rate, which generally requires the use of ground-based rain-gauge network [89–91] as reference and (2) the time needed to transfer data from the observing systems to the meteorological services and then to operational hydrological models used within an early warning system for FFs [92].

The ability of radars to characterize with good accuracy the temporal/spatial variability of rainfall, even though likely affected by large error, contrasts with the relatively more accurate, but spatially limited, rainfall estimate that can be obtained by rain gauges. This makes rainfall estimates provided by radars and rain gauges particularly complementary. In this respect, several techniques using weather radar/rain-gauge adjustment have been developed. They can be classified into two categories [80]: (i) mean bias reduction techniques [70] and (ii) error variance minimization techniques [93–95]. Despite very beneficial for medium/large catchments, the suitability of these techniques for smaller spatial and temporal scales is still questionable. For instance, Morin et al. [96] used weather radars for a FF warning model for two catchments in the Dead Sea region. Despite the large amount of uncertainty, they obtained acceptable model performance with their nowcasting system. However, they highlighted that for smaller catchments the lead time for FFs forecasting might be too short to issue useful warnings and to take effective mitigation actions.

8.3.1.2 Satellite-based rainfall estimates

The steadily improving quality and capabilities of operational and research satellite data, and the enhancement of satellite-based retrieval techniques, are recently leading to the release of products with great potentials for hydrological applications. One important effort in Europe and MA in this direction has been the establishment, on July 3, 2005, of the EUMETSAT Satellite Application Facility on Support to Operational Hydrology and Water Management (H SAF, http://hsaf.meteoam.it/) designed to deliver satellite products of hydrological interest (precipitation, soil moisture, and snow cover parameters) with sufficient time and space resolution to satisfy the needs of operational hydrology [97].

The unprecedented constellation of research and operational microwave (MW) radiometers in the Global Precipitation Measurement (GPM) mission era [98, 99] is specifically conceived to unify and advance precipitation measurements for delivering next-generation global precipitation data products. The GPM constellation consists of a variable and continuously updating number of Low Earth Orbit (LEO) satellites equipped with conical and cross-track scanning radiometers, and ensures 3-hourly global coverage (on average). The main satellite of the constellation, the NASA/JAXA GPM

Core Observatory (GPM-CO), carries an advanced combined MW active/passive sensor package—the GPM Microwave Imager (GMI), acting also as a calibrator for the other MW constellation radiometers, and the Dual-frequency Precipitation Radar (DPR). The GPM-CO is currently considered the reference satellite-borne platform for the quantitative estimation of precipitation (and precipitation microphysics characterization) from space.

Precipitation products in the GPM era are mainly based on the exploitation of all available MW radiometers in the constellation to achieve the best temporal and spatial coverage. Advanced precipitation retrieval algorithms convert multispectral MW radiation measurements, directly related to emission/scattering properties of liquid/solid hydrometeors, into surface precipitation rate estimates (e.g., the official GPM Goddard PROFiling algorithm [100] or the EUMETSAT H SAF algorithms [101–105] among many others). For NRT monitoring or hydrological applications, different approaches have been conceived to combine the MW-based rainfall rate estimates, available at low spatial (10–50 km) resolution and at the local revisit time of each radiometer, with geostationary (GEO) infrared (IR) observations, available at higher spatial (3–5 km) and temporal resolution (10–15 min). In particular, the NASA official GPM MW/IR level 3 product, called Integrated Multi-satellitE Retrievals for GPM (IMERG [106]), is based on this approach (with three different runs characterized by different calibration and latency), and provides global precipitation rate estimates at 0.1°/30 min spatial/temporal resolution. Similarly, EUMETSAT H SAF delivers MW/IR precipitation rate products for NRT applications at 5 km/15 min spatial/temporal resolution.

Use of satellite-based precipitation products for flood modeling

Satellite-based precipitation data have a large potential for hydrological applications and in particular for floods (the reader interested in the topic is referred to Maggioni and Massari [107]) due to the fact that they are readily available, have good coverage, and do not depend upon ground stations density and malfunctions. However, given the relatively coarse spatial resolution and low temporal sampling of these products with respect to the spatial and temporal scales of floods impacting MA their use is limited for flood forecasting. Other reasons are the scarce knowledge on how to efficiently use these data and the associated uncertainties [107].

Few studies addressed the use of satellite precipitation estimates for FFs forecasting while for floods occurring in larger catchments, the number of studies is relevant (see Ref. [107] for a review of these studies). As FFs are generally typical in MA here we only reported few important studies considering FFs.

Nikolopoulos et al. [108] used three different satellite datasets (3B42RT [109], CMORPH [110], and PERSIANN [110]) for analyzing a severe FF occurred in northern Italy. They found that satellite precipitation estimates were able to represent the temporal variability of catchment-averaged rainfall but their bias was too high and could not guarantee optimal performance. Another attempt was carried out by Stampoulis et al. [111], who simulated several FFs events occurred in MA. The authors found it difficult to capture the catchment hydrologic response during the FFs event, even after adjusting the satellite rainfall fields for the mean-field bias. Mei et al. [112] investigated the error propagation of six satellite precipitation products in FFs simulations for a range of mountainous catchment sizes (ranging from 255 to 6967 km^2) in north-eastern Italy during fall and summer. They found better performances in the summer season for moderate to large catchments, with gauge-adjusted satellite precipitation products outperforming satellite-based ones.

The relatively low skills of satellite rainfall estimates for mountainous small-scale catchments and quickly evolving floods, like FFs in MA, are something expected, given the relatively coarse spatial resolution and low temporal sampling of satellite rainfall estimates with respect to the catchment size and the development time of FFs, and the sampling errors associated with these products [113]. In addition, MA, characterized by a complex orography also in proximity of coastal areas, is a challenging region for satellite rainfall estimates [13].

The new generation of satellite-based global precipitation products, derived from the GPM, could provide significant advancements in the estimation of precipitation associated with strong convective systems (such as FFs in MA) [13, 114, 115] and thus could potentially lead to better forecasting skills especially for medium to large catchments as the latency of these products is as low as 4/5 h for the NASA IMERG products and half hour for the EUMETSAF H SAF NRT MW/IR products.

Huffman et al. [106] tested the capabilities of EUMETSAT H SAF regional satellite precipitation products and of the global NASA GPM products to quantitatively estimate precipitation for the very intense and localized event that caused the Livorno FFs, on the coast of Tuscany, Italy, during the night between September 9 and 10, 2017. They found that the satellite products, derived from the GPM constellation of microwave radiometer measurements, offered a wide spectrum of results, showing better performances when precipitation is retrieved by algorithms tailored for the specific region. On the contrary, algorithms designed for global application, such as IMERG [106], were less sensitive to such small-scale thunderstorms.

Attempts to improve satellite rainfall estimates have considered the integration with ground-based observations when available, as proposed by Beck et al., Verdin et al., and He et al. [116–118]. In this respect, many satellite rainfall products use rain gauges to improve satellite rainfall estimates through very different approaches. However, many of the "gauge-corrected" rainfall products are characterized by a latency not suited for operational applications. For example, in the GPM IMERG products portfolio, the so called "final run" uses monthly gauge data to create research-level products and it is released after ~2.5 months. Moreover, spatial resolution of most satellite precipitation products is generally too coarse to resolve the precipitation variability in small catchments, and are thus useless for flood forecasting in the small/medium catchments of MA. In addition, as highlighted by Maggioni and Massari [107], a possible research direction for improving the use of satellite rainfall estimates in flood prediction, even in near real time, is to explore the possibility of merging them with ancillary information like soil moisture [119–122].

8.3.1.3 Numerical weather predictions (NWP)

To extend the forecasting lead time especially for FFs, several studies have investigated the use of rainfall estimates provided by numerical weather prediction (NWP) models. Leading global NWP centers are typically now running a combination of global models at grid resolutions around 16–45 km and high-resolution regional models with grid resolutions between 1.5 and 12 km. The lead times vary from 3 to 6 h (nowcasting, very short-range forecasting) to 12–48 h (short-medium range forecasting), up to 240 h (long range forecasting) [123, 124]. A notable example in Europe is COSMO-2 of the Consortium for Small-scale Modeling, which is a deterministic model with a grid size of 2.2 km and a 24 h lead time computed every 3 h. The increased computing power and improvement of the NWP models have led to considerable advancement in their forecasting skills. However, forecasting precipitation is challenging because it is discontinuous and varies rapidly in space and time, and because precipitation processes depend not only on the synoptic situation but also on dynamical and microphysical processes that are not explicitly resolved [125]. NWP model precipitation forecast skills are still relatively low, especially for very short lead times (e.g., <12 h), for long lead times (e.g., >5 days) and for fine scale weather systems such as local-regional convective systems (e.g., thunderstorms) [126].

Use of NWP for flood and FFs forecasting

One of the most known initiatives in using NWP rainfall estimates for floods and FFs forecasting is the European Flood Awareness System (EFAS [127]). EFAS uses both deterministic (i.e., single realization from one NWP system such as COSMO) and ensemble (i.e., multiple realizations—also called ensemble members—from a single numerical weather prediction system) regional NWP rainfall estimates to run LISFLOOD hydrological model [128] on a 5 km grid across Europe. The system provides local water authorities with probabilistic flood forecasting information up to 10 days in advance (with 5 days for FFs forecasting).

A relatively large number of studies have tested the skills of NWPs, alone and coupled with hydrological models in particular in MA [15, 128–132]. The main difficulties highlighted by these studies are the relatively coarse spatial resolution of NWPs relative to dynamic and scale of FFs in MA ([133], convective systems can develop also below 2–5 km scale), the uncertainties in the initial conditions which can determine significant errors in the quantitative precipitation forecast (QPF [128, 134, 135]) and the difficulties from NWPs to simulate deep convection [134].

For instance, Davolio et al. [131] tested the performance of the BOLogna Limited Area Model (BOLAM) and the MOdello LOCale in Hybrid coordinates (MOLOCH) NWP (by varying the grids and the initial conditions) coupled with the operational hydrological forecasting chain of the Ligurian Hydro-Meteorological Functional Centre to simulate two major floods that occurred during autumn 2011 in Liguria (northern Italy). The results showed an improvement in flood prediction when a high-resolution grid was employed for atmospheric simulations but also found an underestimation of the severity of both floods.

Alfieri et al. [135] analyzed the performance of an NWP-driven FF alert system in Southern Switzerland highlighting (1) the quick reaction of the catchment as an important limiting factor to operate a hydrological model and issue the alert, and (2) the low accuracy of NWP in the estimation of precipitation associated with rapidly evolving, and localized convective events often leading to FFs. Difficulties in the operational forecasting of such events are often due to the small spatial and temporal scales that are involved, while the correct timing and position of the relevant rain amounts are crucial.

To overcome these limitations, different studies (Rossa et al. [136], September 26, 2007 Venice FFs in North East Italy, Federico et al. [137] for the Livorno FFs in September 9–10, 2017, and Liechti et al. [76], FFs in Southern Swiss Alps) proposed to assimilate radar data in NWP models, finding out benefits for discharge forecasts. Federico et al. [137], for instance, assimilated radar reflectivity and lighting activity into the RAMS@ISAC model (Regional Atmospheric Modeling System at Institute for Atmospheric Sciences and Climate of the Italian National Research Council) showing that the assimilation of these variables, especially when used together, has a significant and positive impact on precipitation forecast.

8.3.2 Soil moisture

As depicted in Section 8.1, the catchment initial wetness conditions are very important to discriminate between small and catastrophic floods. Several studies explored the importance of soil moisture in flood forecasting by, for example, attempting to relate the rainfall-runoff model initial conditions with different external indicators of soil moisture [59, 63, 138–145]. All these studies demonstrated that soil moisture is an extremely important variable for improving the capability of flood forecasting models as it is very much related to the catchment initial wetness conditions than other commonly used indices (e.g., Antecedent Precipitation Index [145]). Therefore the monitoring, modeling and ingestion of soil moisture observations into models of is paramount in flood forecasting.

8.3.2.1 Ground-based soil moisture observations

Different ground-based solutions exist to monitor soil moisture at the catchment scale. The most direct method is the use of ground observations based on gravimetric measurements. Less direct but viable solutions are those exploiting the dissipation of an electric field or heat source within the soil medium and then relating this dissipation to the presence of water in the soil. Most of the stations measuring soil moisture are based on the latter method. Despite the recent progress in the systematic collection and dissemination of ground-based soil moisture datasets derived from these stations [146], an inherent problem with in situ measurements is the very small sample volume of soil they represent (typically of the order of 1–10 cm^3). Moreover, it must be considered that soil moisture commonly exhibits very high unstructured spatial variability and then it is often difficult to upscale isolated, ground-based soil moisture measurements to larger geographic extent although Vachaud et al. [147] first, followed by other studies (e.g., [148, 149]) based on the concept of temporal stability of soil moisture spatial patterns, demonstrated that under certain conditions even few stations can be representative for soil moisture of larger areas.

To circumvent the problem of the small spatial representativeness of in situ measurements, in the last decade, some novel technologies have been developed for providing soil moisture measurements over larger areas [150]. Currently, cosmic ray neutron sensors, and geophysical measurements (e.g., electrical resistivity and electromagnetic induction) are the most promising techniques [151–154]. The new techniques are able to provide estimates with a spatial support much larger than the few cubic centimeters of classical in situ measurements.

8.3.2.2 Land surface modeling

Alternative to in situ stations is the modeling of soil moisture based on meteorological inputs (e.g., rainfall rates, near-surface air temperature, wind speed, and relative humidity), and various parameterizations of soil and vegetation water flux processes, using a soil water balance approach. However, the accuracy of modeled soil moisture data is strongly dependent on the employed model, the quality of meteorological observations used as input data, which is related to the issues about precipitation described above, and the simplification of the necessary model parameterizations of the soil and land cover (see Ref. [155] for further details). For instance, high-resolution modeled data (e.g., 500 m) are often obtained by using coarse scale meteorological forcing (e.g., precipitation data at 10–25 km resolution) as inputs suggesting that the real spatial resolution is much larger.

Last but not least, there is the need of continuous hydrometeorological forcing (rainfall, evapotranspiration) to run a model with a sufficient accuracy and this is often not possible within MA [65]. A possible viable solution to overcome these problems is the use of satellite-based soil moisture observations [156–159], which have demonstrated large potential for improving flood forecasting even in small catchments [67, 160].

8.3.2.3 Satellite-based soil moisture observations

In the last decades, two important satellite missions have been specifically launched for the measurement of soil moisture from space, namely the Soil Moisture Ocean Salinity (SMOS [161]) and the Soil Moisture Active and Passive (SMAP [158]) missions. Nevertheless, other active and passive sensors on board of operational and nonoperational satellites have been (and are still) used to retrieve soil moisture. Notable examples are the Advanced Scatterometer (ASCAT [162]) and the Advanced Microwave Scanning Radiometer for Earth observation science (AMSR-E [163]) and its successor AMSR2 [164]. These sensors offer the possibility to measure soil moisture continuously in all-sky conditions and thus are extremely useful for flood forecasting purposes.

Despite these great advantages, the small penetration depth (i.e., smaller than 5 cm) and the large spatial support (i.e., >25 km) of these observations pose important challenges in flood and FFs forecasting. Indeed, unless the runoff generation is mainly driven by fast component and thus controlled by upper soil layers, root zone soil moisture (i.e., >5 cm depth depending on the vegetation type) has often a large influence on the runoff generation [165]. Additionally, the evident contrast between the small scale of catchments and the coarse spatial resolution of products suggests that the utility of satellite soil moisture observations for flood forecasting in medium to small (<500–1000 km^2) catchments is limited. To circumvent this problem, various downscaling/data fusion methods of satellite soil moisture products [166–170], incorporating geophysical variables such as land surface temperature and vegetation, have been studied for improving their spatial resolution. Moreover, the recent launch of Sentinel-1 (S1) satellites has provided new information for potentially retrieving soil moisture from Synthetic Aperture Radar (SAR) at plot scale (<100 m). However, the noise associated with the retrieval of soil moisture has limited the spatial resolution at values larger than 1 km while temporal resolution varies from 1.5 to 4 days in Europe [171]. Note that other techniques exist to retrieve soil moisture from optimal and thermal bands; for these, the reader is referred to Rahimzadeh-Bajgiran and Berg [172].

8.3.2.4 Use of soil moisture observations for improving flood predictions: Data assimilation

The evidences about the importance of initial soil moisture conditions have pushed many researchers to explore the benefit of ingesting soil moisture observations derived both from in situ and satellite sensors into RR models via data assimilation [67, 160, 173–182]. The idea is that an improvement of the initial model state can provide benefit in flood simulations and forecast.

Despite very promising, the topic has demonstrated to be very challenging for the number of assumptions and preprocessing steps needed prior to the ingestion of observations into the models [67]. For instance, some studies showed that the RR model performance did not generally improve [81] after the assimilation of satellite soil moisture observations, whereas in other cases, the improvements were significant [181]. The little improvements obtained were explained by the strong spatial mismatch between satellite soil moisture observations and catchment scales, and by the depth sensed by satellites (2–5 cm), that does not match the soil depth (1–2 m), usually simulated through RR models [181].

A recent example demonstrating the complexity of the topic has been reported by Loizu et al. [183], where two different rainfall-runoff models (MISDc and TOPLATS [183, 184]) were used in two catchments in Spain and Italy. The comparison revealed that the performance of the two assimilation schemes varied as a function of the chosen model. These results are in line with what highlighted by Massari et al. [67], who suggested the need of additional studies about this topic to better understand the real benefit of the ingestion of satellite soil moisture observations into hydrological models.

Despite these evidences, Massari et al. [65, 66] developed a simple RR event-based hydrological model initialized by satellite soil moisture observations and applied it to a small catchment in Greece. The authors found that their model performed similarly to, and in some cases better than, a state-of-the art continuous hydrological model (i.e., MISDc [181]). In the framework of the H SAF project, the model was also extensively applied to 35 catchments in Italy by using soil moisture observations based on ASCAT and a product obtained from the assimilation of ASCAT in the ECMWF Land Data Assimilation System. The outcome of the study confirmed the reliability of the approach and its potential application to forecast floods without being subjected on the knowledge of long-time series of temperature, potential evapotranspiration, and rainfall to simulate the model state.

The controversies of these studies require more investigation in the topic. For instance, objective assessments of satellite soil moisture products are needed to clarify the impact of satellite soil moisture limitations in flood forecasting via more investigations on the role of the antecedent wetness conditions. Recently, Crow et al. [185, 186] sampled the rank correlation between prestorm antecedent wetness conditions (AWCs) estimates, acquired from a range of soil moisture products, and the storm-scale runoff coefficients in the United States. They found that prestorm 9-km surface (0–5 cm) soil moisture estimates, extracted from the SMAP Level 4 Surface and Root-Zone (SMAP L4SM) soil moisture product, are a better predictor of storm-scale runoff coefficient than surface soil moisture products derived from land surface and large-scale hydrological models. This suggests that many existing models may underrepresent the role of AWC on subsequent catchment-scale runoff response. These results challenge the common wisdom that antecedent soil moisture conditions are of little importance in determining the magnitude of extreme floods.

The importance of soil moisture does not only lie in the determination of initial wetness conditions. Different studies have recently demonstrated that soil moisture data can be used to improve rainfall estimates over land (e.g., [187–189]). Based on this concept, some scientists started to integrate precipitation estimates with soil moisture observations for flood modeling (Crow et al. [180] in a synthetic study, Massari et al. [190] in a small catchment in France with real in situ data). Based on these results, attention was paid to satellite-based observations [165, 182, 191–193]. In particular, Ciabatta et al. [192] merged soil moisture-derived rainfall data obtained with the SM2RAIN algorithm

[194] with in situ and satellite rainfall data, in order to obtain a higher quality rainfall product to be used for flood prediction in four catchments in Italy. Results revealed that the use of soil moisture to estimate precipitation improves both rainfall estimation and flood simulation. In particular, the improved performances were ascribed to the higher stability over time of the soil moisture corrected precipitation product with respect to the original TMPA-3B42 RT [109]. Similar results were found by Camici et al. [193] in 15 catchments in MA.

Massari et al. [120] compared the performance of state correction (through data assimilation) versus rainfall correction (through SM2RAIN-derived rainfall) in flood simulations for 15 catchments located in MA, by using MISDc model forced with TMPA-RT satellite-based rainfall dataset. The authors found that state correction was able to improve to the simulation of low flows and only slightly improved high flows. By contrast rainfall correction provided larger benefit for high-flow simulations (close to the simulation obtained with a ground-based dataset). Similar results [182, 191] were also found by other studies outside the MA regions.

8.4 Flash flood guidance (FFG)

To overcome the problems of relying on a very short latency products and complex modeling chains, many authors have proposed the use of the Flash Flood Guidance (FFG) method to issue warning and alert about FFs occurrence. FFG consists of determining a priori, based on RR modeling, the amount of rainfall needed to trigger a flood at the outlet of a specific catchment [195–198]. For a meaningful application, FFG must rely upon the initial wetness conditions of the catchment as they influence the different catchment responses to the rainfall inputs. Even though the occurrence, location, and (or) timing of the FFs are still uncertain with FFG, this method may provide enough lead time so that FFs mitigation measures can be planned and managed in an anticipatory rather than responsive manner [198]. However, the determination of the initial wetness conditions of the catchment is not always straightforward and requires the use of hydrological models with resulting delays in the emission of the warning. To overcome this issue, Alfieri et al. [199] proposed the use of ensemble early warning systems relying on precipitation information only, via the European Precipitation Index Based on Simulated Climatology (EPIC), which is an indicator aimed to monitor the European domain for upcoming severe storms possibly leading to FFs. Corral et al. [200] adopted a FFs forecasting system in Catalonia [201] based on rainfall estimates and forecasts generated from weather radar observations. However, more solid alternatives exist to characterize wetness initial conditions of the catchments like satellite soil moisture derived from satellites described above. The latter could be a viable alternative to models but the research in the topic is to our knowledge still scarce.

8.5 Conclusions

In this chapter, we presented a summary of the research studies related to floods in MA as well as the role of the precipitation and soil moisture in the triggering and evolution of these phenomena. We revised the main techniques to monitor these two important variables and discussed the challenges and opportunities of these techniques as well as their potential synergy.

MA is characterized by a great variety of landscape, with complex and contrasting topography, where floods are triggered by convective and very intense precipitations in small headwater catchments as well as frontal precipitation systems in medium/large catchments. In general, the most destructive floods in MA are the former, known as FFs, which are also the most difficult to predict, given the very short time response of the catchments and the difficulties to observe and model very intense-convective type of events by current monitoring and prediction techniques. In particular, the small spatial and temporal scales of these floods, relative to the sampling characteristics of conventional networks for rain and discharge measurement, have considerable impact on the observability of these events. Very often, the flood itself is the cause of instrument malfunctioning, which makes the study of these types of events very challenging.

The specific literature clearly demonstrates that two hydroclimatic variables take part in the triggering and evolution of floods and FFs, that is, precipitation and soil moisture (in addition to the catchment geology, geomorphology, and land cover). In particular, the different runoff generation mechanisms result from complex interactions between precipitation intensity, initial soil moisture conditions, and catchment characteristics. As a result, the monitoring of these two variables is paramount for understanding and predicting these events.

Although the monitoring of precipitation has been carried out by convectional rain gauge networks and weather radars, these instruments have demonstrated to be not always suitable for a prompt and accurate warning. As a matter of fact, rain gauges are subject to limitations due to the scarce representativeness of point measurements, in case of extreme convective events characterized by a high spatial variability of rainfall. Moreover, weather radar precipitation estimates can be affected by large biases in mountainous terrains. For these reasons, a considerable amount of research has considered the fusion of both rain gauges and radar observations with NWP models and/or satellite-based rainfall estimates. The latter seems to be a promising technique, as it can offer continuous monitoring of precipitation from space, do not need maintenance, and is free for users. However, its capability to improve forecasting (and nowcasting) in complex areas, such as MA, is still questionable, particularly for globally calibrated precipitation products. Thus, efforts should be undertaken to develop satellite products tailored for these areas (working on algorithm calibration), and to provide the error structure of the products (working on validation).

In summary, what emerges is the importance of further research in the topic, that is, developing and studying procedures of data fusion for integrating different sources of information about rainfall, such as soil moisture, NWP, radars, and rain gauge in order to build the best rainfall estimate able to detect and characterize the intense rainfall events typical of MA.

While precipitation has been commonly measured from all over the world since the 20th century, soil moisture measurements are pretty new and today rely on three main alternatives, that is, ground-based observations, models, and remote sensing. Ground observations of soil moisture are very scarce, whereas soil moisture estimated by models, despite fundamental, can be subjected to limitations due to the difficulty in optimal parameterizations of soil and landscapes and the low quality of forcing [202]. Even when this can be carried out with a sufficient accuracy, the times required to run models can be not compatible with those to issue the warning. Satellite-based soil moisture observations are a viable tool for characterizing the prestorm conditions and can significantly help in forecasting extreme flood events. Indeed, many studies have highlighted the utility of these products in flood forecasting not only as predictors of the catchment initial conditions but also as ancillary information for improving rainfall accumulation amounts. This potential is expected to increase thanks to the higher spatial resolutions achieved by new products. For instance, the recent launch of Sentinel-1 satellites, with two satellites currently in orbit (launched in April 2014 and April 2016), opens new challenges in the use of this product in flood prediction. The added value of this new kind of soil moisture measurements is still under investigation, as the first large-scale Sentinel-1 soil moisture products have become available only in 2018 [68, 203]. For instance, Azimi et al. [203] highlighted the benefit of a more frequent revisit time with respect to the higher spatial resolution. In this respect, the current constellation of Metop A, B, and C satellites, carrying ASCAT onboard, in addition to the future SCA Scatterometer [204] or with potential availability of geosynchronous C-band radars will make it possible to collect multiple satellite soil moisture observations per day, which could be extremely important for flood and FFs forecasting in MA.

Acknowledgments

The financial support by the EUMETSAT "Satellite Application Facility on Support to Operational Hydrology and Water Management" (H SAF), Third Continuous Development and Operation Phase (CDOP-3), and by the Italian Research Project of National Interest 2015 (PRIN 2015) 4WX5NA is acknowledged. The authors would like to thank the NASA PMM Research Program for supporting scientific collaboration between GPM and H SAF.

References

[1] B. Jongman, S. Hochrainer-Stigler, L. Feyen, J.C.J.H. Aerts, R. Mechler, W.J.W. Botzen, L.M. Bouwer, G. Pflug, R. Rojas, P.J. Ward, Increasing stress on disaster-risk finance due to large floods, Nat. Clim. Chang. 4 (4) (2014) 264.

[2] J. Pollner, J. Kryspin-Watson, S. Niewwejaar, Disaster Risk Management and Climate Change Adaptation in Europe and Central Asia, The World Bank, 2010.
[3] S.N. Willner, C. Otto, A. Levermann, Global economic response to river floods, Nat. Clim. Chang. 8 (7) (2018) 594.
[4] D. Paprotny, A. Sebastian, O. Morales-Nápoles, S.N. Jonkman, Trends in flood losses in Europe over the past 150 years, Nat. Commun. 9 (1) (2018) 1985.
[5] A. Manzato, S. Davolio, M.M. Miglietta, A. Pucillo, M. Setvák, 12 September 2012: a supercell outbreak in NE Italy? Atmos. Res. 153 (2014) 98–118.
[6] E. Ricciardelli, F. Di Paola, S. Gentile, A. Cersosimo, D. Cimini, D. Gallucci, F. Romano, Analysis of Livorno heavy rainfall event: examples of satellite-based observation techniques in support of numerical weather prediction, Remote Sens. 10 (10) (2018) 1549.
[7] F. Di Paola, E. Ricciardelli, D. Cimini, F. Romano, M. Viggiano, V. Cuomo, Analysis of Catania flash flood case study by using combined microwave and infrared technique, J. Hydrometeorol. 15 (5) (2014) 1989–1998.
[8] E. Gaume, M. Borga, M.C. Llassat, S. Maouche, M. Lang, M. Diakakis, Mediterranean extrem floods and flash floods, in: S. Thièbault, J.-P. Moatti (Eds.), Mediterranean Region under Climate Change. A Scientific Update, IRD editions, 2016, pp. 133–144.
[9] L. Marchi, M. Borga, E. Preciso, E. Gaume, Characterisation of selected extreme flash floods in Europe and implications for flood risk management, J. Hydrol. 394 (1–2) (2010) 118–133.
[10] O.G. Terranova, S.L. Gariano, Rainstorms able to induce flash floods in a Mediterranean-climate region (Calabria, southern Italy), Nat. Hazards Earth Syst. Sci. 14 (9) (2014) 2423–2434.
[11] E. Gaume, V. Bain, P. Bernardara, O. Newinger, M. Barbuc, A. Bateman, L. Blaškovičová, G. Blöschl, M. Borga, A. Dumitrescu, I. Daliakopoulos, J. Garcia, A. Irimescu, S. Kohnova, A. Koutroulis, L. Marchi, S. Matreata, V. Medina, E. Preciso, D. Sempere-Torres, G. Stancalie, J. Szolgay, I. Tsanis, D. Velasco, A. Viglione, A compilation of data on European flash floods, J. Hydrol. 367 (1–2) (2009) 70–78.
[12] R. Romero, C. Ramis, V. Homar, On the severe convective storm of 29 October 2013 in the Balearic Islands: observational and numerical study, Q. J. R. Meteorol. Soc. 141 (2015) 1208–1222.
[13] G. Panegrossi, D. Casella, S. Dietrich, A.C. Marra, P. Sanò, A. Mugnai, L. Baldini, N. Roberto, E. Adirosi, R. Cremonini, R. Bechini, G. Vulpiani, M. Petracca, F. Porcù, Use of the GPM constellation for monitoring heavy precipitation events over the Mediterranean region, IEEE J. Sel. Top. Appl. Earth Obs. Remote Sens. 9 (2016) 2733–2753.
[14] A. Buzzi, S. Davolio, P. Malguzzi, O. Drofa, D. Mastrangelo, Heavy rainfall episodes over Liguria in autumn 2011: numerical forecasting experiments, Nat. Hazards Earth Syst. Sci. 14 (2014) 1325–1340.
[15] F. Silvestro, N. Rebora, F. Giannoni, A. Cavallo, L. Ferraris, The flash flood of the Bisagno Creek on 9th October 2014: an "unfortunate" combination of spatial and temporal scales, J. Hydrol. 541 (10) (2016) 50–62.
[16] V. Kotroni, K. Lagouvardos, E. Defer, S. Dietrich, F. Porcù, C.M. Medaglia, M. Demirtas, The Antalya 5 December 2002 storm: observations and model analysis, J. Appl. Meteorol. Climatol. 45 (2006) 576–590.
[17] F. Porcù, A. Carrassi, C.M. Medaglia, F. Prodi, A. Mugnai, A study on cut-off low vertical structure and precipitation in the Mediterranean region, Meteorog. Atmos. Phys. 96 (2007) 121–140.
[18] G.J. Tripoli, C.M. Medaglia, S. Dietrich, A. Mugnai, G. Panegrossi, S. Pinori, E.A. Smith, The 9–10 November 2001 Algerian flood: a numerical study, Bull. Amer. Meteorol. Soc. 86 (2005) 1229–1235.
[19] E. Flaounas, K. Lagouvardos, V. Kotroni, C. Claud, J. Delanoë, C. Flamant, E. Madonna, H. Wernli, Processes leading to heavy precipitation associated with two Mediterranean cyclones observed during the HyMeX SOP1, Q. J. R. Meteorol. Soc. 142 (2016) 275–286.
[20] O. Reale, R. Atlas, Tropical cyclone–like vortices in the extratropics: observational evidence and synoptic analysis, Weather Forecast. 16 (2001) 7–34.
[21] K. Emanuel, Genesis and maintenance of "Mediterranean hurricanes", Adv. Geosci. 2 (2005) 217–220.
[22] M.M. Miglietta, A. Moscatello, D. Conte, G. Mannarini, G. Lacorata, R. Rotunno, Numerical analysis of a Mediterranean 'hurricane' over south-eastern Italy: sensitivity experiments to sea surface temperature, Atmos. Res. 101 (2011) 412–426.

[23] A.C. Marra, S. Federico, M. Montopoli, E. Avolio, L. Baldini, D. Casella, L.P. D'Adderio, S. Dietrich, P. Sanò, R.C. Torcasio, G. Panegrossi, The precipitation structure of the Mediterranean tropical-like cyclone numa: analysis of GPM observations and numerical weather prediction model simulations, Remote Sens. (Basel) 11 (2019) 1690.
[24] C. Massari, A. Tarpanelli, T. Moramarco, A fast simplified model for predicting river flood inundation probabilities in poorly gauged areas, Hydrol. Process. 29 (10) (2015) 2275–2289.
[25] Z.W. Kundzewicz, I. Pińskwar, G.R. Brakenridge, Large floods in Europe, 1985–2009, Hydrol. Sci. J. 58 (1) (2013) 1–7.
[26] J. Hall, B. Arheimer, M. Borga, R. Brázdil, P. Claps, A. Kiss, T.R. Kjeldsen, J. Kriaučiūnienė, Z.W. Kundzewicz, M. Lang, M.C. Llasat, N. Macdonald, N. McIntyre, L. Mediero, B. Merz, R. Merz, P. Molnar, A. Montanari, C. Neuhold, J. Parajka, R.A.P. Perdigão, L. Plavcová, M. Rogger, J.L. Salinas, E. Sauquet, C. Schär, J. Szolgay, A. Viglione, G. Blöschl, Understanding flood regime changes in Europe: a state-of-the-art assessment, Hydrol. Earth Syst. Sci. 18 (2014) 2735–2772.
[27] G. Blöschl, J. Hall, J. Parajka, R.A.P. Perdigão, B. Merz, B. Arheimer, G.T. Aronica, A. Bilibashi, O. Bonacci, M. Borga, I. Čanjevac, A. Castellarin, G.B. Chirico, P. Claps, K. Fiala, N. Frolova, L. Gorbachova, A. Gül, J. Hannaford, S. Harrigan, M. Kireeva, A. Kiss, T.R. Kjeldsen, S. Kohnová, J.J. Koskela, O. Ledvinka, N. Macdonald, M. Mavrova-Guirguinova, L. Mediero, R. Merz, P. Molnar, A. Montanari, C. Murphy, M. Osuch, V. Ovcharuk, I. Radevski, M. Rogger, J.L. Salinas, E. Sauquet, M. Šraj, J. Szolgay, A. Viglione, E. Volpi, D. Wilson, K. Zaimi, N. Živković, Changing climate shifts timing of European floods, Science 357 (6351) (2017) 588–590.
[28] W. Mangini, A. Viglione, J. Hall, Y. Hundecha, S. Ceola, A. Montanari, M. Rogger, J.L. Salinas, I. Borzì, J. Parajka, Detection of trends in magnitude and frequency of flood peaks across Europe, Hydrol. Sci. J. 63 (4) (2018) 493–512.
[29] L. Mediero, D. Santillán, L. Garrote, A. Granados, Detection and attribution of trends in magnitude, frequency and timing of floods in Spain, J. Hydrol. 517 (2014) 1072–1088.
[30] M.C. Llasat, R. Marcos, M. Turco, J. Gilabert, M. Llasat-Botija, Trends in flash flood events versus convective precipitation in the Mediterranean region: the case of Catalonia, J. Hydrol. 541 (2016) 24–37.
[31] I. Giuntoli, B. Renard, M. Lang, Floods in France, in: Z.W. Kundzewicz (Ed.), Changes in Flood Risk in Europe, CRC Press, 2012, pp. 199–211.
[32] J. Knighton, S.M. Saia, C.K. Morris, J.A. Archiblad, M.T. Walter, Ecohydrologic considerations for modeling of stable water isotopes in a small intermittent watershed, Hydrol. Process. 31 (13) (2017) 2438–2452.
[33] A. Barrera-Escoda, M.C. Llasat, Evolving flood patterns in a Mediterranean region (1301–2012) and climatic factors—the case of Catalonia, Hydrol. Earth Syst. Sci. 19 (1) (2015) 465–483.
[34] Y. Tramblay, L. Mimeau, L. Neppel, F. Vinet, E. Sauquetm, Detection and attribution of flood trends in Mediterranean basins, Hydrol. Earth Syst. Sci. Discuss. 23 (2019) 4419–4431, https://doi.org/10.5194/hess-2019-307 (under review).
[35] Y. Hirabayashi, R. Mahendran, S. Koirala, L. Konoshima, D. Yamazaki, S. Watanabe, H. Kim, S. Kanae, Global flood risk under climate change, Nat. Clim. Chang. 3 (9) (2013) 816.
[36] S. Camici, L. Brocca, F. Melone, T. Moramarco, Impact of climate change on flood frequency using different climate models and downscaling approaches, J. Hydrol. Eng. 19 (8) (2013) 04014002.
[37] C.P. Wake, E.A. Burakowski, P. Wilkinson, K. Hayhoe, A. Stoner, C. Keeley, J. LaBranche, Climate Change in Southern New Hampshire: Past, Present and Future, vol. 2, The Sustainability Institute, 2014. Available at https://scholars.unh.edu/sustainability/2.
[38] L. Alfieri, B. Bisselink, F. Dottori, G. Naumann, A. de Roo, P. Salamon, K. Wyser, L. Feyen, Global projections of river flood risk in a warmer world, Earth's Future 5 (2) (2017) 171–182.
[39] M.C. Llasat, M. Llasat-Botija, M.A. Prat, F. Porcú, C. Price, A. Mugnai, K. Lagouvardos, V. Kotroni, D. Katsanos, S. Michaelides, Y. Yair, K. Savvidou, K. Nicolaides, High-impact floods and flash floods in Mediterranean countries: the FLASH preliminary database, Adv. Geosci. 23 (2010) 47–55.
[40] D. Penna, M. Borga, D. Zoccatelli, R. Marston, M. Stoffel, Analysis of Flash-Flood Runoff Response, with Examples from Major European Events. (2013). https://doi.org/10.1016/B978-0-12-374739-6.00153-6.

[41] R.E. Horton, The role of infiltration in the hydrologic cycle, EOS Trans. Am. Geophys. Union 14 (1) (1933) 446–460.
[42] T. Dunne, R.D. Black, Partial area contributions to storm runoff in a small New England watershed, Water Resour. Res. 6 (5) (1970) 1296–1311.
[43] K. Beven, P. Germann, Macropores and water flow in soils, Water Resour. Res. 18 (5) (1982) 1311–1325.
[44] S.W. Lyon, S.L. Desilets, P.A. Troch, Characterizing the response of a catchment to an extreme rainfall event using hydrometric and isotopic data, Water Resour. Res. 44 (6) (2008) W06413.
[45] T. Dunne, Field studies of hillsope flow processes. in: M.J. Kirkby (Ed.), Hillslope Hydrology, 1978, pp. 227–293, https://doi.org/10.18172/cig.1099.
[46] D. Penna, H.J. Tromp-van Meerveld, A. Gobbi, M. Borga, G. Dalla Fontana, The influence of soil moisture on threshold runoff generation processes in an alpine headwater catchment, Hydrol. Earth Syst. Sci. 15 (3) (2011) 689–702.
[47] A.W. Western, R.B. Grayson, The Tarrawarra data set: soil moisture patterns, soil characteristics, and hydrological flux measurements, Water Resour. Res. 34 (10) (1998) 2765–2768.
[48] J. Tromp-van Meerveld, J.J. McDonnell, Threshold relations in subsurface stormflow: 2. The fill and spill hypothesis, Water Resour. Res. 42 (2) (2006).
[49] J. Latron, F. Gallart, Runoff generation processes in a small Mediterranean research catchment (Vallcebre, Eastern Pyrenees), J. Hydrol. 358 (3–4) (2008) 206–220.
[50] E. Zehe, T. Graeff, M. Morgner, A. Bauer, A. Bronstert, Plot and field scale soil moisture dynamics and subsurface wetness control on runoff generation in a headwater in the Ore Mountains, Hydrol. Earth Syst. Sci. 14 (6) (2010) 873.
[51] S. Scherrer, F. Naef, A.O. Faeh, I. Cordery, Formation of runoff at the hillslope scale during intense precipitation, Hydrol. Earth Syst. Sci. Discuss. 11 (2) (2007) 907–922.
[52] M. Sivapalan, Process complexity at hillslope scale, process simplicity at the watershed scale: is there a connection? Hydrol. Process. 17 (5) (2003) 1037–1041.
[53] R. Merz, G. Blöschl, J. Parajka, Spatio-temporal variability of event runoff coefficients, J. Hydrol. 331 (3–4) (2006) 591–604.
[54] W.R. Berghuijs, S.T. Allen, S. Harrigan, J.W. Kirchner, Growing spatial scales of synchronous river flooding in Europe, Geophys. Res. Lett. 46 (3) (2019) 1423–1428.
[55] P. Sturdevant-Rees, J.A. Smith, J. Morrison, M.L. Baeck, Tropical storms and the flood hydrology of the central Appalachians, Water Resour. Res. 37 (8) (2001) 2143–2168.
[56] M. Borga, P. Boscolo, F. Zanon, M. Sangati, Hydrometeorological analysis of the 29 August 2003 flash flood in the eastern Italian Alps, J. Hydrometeorol. 8 (5) (2007) 1049–1067.
[57] E. Meyles, A. Williams, L. Ternan, J. Dowd, Runoff generation in relation to soil moisture patterns in a small Dartmoor catchment, Southwest England, Hydrol. Process. 17 (2) (2003) 251–264.
[58] A. Longobardi, P. Villani, R.B. Grayson, A.W. Western, On the relationship between runoff coefficient and catchment initial conditions, in: In Proceedings of MODSIM, 2003, pp. 867–872.
[59] L. Brocca, G. Calice, T. Lacava, F. Melone, T. Moramarco, N. Pergola, V. Tramutoli, Soil moisture estimation through the AMSU-based soil wetness variation index (SWVI) for hydrological applications, in: Proceedings from the 33rd International Symposia on Remote Sensing of Environment, Sustaining the Millennium Development Goals, Volumes I & II, Omni-Press, Madison WI (United States of America), 2009, pp. 981–984.
[60] M. Huang, J. Gallichand, C. Dong, Z. Wang, M. Shao, Use of soil moisture data and curve number method for estimating runoff in the Loess Plateau of China, Hydrol. Process. 21 (11) (2007) 1471–1481.
[61] L. Pfister, G. Drogue, A. El Idrissi, J. Humbert, J.F. Iffly, P. Matgen, L. Hoffmann, Predicting peak discharge through empirical relationships between rainfall, groundwater level and catchment humidity in the Alzette River catchment (Grand-Duchy of Luxembourg), J. Hydrol. Hydromech. 51 (3) (2009) 210–220.
[62] Y. Tramblay, C. Bouvier, C. Martin, J.F. Didon-Lescot, D. Todorovik, J.M. Domergue, Assessment of initial soil moisture conditions for event-based rainfall–runoff modelling, J. Hydrol. 387 (3–4) (2010) 176–187.

[63] H.E. Beck, R.A.M. de Jeu, L.A. Bruijnzeel, J. Schellekens, A.I.J.M. van Dijk, Improving curve number based storm runoff estimates using soil moisture proxies, IEEE J. Sel. Top. Appl. Earth Obs. Remote Sens. 2 (4) (2009) 250–259.
[64] V.T. Chow, D.R. Maidment, L.W. Mays, Applied Hydrology, McGraw-Hill, New York, 1998.
[65] C. Massari, L. Brocca, S. Barbetta, C. Papathanasiou, M. Mimikou, T. Moramarco, Using globally available soil moisture indicators for flood modelling in Mediterranean catchments, Hydrol. Earth Syst. Sci. 18 (2014) 839–853.
[66] C. Massari, L. Brocca, L. Ciabatta, T. Moramarco, S. Gabellani, C. Albergel, P. de Rosnay, S. Puca, W. Wagner, The use of H-SAF soil moisture products for operational hydrology: flood modelling over Italy, Hydrology 2 (1) (2015) 2–22.
[67] C. Massari, L. Brocca, A. Tarpanelli, T. Moramarco, Data assimilation of satellite soil moisture into rainfall-runoff modelling: a complex recipe? Remote Sens. 7 (9) (2015) 11403–11433.
[68] L. Cenci, L. Pulvirenti, G. Boni, M. Chini, P. Matgen, S. Gabellani, G. Squicciarino, N. Pierdicca, An evaluation of the potential of sentinel 1 for improving flash flood predictions via soil moisture–data assimilation, Adv. Geosci. 44 (2017) 89–100.
[69] G. Villarini, P.V. Mandapaka, W.F. Krajewski, R.J. Moore, Rainfall and sampling uncertainties: a rain gauge perspective, J. Geophys. Res.: Atmos. 113 (D11102) (2008) 2156–2202.
[70] S.J. Cole, R.J. Moore, Hydrological modelling using raingauge-and radar-based estimators of areal rainfall, J. Hydrol. 358 (3–4) (2008) 159–181.
[71] D.B. Michelson, Systematic correction of precipitation gauge observations using analyzed meteorological variables, J. Hydrol. 290 (3–4) (2004) 161–177.
[72] B. Sevruk, Adjustment of tipping-bucket precipitation gauge measurements, Atmos. Res. 42 (1996) 237–246.
[73] G.Q. Tabios III, J.D. Salas, A comparative analysis of techniques for spatial interpolation of precipitation, J. Amer. Water Resour. Assoc. 21 (3) (1985) 365–380.
[74] K.H. Syed, D.C. Goodrich, D.E. Myers, S. Sorooshian, Spatial characteristics of thunderstorm rainfall fields and their relation to runoff, J. Hydrol. 271 (1–4) (2003) 1–21.
[75] C.A. Velasco-Forero, D. Sempere-Torres, E.F. Cassiraga, J.J. Gómez-Hernández, A non-parametric automatic blending methodology to estimate rainfall fields from rain gauge and radar data, Adv. Water Resour. 32 (7) (2009) 986–1002.
[76] K. Liechti, L. Panziera, U. Germann, M. Zappa, The potential of radar-based ensemble forecasts for flash-flood early warning in the southern Swiss Alps, Hydrol. Earth Syst. Sci. 17 (10) (2013) 3853–3869.
[77] U. Germann, G. Galli, M. Boscacci, M. Bolliger, Radar precipitation measurement in a mountainous region, Q. J. R. Meteorol. Soc. 132 (618) (2006) 1669–1692.
[78] J. Szturc, A. Jurczyk, L. Jelonek, Concept of dealing with uncertainty in radar-based data for hydrological purpose, Nat. Hazards Earth Syst. Sci. 8 (2) (2008) 267–279.
[79] M. Werner, M. Cranston, Understanding the value of radar rainfall nowcasts in flood forecasting and warning in flashy catchments, Meteorol. Appl. 16 (1) (2009) 41–55.
[80] L.P. Wang, S. Ochoa-Rodríguez, N.E. Simões, C. Onof, Č. Maksimović, Radar–raingauge data combination techniques: a revision and analysis of their suitability for urban hydrology, Water Sci. Technol. 68 (4) (2013) 737–747.
[81] T. Einfalt, M. Jessen, B. Mehlig, Comparison of radar and raingauge measurements during heavy rainfall, Water Sci. Technol. 51 (2) (2005) 195–201.
[82] Q. Zeng, H. Chen, C.Y. Xu, M.X. Jie, J. Chen, S.L. Guo, J. Liu, The effect of rain gauge density and distribution on runoff simulation using a lumped hydrological modelling approach, J. Hydrol. 563 (2018) 106–122.
[83] A. Bárdossy, T. Das, Influence of rainfall observation network on model calibration and application, Hydrol. Earth Syst. Sci. 12 (2008) 77–89.
[84] R.L. Bras, I. Rodríguez-Iturbe, Network design for the estimation of areal mean of rainfall events, Water Resour. Res. 12 (6) (1976) 1185–1195.
[85] M. Girons Lopez, H. Wennerström, L.Å. Nordén, J. Seibert, Location and density of rain gauges for the estimation of spatial varying precipitation, Geogr. Ann. Ser. A Phys. Geogr. 97 (1) (2015) 167–179.

[86] L. Moulin, E. Gaume, C. Obled, Uncertainties on mean areal precipitation: assessment and impact on streamflow simulations, Hydrol. Earth Syst. Sci. 13 (2009) 2067–2110.

[87] S. Liguori, M.A. Rico-Ramirez, A.N.A. Schellart, A.J. Saul, Using probabilistic radar rainfall nowcasts and NWP forecasts for flow prediction in urban catchments, Atmos. Res. 103 (2012) 80–95.

[88] A.N.A. Schellart, W.J. Shepherd, A.J. Saul, Influence of rainfall estimation error and spatial variability on sewer flow prediction at a small urban scale, Adv. Water Resour. 45 (2012) 65–75.

[89] R.A. Fulton, J.P. Breidenbach, D.J. Seo, D.A. Miller, T. O'Bannon, The WSR-88D rainfall algorithm, Weather Forecast. 13 (2) (1998) 377–395.

[90] D.J. Seo, J.P. Breidenbach, E.R. Johnson, Real-time estimation of mean field bias in radar rainfall data, J. Hydrol. 2231 (1999) 131–147.

[91] D.L. Harrison, M. Kitchen, High-resolution precipitation estimates for hydrological uses, Proc. Inst. Civ. Eng. Water Manag. 162 (2) (2009) 125–135.

[92] L. Alfieri, J. Thielen, A European precipitation index for extreme rain-storm and flash flood early warning, Meteorol. Appl. 22 (1) (2015) 3–13.

[93] W.F. Krajewski, Cokriging radar-rainfall and rain gage data, J. Geophys. Res. 92 (D8) (1987) 9571–9580.

[94] E. Todini, Influence of parameter estimation uncertainty in kriging: Part 1—theoretical development, Hydrol. Earth Syst. Sci. Discuss. 5 (2) (2001) 215–223.

[95] E. Todini, F. Pellegrini, C. Mazzetti, Influence of parameter estimation uncertainty in kriging: Part 2—test and case study applications, Hydrol. Earth Syst. Sci. Discuss. 5 (2) (2001) 225–232.

[96] E. Morin, Y. Jacoby, S. Navon, E. Bet-Halachmi, Towards flash-flood prediction in the dry Dead Sea region utilizing radar rainfall information, Adv. Water Resour. 32 (7) (2009) 1066–1076.

[97] A. Mugnai, D. Casella, E. Cattani, S. Dietrich, S. Laviola, V. Levizzani, G. Panegrossi, P. Sanò, D. Biron, L. De Leonibus, D. Melfi, P. Rosci, A. Vocino, F. Zauli, S. Puca, A. Rinollo, L. Milani, F. Porcù, F. Gattari, Precipitation products from the hydrology SAF, Nat. Hazards Earth Syst. Sci. 13 (2013) 1959–1981.

[98] A.Y. Hou, R.K. Kakar, S. Neeck, A.A. Azarbarzin, C.D. Kummerow, M. Kojima, R. Oki, K. Nakamura, T. Iguchi, The global precipitation measurement mission, Bull. Am. Meteorol. Soc. 95 (5) (2014) 701–722.

[99] G. Skofronick-Jackson, W.A. Petersen, W. Berg, C. Kidd, E.F. Stocker, R. Kakar, S.A. Braun, T. Iguchi, P.E. Kirstetter, C. Kummerow, R. Oki, W.S. Olson, Y.N. Takayabu, K. Furukawa, T. Wilheit, The global precipitation measurement (GPM) mission for science and society, Bull. Am. Meteorol. Soc. 98 (8) (2017) 1679–1695.

[100] C.D. Kummerow, D.L. Randel, M. Kulie, N.Y. Wang, R. Ferraro, S. Joseph Munchak, V. Petkovic, The evolution of the Goddard profiling algorithm to a fully parametric scheme, J. Atmos. Ocean. Technol. 32 (2015) 2265–2280.

[101] D. Casella, G. Panegrossi, P. Sanò, A. Mugnai, E.A. Smith, G.J. Tripoli, S. Dietrich, M. Formenton, F. Di Paola, H.W.-Y. Leung, A.V. Mehta, Transitioning from CRD to CDRD in Bayesian retrieval of rainfall from satellite passive microwave measurements, Part 2: overcoming database profile selection ambiguity by consideration of meteorological control on microphysics, IEEE Trans. Geosci. Remote Sens. 51 (9) (2013) 4650–4671.

[102] D. Casella, L.M. Amaral, S. Dietrich, A.C. Marra, P. Sanò, G. Panegrossi, The cloud dynamics and radiation database algorithm for AMSR2: exploitation of the GPM observational dataset for operational applications, IEEE J. Sel. Topics Appl. Earth Obs. Remote Sens. 10 (9) (2017) 3985–4001.

[103] P. Sanò, G. Panegrossi, D. Casella, A.C. Marra, L.P. D'Adderio, J.-F. Rysman, S. Dietrich, The passive microwave neural network precipitation retrieval (PNPR) algorithm for the conical scanning GMI radiometer, Remote Sens. 10 (2018) 1122.

[104] P. Sanò, G. Panegrossi, D. Casella, A.C. Marra, F. Di Paola, S. Dietrich, The new passive microwave neural network precipitation retrieval (PNPR) algorithm for the cross-track scanning ATMS radiometer: description and verification study over Europe and Africa using GPM and TRMM spaceborne radars, Atmos. Meas. Tech. 9 (2016) 5441–5460.

[105] P. Sanò, G. Panegrossi, D. Casella, F. Di Paola, L. Milani, A. Mugnai, M. Petracca, S. Dietrich, The passive microwave neural network precipitation retrieval (PNPR) algorithm for AMSU/MHS observations: description and application to European case studies, Atmos. Meas. Tech. 8 (2015) 837–857.

[106] G.J. Huffman, D.T. Bolvin, E.J. Nelkin, Integrated multi-satellitE retrievals for GPM (IMERG) technical documentation, IMERG Tech Doc. 612 (47) (2017) 2019.
[107] V. Maggioni, C. Massari, On the performance of satellite precipitation products in riverine flood modeling: a review, J. Hydrol. 558 (2018) 214–224.
[108] E.I. Nikolopoulos, E.N. Anagnostou, M. Borga, Using high-resolution satellite rainfall products to simulate a major flash flood event in northern Italy, J. Hydrometeorol. 14 (1) (2013) 171–185.
[109] G.J. Huffman, R.F. Adler, D.T. Bolvin, G. Gu, E.J. Nelkin, K.P. Bowman, Y. Hong, E.F. Stocker, D.B. Wolff, The TRMM multisatellite precipitation analysis (TMPA): quasi-global, multiyear, combined-sensor precipitation estimates at fine scales, J. Hydrometeorol. 8 (1) (2007) 38–55.
[110] R.J. Joyce, J.E. Janowiak, P.A. Arkin, P. Xie, CMORPH: a method that produces global precipitation estimates from passive microwave and infrared data at high spatial and temporal resolution, J. Hydrometeorol. 5 (3) (2004) 487–503.
[111] D. Stampoulis, E.N. Anagnostou, E.I. Nikolopoulos, Assessment of high-resolution satellite-based rainfall estimates over the Mediterranean during heavy precipitation events, J. Hydrometeorol. 14 (2013) 1500–1514.
[112] Y. Mei, E. Nikolopoulos, E. Anagnostou, D. Zoccatelli, M. Borga, Error analysis of satellite precipitation-driven modeling of flood events in complex alpine terrain, Remote Sens. 8 (4) (2016) 293.
[113] A. Behrangi, Y. Wen, On the spatial and temporal sampling errors of remotely sensed precipitation products, Remote Sens. 9 (11) (2017) 1127.
[114] A.C. Marra, F. Porcù, L. Baldini, M. Petracca, D. Casella, S. Dietrich, A. Mugnai, P. Sanò, G. Vulpiani, G. Panegrossi, Observational analysis of an exceptionally intense hailstorm over the Mediterranean area: role of the GPM Core Observatory, Atmos. Res. 192 (2017) 72–90.
[115] G. Panegrossi, A.C. Marra, P. Sanò, L. Baldini, D. Casella, F. Porcù, Heavy precipitation systems in the Mediterranean area: the role of GPM, in: V. Levizzani, C. Kidd, D.B. Kirschbaum, C.D. Kummerow, K. Nakamura, F.J. Turk (Eds.), Satellite Precipitation Measurement, Springer, Dordrecht, 2019.
[116] H.E. Beck, A.I.J.M. Van Dijk, V. Levizzani, J. Schellekens, D.G. Miralles, B. Martens, A. De Roo, MSWEP: 3-hourly 0.25° global gridded precipitation (1979–2015) by merging gauge, satellite, and reanalysis data, Hydrol. Earth Syst. Sci. 21 (1) (2017) 589–615.
[117] A. Verdin, B. Rajagopalan, W. Kleiber, C. Funk, A Bayesian kriging approach for blending satellite and ground precipitation observations, Water Resour. Res. 51 (2) (2015) 908–921.
[118] Y. He, Y. Zhang, R. Kuligowski, R. Cifelli, D. Kitzmiller, Incorporating satellite precipitation estimates into a radar-gauge multi-sensor precipitation estimation algorithm, Remote Sens. 10 (1) (2018) 106.
[119] L. Ciabatta, L. Brocca, C. Massari, T. Moramarco, S. Gabellani, S. Puca, W. Wagner, Rainfall-runoff modelling by using SM2RAIN-derived and state-of-the-art satellite rainfall products over Italy, Int. J. Appl. Earth Observ. Geoinform. 48 (2016) 163–173.
[120] C. Massari, S. Camici, L. Ciabatta, L. Brocca, Exploiting satellite-based surface soil moisture for flood forecasting in the Mediterranean area: state update versus rainfall correction, Remote Sens. 10 (2) (2018) 292.
[121] C. Massari, V. Maggioni, S. Barbetta, L. Brocca, L. Ciabatta, S. Camici, T. Moramarco, G. Coccia, E. Todini, Complementing near-real time satellite rainfall products with satellite soil moisture-derived rainfall through a Bayesian inversion approach, J. Hydrol. 573 (2019) 341–351.
[122] C. Massari, L. Brocca, T. Pellarin, G. Abramowitz, F. Filippucci, L. Ciabatta, V. Maggioni, Y. Kerr, P.F. Prieto, A daily/25km short-latency rainfall product for data scarce regions based on the integration of the GPM IMERG Early Run with multiple satellite soil moisture products. Hydrol. Earth Syst. Sci. Discuss. (2019) https://doi.org/10.5194/hess-2019-387 in review.
[123] M. Zappa, M.W. Rotach, M. Arpagaus, M. Dorninger, C. Hegg, A. Montani, R. Ranzi, F. Ament, U. Germann, G. Grossi, S. Jaun, A. Rossa, S. Vogt, A. Walser, J. Wehrhan, C. Wunram, MAP D-PHASE: real-time demonstration of hydrological ensemble prediction systems, Atmos. Sci. Lett. 9 (2) (2008) 80–87.
[124] A. Montani, D. Cesari, C. Marsigli, T. Paccagnella, Seven years of activity in the field of mesoscale ensemble forecasting by the COSMO-LEPS system: main achievements and open challenges, Tellus Ser. A Dyn. Meteorol. Oceanogr. 63 (3) (2011) 605–624.

[125] U. Damrath, G. Doms, D. Fruhwald, E. Heise, B. Richter, J. Steppeler, Operational quantitative precipitation forecasting at the German weather service, J. Hydrol. 239 (2000) 260–285.

[126] L. Cuo, T.C. Pagano, Q. Wang, A review of quantitative precipitation forecasts and their use in short- to medium range streamflow forecasting, J. Hydrometeorol. 12 (2011) 713–728.

[127] A.P.J. De Roo, B. Gouweleeuw, J. Thielen, J. Bartholmes, P. Bongioannini-Cerlini, E. Todini, P.D. Bates, M. Horritt, N. Hunter, K. Beven, F. Pappenberger, E. Heise, G. Rivin, M. Hils, A. Hollingsworth, B. Holst, J. Kwadijk, P. Reggiani, M. Van Dijk, K. Sattler, E. Sprokkereef, Development of a European flood forecasting system, Int. J. River Basin Manag. 1 (1) (2003) 49–59.

[128] S. Anquetin, E. Yates, V. Ducrocq, S. Samouillan, K. Chancibault, S. Davolio, C. Accadia, M. Casaioli, S. Mariani, G. Ficca, B. Gozzini, F. Pasi, M. Pasqui, A. Garcia, M. Martorell, R. Romero, P. Chessa, The 8 and 9 September 2002 flash flood event in France: a model intercomparison, Nat. Hazards Earth Syst. Sci. 5 (2005) 741–754.

[129] B. Vincendon, V. Ducrocq, O. Nuissier, B. Vié, Perturbation of convection-permitting NWP forecasts for flash-flood ensemble forecasting, Nat. Hazards Earth Syst. Sci. 11 (5) (2011) 1529–1544.

[130] E. Fiori, A. Comellas, L. Molini, N. Rebora, F. Siccardi, D.J. Gochis, S. Tanelli, A. Parodi, Analysis and hindcast simulations of an extreme rainfall event in the Mediterranean area: the Genoa 2011 case, Atmos. Res. 138 (2014) 13–29.

[131] S. Davolio, F. Silvestro, P. Malguzzi, Effects of increasing horizontal resolution in a convection-permitting model on flood forecasting: the 2011 dramatic events in Liguria, Italy, J. Hydrometeorol. 16 (4) (2015) 1843–1856.

[132] F. Silvestro, L. Rossi, L. Campo, A. Parodi, F. Fiori, R. Rudari, L. Ferraris, Impact-based flash-flood forecasting system: sensitivity to high resolution numerical weather prediction systems and soil moisture, J. Hydrol. 572 (2019) 388–402.

[133] N.M. Roberts, S.J. Cole, R.M. Forbes, R.J. Moore, D. Boswell, Use of high-resolution NWP rainfall and river flow forecasts for advance warning of the Carlisle flood, north-west England, Meteorol. Appl. 16 (1) (2009) 23–34.

[134] K. Chancibault, S. Anquetin, V. Ducrocq, G.M. Saulnier, Hydrological evaluation of high-resolution precipitation forecasts of the Gard flash-flood event (8–9 September 2002), Q. J. R. Meteorol. Soc. 132 (617) (2006) 1091–1117.

[135] L. Alfieri, J. Thielen, F. Pappenberger, Ensemble hydro-meteorological simulation for flash flood early detection in southern Switzerland, J. Hydrol. 424 (2012) 143–153.

[136] A.M. Rossa, G. Cenzon, M. Monai, Quantitative comparison of radar QPE to rain gauges for the 26 September 2007 Venice Mestre flood, Nat. Hazards Earth Syst. Sci. 10 (2) (2010) 371–377.

[137] S. Federico, R.C. Torcasio, E. Avolio, O. Caumont, M. Montopoli, L. Baldini, G. Vulpiani, S. Dietrich, The impact of lightning and radar data assimilation on the performance of very short term rainfall forecast for two case studies in Italy, Nat. Hazards Earth Syst. Sci. 19 (2019) 1839–1864.

[138] L. Brocca, F. Melone, T. Moramarco, On the estimation of antecedent wetness conditions in rainfall-runoff modelling, Hydrol. Process. 22 (5) (2008) 629–642.

[139] L. Brocca, F. Melone, T. Moramarco, R. Morbidelli, Antecedent wetness conditions based on ERS scatterometer data, J. Hydrol. 364 (1–2) (2009) 73–87.

[140] L. Brocca, F. Melone, T. Moramarco, W. Wagner, What perspective in remote sensing of soil moisture for hydrological applications, in: C.M.U. Neale, A. Maltese, K. Richter (Eds.), Remote Sensing for Agriculture, Ecosystems, and Hydrology XIII, Proc. of SPIE, vol. 8174, 2011, 81740A.

[141] Y. Tramblay, C. Bouvier, P.-A. Ayral, A. Marchandise, Impact of rainfall spatial distribution on rainfall-runoff modeling efficiency and initial soil moisture conditions estimation, Nat. Hazards Earth Syst. Sci. 11 (2011) 157–170.

[142] Y. Tramblay, R. Bouaicha, L. Brocca, W. Dorigo, C. Bouvier, S. Camici, E. Servat, Estimation of antecedent wetness conditions for flood modelling in northern Morocco, Hydrol. Earth Syst. Sci. 16 (11) (2012) 4375–4386.

[143] M. Coustau, C. Bouvier, V. Borrell-Estupina, H. Jourde, Flood modelling with a distributed event-based parsimonious rainfall-runoff model: case of the karstic Lez river catchment, Nat. Hazard Earth Sys. 12 (4) (2012) 1119–1133.

[144] T. Graeff, E. Zehe, T. Blume, T. Francke, B. Schröder, Predicting event response in a nested catchment with generalized linear models and a distributed watershed model, Hydrol. Process. 26 (2012) 3749–3769.
[145] SCS, Soil Conservation Service: Hydrology, National Engineering Handbook, Supplement A, Section 4, Soil Conservation Service, USDA, Washington, DC, 1993.
[146] W.A. Dorigo, W. Wagner, R. Hohensinn, S. Hahn, C. Paulik, A. Xaver, A. Gruber, M. Drusch, S. Mecklenburg, P. Van Oevelen, A. Robock, T. Jackson, The international soil moisture network: a data hosting facility for global in situ soil moisture measurements, Hydrol. Earth Syst. Sci. 15 (5) (2011) 1675–1698.
[147] G.A. Vachaud, A. Passerat de Silans, P. Balabanis, M. Vauclin, Temporal stability of spatially measured soil water probability density function, Soil Sci. Soc. Am. J. 49 (1985) 822–828.
[148] L. Brocca, F. Melone, T. Moramarco, Storm runoff estimation based on the soil conservation service—curve number method with soil moisture data assimilation, in: L. Pfister, L. Hoffmann (Eds.), Uncertainties in the 'Monitoring-Conceptualisation-Modelling' Sequence of Catchment Research, IHP-VI, Technical Documents in Hydrology, vol. 81, UNESCO, Paris, 2007, pp. 141–148. Working Series SC-2007/WS/54.
[149] D. Penna, L. Brocca, M. Borga, G. Dalla Fontana, Soil moisture temporal stability at different depths on two alpine hillslopes during wet and dry periods, J. Hydrol. 477 (2013) 55–71.
[150] H.R. Bogena, J.A. Huisman, A. Güntner, C. Hübner, J. Kusche, F. Jonard, S. Vey, H. Vereecken, Emerging methods for noninvasive sensing of soil moisture dynamics from field to catchment scale: a review, Wiley Interdiscip. Rev. Water 2 (2015) 635–647.
[151] K.M. Larson, E.E. Small, E.D. Gutmann, A.L. Bilich, J.J. Braun, V.U. Zavorotny, Use of GPS receivers as a soil moisture network for water cycle studies, Geophys. Res. Lett. 35 (2008) L24405.
[152] H.R. Bogena, M. Herbst, J.A. Huisman, U. Rosenbaum, A. Weuthen, H. Vereecken, Potential of wireless sensor networks for measuring soil water content variability, Vadose Zone J. 9 (2010) 1002–1013.
[153] H.R. Bogena, J.A. Huisman, B. Schilling, A. Weuthen, H. Vereecken, Effective calibration of low-cost soil water content sensors, Sensors (Switzerland) 17 (2017) 208.
[154] T.E. Franz, T. Wang, W. Avery, C. Finkenbiner, L. Brocca, Combined analysis of soil moisture measurements from roving and fixed cosmic ray neutron probes for multiscale real-time monitoring, Geophys. Res. Lett. 42 (2015) 3389–3396.
[155] L. Brocca, L. Ciabatta, C. Massari, S. Camici, A. Tarpanelli, Soil moisture for hydrological applications: open questions and new opportunities, Water (Switzerland) 9 (2) (2017) 140.
[156] W. Wagner, G. Lemoine, H. Rott, A method for estimating soil moisture from ERS scatterometer and soil data, Remote Sens. Environ. 70 (1999) 191–207.
[157] Y.H. Kerr, P. Waldteufel, J.P. Wigneron, S. Delwart, F. Cabot, J. Boutin, M.J. Escorihuela, J. Font, N. Reul, C. Gruhier, S.E. Juglea, M.R. Drinkwater, A. Hahne, M. Martin-Neira, S. Mecklenburg, The SMOS L: new tool for monitoring key elements of the global water cycle, Proc. IEEE 98 (5) (2010) 666–687.
[158] D. Entekhabi, E.G. Njoku, P.E. Neill, K.H. Kellogg, W.T. Crow, W.N. Edelstein, et al., The soil moisture active passive (SMAP) mission, Proc. IEEE 98 (5) (2010) 704–716.
[159] W. Dorigo, W. Wagner, C. Albergel, F. Albrecht, G. Balsamo, L. Brocca, et al., ESA CCI soil moisture for improved earth system understanding: state-of-the art and future directions, Remote Sens. Environ. 203 (2017) 185215.
[160] H. Lievens, S.K. Tomer, A. Al Bitar, G.J.M. De Lannoy, M. Drusch, G. Dumedah, H.-J. Hendricks Franssen, Y.H. Kerr, B. Martens, M. Pan, J.K. Roundy, H. Vereecken, J.P. Walker, E.F. Wood, N.E.C. Verhoest, V.R.N. Pauwels, SMOS soil moisture assimilation for improved hydrologic simulation in the Murray Darlng catchment, Australia, Remote Sens. Environ. 168 (2015) 146–162.
[161] Y.H. Kerr, P. Waldteufel, J.P. Wigneron, J.M. Martinuzzi, J. Font, M. Berger, Soil moisture retrieval from space: the soil moisture and ocean salinity (SMOS) mission, IEEE Trans. Geosci. Remote Sens 39 (2001) 1729–1735.
[162] W. Wagner, S. Hahn, R. Kidd, T. Melzer, Z. Bartalis, S. Hasenauer, et al., The ASCAT soil moisture product: a review of its specifications, validation results, and emerging applications, Meteorol. Z. 22 (1) (2013) 5–33.

[163] E.G. Njoku, T.J. Jackson, V. Lakshmi, T.K. Chan, S.V. Nghiem, Soil moisture retrieval from AMSR-E, IEEE Trans. Geosci. Remote Sens. 41 (2) (2003) 215–229.

[164] H. Lu, T. Koike, H. Fujii, T. Ohta, T. Tamagawa, Development of a physically based soil moisture retrieval algorithm for space borne passive microwave radiometers and its application to AMSR-E, Int. J. Remote Sens. 29 (2009) 253–262.

[165] Y. Mao, W.T. Crow, B. Nijssen, Dual state/rainfall correction via soil moisture assimilation for improved streamflow simulation: evaluation of a large-scale implementation with Soil Moisture Active Passive (SMAP) satellite data. Hydrol. Earth Syst. Sci. 24 (2020) 615–631, https://doi.org/10.5194/hess-24-615-2020.

[166] X. Zhan, P.R. Houser, J.P. Walker, W.T. Crow, A method for retrieving high-resolution surface soil moisture from hydros L-band radiometer and radar observations, IEEE Trans. Geosci. Remote Sens. 44 (2006) 1534–1544.

[167] M. Piles, A. Camps, M. Vall-Llossera, I. Corbella, R. Panciera, C. Rudiger, Y.H. Kerr, J. Walker, Downscaling SMOS-derived soil moisture using MODIS visible/infrared data, IEEE Trans. Geosci. Remote Sens. 49 (2011) 3156–3166.

[168] O. Merlin, C. Rudiger, A. Al Bitar, P. Richaume, J.P. Walker, Y.H. Kerr, Disaggregation of SMOS soil moisture in Southeastern Australia, IEEE Trans. Geosci. Remote Sens. 50 (2012) 1556–1571.

[169] Y. Malbéteau, O. Merlin, B. Molero, C. Rüdiger, S. Bacon, DisPATCh as a tool to evaluate coarse-scale remotely sensed soil moisture using localized in situ measurements: application to SMOS and AMSR-E data in Southeastern Australia, Int. J. Appl. Earth Obs. Geoinf. 45 (2016) 221–234.

[170] B. Bauer-Marschallinger, C. Paulik, S. Hochstöger, T. Mistelbauer, S. Modanesi, L. Ciabatta, C. Massari, L. Brocca, W. Wagner, Soil moisture from fusion of scatterometer and SAR: closing the scale gap with temporal filtering, Remote Sens. 10 (7) (2018) 1030.

[171] B. Bauer-Marschallinger, V. Freeman, S. Cao, C. Paulik, S. Schaufler, T. Stachl, S. Modanesi, C. Massari, L. Ciabatta, L. Brocca, W. Wagner, Toward global soil moisture monitoring with Sentinel-1: harnessing assets and overcoming obstacles, IEEE Trans. Geosci. Remote Sens. 57 (2018) 520–539.

[172] P. Rahimzadeh-Bajgiran, A. Berg, Soil moisture retrievals using optical/TIR methods, in: Satellite Soil Moisture Retrieval, Elsevier, 2016, pp. 47–72.

[173] C. Loumagne, M. Normand, M. Riffard, A. Weisse, A. Quesney, S. Le Hegarat-Mascle, F. Alem, Methodology for integration of remote sensing data into hydrological models for reservoir management purposes, Hydrol. Sci. J. 46 (1) (2001) 89–102.

[174] V.R.N. Pauwels, R. Hoeben, N.E.C. Verhoest, F.P. DeTroch, The importance of the spatial patterns of remotely sensed soil moisture in the improvement of discharge predictions for small scale catchments through data assimilation, J. Hydrol. 251 (2001) 88–102.

[175] R.N. Pauwels, R. Hoeben, N.E.C. Verhoest, F.P. De Troch, P.A. Troch, Improvements of TOPLATS-based discharge predictions through assimilation of ERS-based remotely-sensed soil moisture values, Hydrol. Process. 16 (2002) 995–1013.

[176] C. Francois, A. Quesney, C. Ottlé, Sequential assimilation of ERS-1 SAR data into a coupled land surface-hydrological model using an extended Kalman filter, J. Hydrometeorol. 4 (2) (2003) 473–487.

[177] W.T. Crow, R. Bindlish, T.J. Jackson, The added value of spaceborne passive microwave soil moisture retrievals for fore-casting rainfall-runoff ratio partitioning, Geophys. Res. Lett. 32 (2005) L18401.

[178] P.J. Matgen, B. Henry, L. Hoffmann, L. Pfister, Assimilation of remotely sensed soil saturation levels in conceptual rainfall-runoff models, in IAHS red book series: prediction in ungauged basins: promise and progress, IAHS Publ. 303 (2006) 226–234.

[179] F. Anctil, N. Lauzon, M. Filion, Added gains of soil moisture content observations for streamflow predictions using neural networks, J. Hydrol. 359 (3–4) (2008) 225–234.

[180] W.T. Crow, D. Ryu, A new data assimilation approach for improving runoff prediction using remotely-sensed soil moisture retrievals, Hydrol. Earth Syst. Sci. 13 (1) (2009) 1–16.

[181] L. Brocca, T. Moramarco, F. Melone, W. Wagner, S. Hasenauer, S. Hahn, Assimilation of surface- and root-zone ASCAT soil moisture products into rainfall-runoff modeling, IEEE Trans. Geosci. Remote Sens. 50 (7) (2012) 2542–2555.

[182] C. Alvarez-Garreton, D. Ryu, A.W. Western, W.T. Crow, C.H. Su, D.R. Robertson, Dual assimilation of satellite soil moisture to improve streamflow prediction in data-scarce catchments, Water Resour. Res. 52 (7) (2016) 5357–5375.

[183] J. Loizu, C. Massari, J. Alvarez-Mozos, A. Tarpanelli, L. Brocca, J. Casali, On the assimilation set-up of ASCAT soil moisture data for improving streamflow catchment simulation, Adv. Water Resour. 111 (2018) 86–104.

[184] L. Brocca, F. Melone, T. Moramarco, Distributed rainfall-runoff modelling for flood frequency estimation and flood forecasting, Hydrol. Process. 25 (18) (2011) 2801–2813.

[185] W.T. Crow, F. Chen, R.H. Reichle, Q. Liu, L band microwave remote sensing and land data assimilation improve the representation of prestorm soil moisture conditions for hydrologic forecasting, Geophys. Res. Lett. 44 (11) (2017) 5495–5503.

[186] W.T. Crow, F. Chen, R.H. Reichle, Y. Xia, Q. Liu, Exploiting soil moisture, precipitation, and streamflow observations to evaluate soil moisture/runoff coupling in land surface models, Geophys. Res. Lett. 45 (2018) 4869–4878.

[187] W.T. Crow, M.J. van Den Berg, G.J. Huffman, T. Pellarin, Correcting rainfall using satellite-based surface soil moisture retrievals: the Soil Moisture Analysis Rainfall Tool (SMART), Water Resour. Res. 47 (8) (2011).

[188] T. Pellarin, S. Louvet, C. Gruhier, G. Quantin, C. Legout, A simple and effective method for correcting soil moisture and precipitation estimates using AMSR-E measurements, Remote Sens. Environ. 136 (2013) 28–36.

[189] L. Brocca, R. Morbidelli, F. Melone, T. Moramarco, Soil moisture spatial variability in experimental areas of Central Italy, J. Hydrol. 333 (2–4) (2007) 356–373.

[190] C. Massari, L. Brocca, T. Moramarco, Y. Tramblay, J.F. Didon Lescot, Potential of soil moisture observations in flood modelling: estimating initial conditions and correcting rainfall, Adv. Water Resour. 74 (2014) 44–53.

[191] F. Chen, T.W. Crow, D. Ryu, Dual forcing and state correction via soil moisture assimilation for improved rainfall–runoff modeling, J. Hydrometeorol. 15 (5) (2014) 1832–1848.

[192] L. Ciabatta, L. Brocca, C. Massari, T. Moramarco, S. Puca, A. Rinollo, S. Gabellani, W. Wagner, Integration of satellite soil moisture and rainfall observations over the Italian territory, J. Hydrometeorol. 16 (3) (2015) 1341–1355.

[193] S. Camici, L. Ciabatta, C. Massari, L. Brocca, How reliable are satellite precipitation estimates for driving hydrological models: a verification study over the Mediterranean area, J. Hydrol. 563 (2018) 950–961.

[194] L. Brocca, L. Ciabatta, C. Massari, T. Moramarco, S. Hahn, S. Hasenauer, R. Kidd, W. Dorigo, W. Wagner, V. Levizzani, Soil as a natural rain gauge: estimating global rainfall from satellite soil moisture data, J. Geophys. Res. 119 (9) (2014) 5128–5141.

[195] K.P. Georgakakos, A generalized stochastic hydrometeorological model for flood and flash-flood forecasting: 1. Formulation, Water Resour. Res. 22 (13) (1986) 2083–2095.

[196] K.P. Georgakakos, Analytical results for operational flash flood guidance, J. Hydrol. 317 (2006) 81–103.

[197] C.G. Collier, Flash flood forecasting: what are the limits of predictability? Q. J. Roy. Meteorol. Soc. 133 (622) (2007) 3–23.

[198] D. Norbiato, M. Borga, R. Dinale, Flash flood warning in ungauged catchments by use of the flash flood guidance and model-based runoff thresholds, Meteorol. Appl. 16 (1) (2009) 65–75.

[199] L. Alfieri, M. Berenguer, V. Knechtl, K. Liechti, D. Sempere-Torres, M. Zappa Flash, Flood forecasting based on rainfall thresholds, in: Q. Duan, F. Pappenberger, J. Thielen, A. Wood, H. Cloke, J. Schaake (Eds.), Handbook of Hydrometeorological Ensemble Forecasting, Springer, Berlin, Heidelberg, 2015.

[200] C. Corral, D. Velasco, D. Forcadell, D. Sempere-Torres, E. Velasco, Advances in Radar-Based Flood Warning Systems. The EHIMI System and the Experience in the Besòs Flash-Flood Pilot Catchment, Taylor & Francis Group, 2009.

[201] P.A. Versini, M. Berenguer, C. Corral, D. Sempere-Torres, An operational flood warning system for poorly gauged catchments: demonstration in the Guadalhorce catchment (Spain), Nat. Hazards 71 (3) (2014) 1355–1378.

[202] L. Brocca, W.T. Crow, L. Ciabatta, C. Massari, P. De Rosnay, M. Enenkel, S. Hahn, G. Amarnath, S. Camici, A. Tarpanelli, W. Wagner, A review of the applications of ASCAT soil moisture products, IEEE J. Sel. Top. Appl. Earth Obs. Remote Sens. 10 (5) (2017) 2285–2306.
[203] S. Azimi, A. Dariane, S. Modanesi, B. Bauer-Marschallinger, R. Bindish, W. Wagner, C. Massari, Assimilation of Sentinel 1 and SMAP-based satellite soil moisture retrievals into SWAT hydrological model: the impact of satellite revisit time and product spatial resolution on flood simulations in small basins, J. Hydrol. 581 (2020) 124367.
[204] F. Rostan, D. Ulrich, S. Riegger, A. Østergaard, MetoP-SG SCA wind scatterometer design and performance, in: 2016 IEEE International Geoscience and Remote Sensing Symposium (IGARSS), IEEE, 2016, pp. 7366–7369.

Further reading

S. Sorooshian, K.L. Hsu, X. Gao, H.V. Gupta, B. Imam, D. Braithwaite, Evaluation of PERSIANN system satellite-based estimates of tropical rainfall, Bull. Am. Meteorol. Soc. 81 (9) (2000) 2035–2046.
J.M. Van Der Knijff, J. Younis, A.P.J. De Roo, LISFLOOD: a GIS-based distributed model for river catchment scale water balance and flood simulation, Int. J. Geogr. Inf. Sci. 24 (2) (2010) 189–212.

CHAPTER 9

Droughts

Sergio M. Vicente-Serrano[a], Fernando Domínguez-Castro[a], J. Julio Camarero[a], Ahmed El Kenawy[b,c]

[a]Instituto Pirenaico de Ecología, Spanish National Research Council (IPE-CSIC), Zaragoza, Spain
[b]Department of Geography, Mansoura University, Mansoura, Egypt
[c]Department of Geography, Sultan Qaboos University, Muscat, Oman

9.1 Introduction

Drought is one of the most difficult natural hazards to quantify, monitor, and predict [1, 2]. This is simply because drought cannot directly be measured using specific instruments in the field. The diverse impacts of drought on natural systems and economic sectors, including meteorological, hydrological, agricultural, and ecological droughts, add further challenges and complications to drought quantification [3, 4]. Given all these constraints, it is quite difficult to determine the onset and cessation of a drought event, as well as its severity and spatial extent.

Drought has a wide spectrum of environmental and socioeconomic impacts in the Mediterranean region, including crops [5], soil moisture [6], surface water resources [7–9], and groundwater [10]. These impacts extend to biosystems, particularly forests [11–14] and wildfires [15, 16]. Several studies confirmed that the drastic impacts of drought on local communities in the Mediterranean have been more evident over the past decades [17, 18].

In the Mediterranean, the dry season (summer) coincides with the warm season, contributing to the severity of drought and its corresponding negative impacts in the region [19]. As such, the Mediterranean ecosystems and human societies have been more vulnerable to water availability, especially with the strong interannual and intraannual variability of precipitation. Given the geographical position of the Mediterranean region, climate is largely influenced by the tropical, extratropical, midlatitudes, and polar configurations. In the pursuit of this influence, drought behavior in the region is usually linked to these complex atmospheric configurations and processes. Atmospheric circulation is usually described as a key driver of drought behavior in the region. Numerous studies attributed the occurrence of several intense drought events in the Mediterranean to atmospheric circulation anomalies over the last two decades [20]. Nonetheless, due to the wide range of circulation mechanisms impacting drought in the Mediterranean (e.g., blocking systems, ridges, anomalies in large-scale atmospheric circulation patterns, latitudinal displacements of the

Water Resources in the Mediterranean Region
https://doi.org/10.1016/B978-0-12-818086-0.00009-1

polar front), it has been difficult to define a unique or dominant mechanism that single handedly can explain the variability of drought [21, 22].

The drastic and adverse impacts of drought in the Mediterranean have made it necessary to develop effective initiatives to mitigate drought consequences and to build society capacity for drought preparedness. With the development of dense networks of reservoirs to store water during wet periods and secure urban and irrigation supply during dry periods, several measures have been developed focusing on surface water resources [23]. Other measures placed more emphasis on the agricultural sector, particularly with the development of agricultural insurance systems [24]. In the last decade, drought plans have been developed by national and regional institutions to improve drought preparedness and monitoring. These plans involved drought monitoring systems [25], as well as early warning and detailed protocols that aim to manage drought conditions once they occur [26].

In the context of the anomalous drought events that trigged the Mediterranean in the past few decades (e.g., 2017), this chapter provides a brief explanation of drought characteristics and implications over the region, stressing the complexity of characterizing and managing this phenomenon.

9.2 State of the art

9.2.1 Importance of droughts in the Mediterranean region from historical point of view

Historically, the Mediterranean witnessed severe drought events that broadly impacted human activities, leading to bad harvests, failure of agricultural crops, increases in food prices, shortages, livestock mortality, hunger, epidemics, and famine. To cope with this situation, early civilizations in the Mediterranean endeavored to mitigate the negative impacts of these events by constructing water infrastructures [27]. The role of severe droughts in mega migrations and decline of ancient civilizations has been documented and evident for the Neo-Assyrian Empire [28] and the Ptolemaic Egypt [29]. Nevertheless, particular caution should be taken when trying to directly and only attribute agricultural and societal crises to dry episodes [30].

Historical dry periods are often identified and characterized using a large variety of natural proxies, such as sediments, corals, or speleothems. Nonetheless, for proper identification of drought events based on these techniques, high temporal resolution proxies (e.g., tree rings) and documentary sources are highly desired. In this regard, while tree rings method allows for generating continuous precipitation or droughts indices records for long periods, its main shortfall is that the reconstructions can be made only for a certain period of the year (mainly in summer or spring across the Mediterranean). On the other hand, documentary sources can provide information on droughts for the whole year at monthly or even higher time resolution. Moreover, documentary sources often provide descriptions of the droughts impacts and sometimes the management decisions adopted to minimize these impacts. Albeit with these advantages, the reconstruction of droughts indices from documentary sources is a challenging task [31, 32].

Documentary sources have been widely used for drought reconstruction in the Mediterranean region. Early drought evidences were provided by Telelis [33], based on analysis of Byzantine documentary sources during 300–1500 AD. According to this study, five periods with severe drought were identified: that is, 560–590, 740–790, 1020–1050, 1070–110, and 1140–1160. Domínguez-Castro et al. [34] analyzed drought events from Islamic chronicles from 711 to 1010 AD, identifying three severe droughts accompanying by severe famines and migrations: that is, 748–754 (southern Spain and Tangier), 812–823 (Al-Andalus) and 867–879 (Al-Andalus, Maghreb, and Egypt). In this context, the pro-pluvia rogation ceremonies, which are catholic rites celebrated in order to beg God for the end of drought, are among the most popular documentary proxies used to reconstruct droughts in the Mediterranean catholic countries. For example, there are 18 series of rogation ceremonies recorded in Spain, that is, Bilbao, Santo Domingo de la Calzada, Huesca, Barbastro, Calahorra, Girona, Vic, Barcelona, Cervera, Tarragona, Tortosa, Zaragoza, Teruel, Zamora, Toledo, Zafra, Sevilla, and Murcia [18, 35–39]. Domínguez-Castro et al. [17, 40] employed the Spanish rogation celebration to explore the spatial distribution of historical droughts in Spain. They reported the most extended droughts of 1664, 1680, 1750–1754, 1779–1783, and 1817. Piervitali and Colacino [41] analyzed the rogation ceremonies from Erice (western Sicily) from 1565 to 1915, highlighting the most severe droughts from 1666 to 1670. Xoplaki et al. [42] also identified severe droughts in the Balkans during 1675–1715 and 1780–1830 using documentary sources.

Drought reconstructions based on tree rings have also been developed in some Mediterranean regions (e.g., Morocco [43, 44], Iberia [45], Turkey [46, 47], and northwestern Africa [48, 49]). Other reconstruction efforts cover most of the Mediterranean area. For example, Brewer et al. [50] reconstructed summer Palmer Drought Severity Index from 1500 to 2000 using tree rings from 136 sites spanning the Mediterranean basin. Similarly, Nicault et al. [51] reconstructed the self-calibrated Palmer Drought Severity Index from April to September for the last 500 years, based on a network of 165 locations. Cook et al. [52] developed the longest summer sPDSI reconstruction of the Mediterranean region from 1100 to 2012. Following this study, the Mediterranean witnessed persistent and extended drought events during 1237–1253, 1312–1329, 1351–1366, 1856–1881, and 1942–1953. Cook et al. [52] stressed that recent droughts must be seen within the natural variability of the last 900 years. Exceptionally, the Levant 1998–2012 drought, mostly represented in Cyprus, Israel, Jordan, Lebanon, Palestine, Syria, and Turkey, is the driest on record over the past 900 years.

9.2.2 Spatio-temporal variability, mechanisms, and recent trends in climatic droughts

Climate in the Mediterranean is highly variable over space and time, due to the influence of latitude, topography, combined with diverse maritime, oceanic, and continental influences. In the era of climate change, the Mediterranean has frequently witnessed several drought events, making it one of the most affected regions worldwide [53]. In their assessment of drought trends in the Northern Hemisphere, Caloiero et al. [54] indicated that the Mediterranean basin and North Africa are among the regions in which more frequent and severe droughts have been recorded in the last decades. Recently, several drastic dry episodes have been detected in the Mediterranean region, mainly in 1991, 1994, 2002, 2003, 2005, 2006, 2012, and 2014, among others [55–57].

Several indicators and techniques have been utilized for quantification of drought conditions in the Mediterranean. Trend assessment shows very different results among the Northern, Eastern, and Southern Mediterranean. This can be expected given that drought occurs over large spatial domains, with different climatic zones [58]. Also, recalling the wide web of environmental, social, and economic impacts of drought, the vulnerability to drought can vary considerably from one region to another and among seasons. Moreover, the response of any region to drought can differ significantly, as a consequence of the time scale at which drought is assessed [59]. Overall, numerous studies indicated that drought variability displays a significant trend in the Western and Northern Mediterranean, mainly the majority of Iberia, besides Greece, Anatolia, and the Balkans [52]. While the increase in drought severity has been verified at the Western and Northern Mediterranean scale [55, 60], it has also been confirmed for specific regions such as the Iberian Peninsula [60, 61], Greece [62], France [63], Turkey [64], and even subregions like the Ebro valley, Spain [65], Southern Italy [66], Apulia region, Italy [67], and central Greece [68]. For example, in the Iberian Peninsula, Greve et al. [69] concluded that the Iberian Peninsula has suffered from severe and long drought events during the last decades.

As Mediterranean droughts are mainly linked to lack of precipitation as well as high atmospheric evaporative demand, a range of studies have assessed the links between drought characteristics (e.g., frequency, duration, intensity) and some relevant climatic drivers, especially rainfall, atmospheric evaporative demand, and air temperature. For example, Vicente-Serrano et al. [60] found that drought severity over the Southern Mediterranean in the past decades can partially be explained by the increase in atmospheric evaporative demand induced by temperature rise, especially during summer season. Also, Teuling et al. [70] linked evapotranspiration with drought episodes in the Western Europe. More recently, Páscoa et al. [61] attributed the strong spatial variability in drought characteristics (e.g., duration, frequency, intensity) over the Iberian Peninsula to variations in precipitation trends, mainly positive trends in northwestern areas of the peninsula and negative trends in southern regions.

Drought variability in the Eastern and Southern Mediterranean is generally poorly understood. Due to data limitations, less information at detailed spatial and temporal scales is available for drought characterization, attribution of its underlying processes, and assessment of associated impacts. In specific, the region suffers from paucity and scarcity of observational climate network, with few high-quality records that extend back several decades. This shortcoming is combined with poor installation and maintenance of meteorological instruments, besides a series of political and socioeconomic instability (e.g., Libya, Algeria, Egypt, Lebanon, and Syria). All these common challenges together constrain a more complete understanding of the Southern and Eastern Mediterranean drought variability. Taken together and in contrast to the Northern Mediterranean, few attempts have been made to describe and analyze the spatial patterns and trends of drought in the Southern Mediterranean. The majority of these studies were conducted at the national scale, employing relatively short instrumental records and unevenly distributed meteorological stations. Based on a comparison of three different drought indices (SPI, SPEI, and the standardized streamflow index–SSI), Turco et al. [71] reported an increase in drought conditions over Israel from 1980 to 2014. A similar finding for Israel is found by Kafle and Bruins [72], who attributed the increase in drought to a statistically significant increase in potential evapotranspiration, combined with a decrease in the average annual precipitation. From 2006 to 2011, Syria experienced severe multiseasonal and multiannual drought episode, with drastic impacts inducing agricultural failure, economic impact, and population displacement [73, 74]. Cook et al. [75] reported that the Levant recorded its most severe drought episode on record during the last two decades (1998–2012). This drying is more linked to warming trend and anthropogenic influences [52, 74].

Temporarily, the frequency and intensity of droughts have increased significantly in the Mediterranean since 1950 [60]. Nonetheless, the Mediterranean showed its highest drought frequency and severity from 1990s onward, with the 2000s recording the most anomalous dry events [52, 57]. This trend can be linked with the location of the Mediterranean in a transitional zone between extra-tropical and subtropical influences. As such, any slight changes in the general circulation patterns can induce significant changes in the Mediterranean climate including drier conditions [76]. For this reason, several studies endeavored to link drought variability with changes in large-scale teleconnection patterns, including changes in the atmospheric waves, storm tracks, jet stream location, and intensity. Numerous studies linked drought variability in the Mediterranean with changes in the North Atlantic Oscillation (NAO) [77]. According to Mariotti et al. [78] and Cook et al. [52], the positive phase of the NAO is favoring for the incidence of drought in vast areas of the Northern Mediterranean (e.g., Spain, Greece, Turkey), while it is directly related to more humid conditions in the Southern Mediterranean (e.g., Libya, Egypt, Levant). Shohami et al. [79] indicated that changes in atmospheric conditions during winter and the transitional seasons support drier conditions in the

Eastern Mediterranean. Other investigations concluded that the dry conditions detected in the Southern Mediterranean could be associated with the Sahel drought, which is related to the response of the African summer monsoon to oceanic forcing, amplified by land-atmosphere interaction [80]. Some studies attributed the increase in drought severity over the Mediterranean to the poleward expansion of subtropical dry zones [81]. Páscoa et al. [61] linked drier conditions in the Mediterranean with global ocean warming, particularly the increase in sea surface temperature (SST), besides other external forces such as anthropogenic greenhouse gas and aerosols. A similar finding is also confirmed by Hoerling et al. [55], but using climate model outputs.

Overall, while drought trends are well established in the northern and western regions of the Mediterranean, the available results on drought behavior in the eastern and southern regions should be seen with particular attention, due to lack of direct observations. In a region with complex climatology like the Mediterranean, the low density of data can hinder any attempt to properly characterize the spatial and temporal complexity of droughts. With the multidimensional character of drought, detailed spatial and temporal information is desired to properly capture these dimensions.

9.2.3 Hydrological droughts: Physical and human drivers

In contrast to climatic droughts that depend only on climate variability, water uses and demands, land transformations, and water regulation strongly determine the severity of hydrological droughts [82]. One of the main objectives of water regulation is to reduce the severity of drought impacts on the economic system and the society by maintaining water supply (for irrigation demands, urban supply, etc.) during drought periods [23].

Hydrological droughts can be measured using different metrics, mostly based on streamflow [83, 84] and, more recently, on groundwater levels [85]. Some authors use daily streamflow records, from which a fixed or variable low flow threshold is selected (i.e., a certain percentile calculated over the entire historical series) [86, 87]. This approach has already been used in the Mediterranean region by Giuntoli et al. [7] and Floriancic et al. [88], among others. Another common approach is to use monthly streamflow data, from which standardized drought index can be calculated. Although these standardized indices are sensitive to the presence of temporal inhomogeneities caused by changes in water regulation and management (e.g., damming), their main advantage is that they allow for direct comparison of drought severity over both space and time. Among the widely recognized standardized indices is the standardized streamflow index (SSI), which has extensively been used in different drought studies across Spain and elsewhere [83, 89, 90].

Climate variability has strong influence on hydrological drought severity in the Mediterranean region. The correlation between the temporal variability of hydrological and climatic droughts is usually high, though being generally higher in near-natural catchments than in managed basins [9]. This finding has been confirmed by López-Moreno et al. [89], who indicated that hydrological droughts in near-natural catchments over the entire Ebro basin (North-East Spain) show stronger and faster response to climate drought variability, compared to basins impacted by large human management and water uses. In few exceptions, there is different response between hydrological and climatic data due to the lithological characteristics [9]. Hydrological droughts respond differently to the variability of different climate variables. This has been recently analyzed by Yang et al. [90] at the global scale, who suggested that streamflow is more sensitive to temporal variability of precipitation than to the atmospheric evaporative demand (AED), albeit with more influence of AED on surface water resources in some regions worldwide. In the Mediterranean region, AED exhibited significant increase in some regions, such as the Iberian Peninsula [91], Romania [92], and Slovenia [93]. This is mostly caused by the increase of vapor pressure deficit driven by temperature rise and stationary atmospheric humidity [94]. In the Mediterranean, recent studies suggested an enhanced severity of drought events during the last decades, mostly associated with changes in the AED [59, 95]. In addition to natural climate change, anthropogenic forcings also contribute significantly to drought severity in the Mediterranean [55, 96]. Vicente-Serrano et al. [60] employed a large network of both near-natural and managed basins to assess the association between climatic and hydrologic droughts in the Iberian Peninsula between 1961 and 2012. Results demonstrate that hydrological droughts mostly respond to the variability of precipitation and—to a less extent—to AED variability and trends. The impact of AED on hydrological droughts is more pronounced during summertime, due to higher AED magnitudes as well as higher water requirements by natural vegetation and irrigated crops.

Hydrological drought trends have already been characterized in different regions of the Mediterranean, including the Iberian Peninsula and France [60, 97]. Fig. 9.1 illustrates an example of the evolution of SSI in two rivers of Southeast France and Northeast Spain. As depicted, there is an increase in the hydrological drought events over the last decades.

Nevertheless, while climatic drivers contribute significantly to variability and trends of hydrological droughts, land processes (e.g., land cover changes, water regulation, and demand) play an important role in explaining recent hydrological drought in the Mediterranean. In specific, the Mediterranean has been impacted by rapid and complex land cover changes during the past decades. These changes are mostly driven by the progressive and generalized abandonment of crops and pasturelands that have conducted to a natural revegetation process, often fueled by the establishment of pine plantations [23]. Due to these changes, shrubs and forests have nowadays become the dominant land cover types in the Mediterranean, with a strong water demand that would generally induce a decrease of streamflow in the region [23, 98]. Moreover, natural vegetation increase is generally consistent with increased AED in the region. This contributes to a stronger

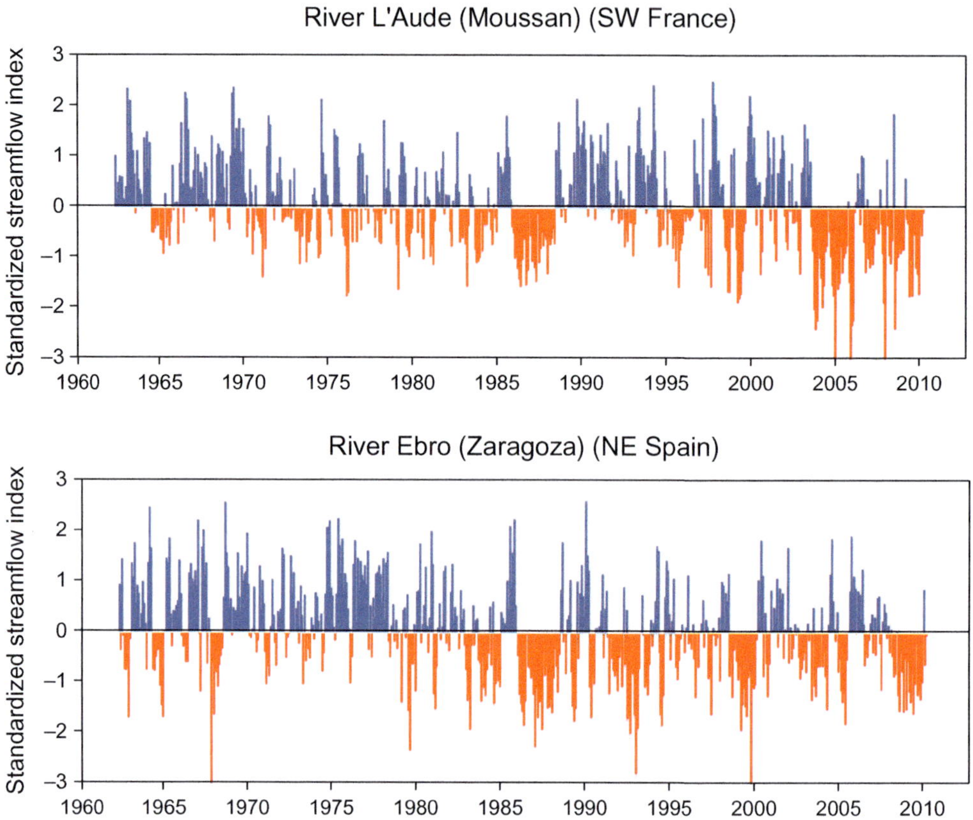

Fig. 9.1 Evolution of a hydrological drought index: the standardized streamflow index in two rivers of Southeast France (L'Aude) and Northeast Spain (Ebro).

actual evapotranspiration, and accordingly an increase of water consumption by vegetation against the surface runoff.

In addition to climate variability and land cover changes, other factors also contribute significantly to changes in hydrological droughts. In particular, water management practices are associated with a higher consumption of water by different sectors (e.g., irrigated agriculture, domestic, industry, etc.). In particular, water withdrawn by irrigated agriculture have been much more important than the consumption by natural forests [99, 100]. Regulation has altered river regimes in a high percentage of basins across the Mediterranean region [101–103]. Reservoirs have secured water supply for irrigated lands and urban/tourist needs. Nevertheless, a reduction of the socioeconomic impacts of droughts can induce an increase in the severity of low flow in the river streams. For example, Vicente-Serrano et al. [100] analyzed the evolution of hydrological droughts in a heavily regulated basin in Northeast Spain (the Segre basin), indicating that hydrological droughts upstream are mostly driven by climate drought variability, but with no obvious trends. In contrast, hydrological droughts exhibited a significant increase downstream of the main reservoirs and over large irrigated areas of the basin. According to this study, the flow routed to the irrigation channels strongly increased over the last decades [99], as compared to the decrease of the river flow. Under current increase of the AED in the region, the direct evaporation from the reservoirs is also of considerable importance [104], as it contributes to the increase of hydrological droughts severity downstream.

In some cases, changes in hydrological drought conditions are mostly determined by water transfers to other basins. A representative example is the headwaters of the Tagus river (central Spain), which are heavily regulated to secure water for irrigation and urban/tourism supply purposes. Hydrological droughts downstream this reservoir have been more frequent and severe, as a consequence of maintaining water amount for transfer purposes [105]. Transboundary issues are also a factor to be considered, with impacts on hydrological drought conditions downstream. The big transboundary basins shared between Spain and Portugal are an excellent example. In particular, the Alcántara dam, the second biggest in Europe located in the border between Spain and Portugal, contributed to the increase of hydrological droughts downstream due to dam management for hydroelectric purposes [106].

Finally, other related climate change processes may have an important influence in recent decades. For example, snow cover has strong hydrological importance in Mediterranean basins [23, 107–109]. The reduction of the snow pack identified in some Mediterranean mountain areas [110–112] and the changes in the snowmelt timing may also affect changes in the hydrological drought frequency and severity since they are affecting not only the timing of the discharges but also their magnitude as observed in recent studies in Spain, Italy, and France [109, 113–115].

Overall, hydrological drought processes are strongly complex over the Mediterranean basin, being impacted by diverse factors, such as climate variability and change, land cover changes, water demands, river regulation, and water management issues. Under future scenarios, these processes are likely to be much more severe. For example, following a process of natural succession, vegetation is expected to increase in the headwaters, with more dense forests [23]. The hydrological influence of these changes can be also enhanced by the projected climate change [11]. Climate droughts are also likely to increase in the future over the Mediterranean basin, as a consequence of the projected decrease in precipitation and an increase of AED [116, 117]. Taken together, coupled climatic and hydrological models that consider vegetation scenarios suggest a generalized increase of the frequency of hydrological droughts, albeit with a degree of uncertainty [118].

9.2.4 Drought planning and management

Drought events caused important losses in the Mediterranean. For example, drought episode in Morocco in 2000 caused losses of US$900 million [119]. As climatic model outputs suggest an increase of droughts in the Mediterranean under future climate simulations, the socioeconomic and environmental costs of frequent and severe droughts could be much higher. This situation stresses the need to improve current strategies and policies for drought planning and management.

Any targeted strategy for drought management in the Mediterranean should be "effective" so that it can minimize the risk of drought events and reduce their corresponding impacts. To archive this objective, Wilhite [120] proposes four key components: (1) the availability of timely and reliable information on which management and policy decisions are made; (2) policies and institutional arrangements that encourage assessment, communication, and application of that information; (3) a suite of appropriate risk management measures for decision makers; and (4) effective and consistent actions by decision makers to support a national drought strategy. We note that the concept of drought risk here is a multiplication of the concepts of hazard × vulnerability × exposure.

The objectives of a drought plan must also take into account the particular characteristics of the analyzed region (e.g., physical, environmental, socioeconomic, and political).

In the Mediterranean, there are different levels of implementation of drought policies, plans, and management initiatives. The northern Mediterranean has common water management politics elaborated by the European Commission (EC). In 2000, the EC adopted the Water Framework Directive (WFD) 2000/60/EC that provides the river basin perspective to the European water policy and management, promoting the development of river basin management plans (RBMPs). In 2007, the Communication "Addressing the challenge of water scarcity and droughts in the European Union" underlined that water saving must be the priority, improving water efficiency prior to increasing water supply, and recommended the development of drought risk management plans (DMPs) [121]. In 2007, the EC also published the "Drought Management Plan Report, including agricultural, drought indicators, and climate change aspects" to guide governments in the development of DMPs. In 2011, the Joint Research Centre of the European Commission created the European Drought Observatory (EDO) with the objective of improving the knowledge of droughts, its monitoring, detection and forecasting. The EDO provides drought indicators from different data sources, such as meteorological stations, satellite, or models to monitor drought. Spain was one of the first European countries in implementing the EC drought policies. It transposed the WFD into the Spanish law in 2003 and in 2016, approving the second cycle of their RBMPs for most of their river basins [26]. DMPs were also approved in 2007 as specific plans of the RBMPs. The Spanish DMPs included a drought diagnosis, a program of related measures associated with each drought status, and a management and follow-up system [122, 123].

Similar to the Northern Mediterranean, the Southern and Eastern Mediterranean countries have a large diversity of water policies, which place more focus on the recovery from drought crisis rather than the prevention of drought impacts. While current global drought risk management policies account for risk emergency, emphasizing on actions taken during drought crisis either for providing food and feed or for creating jobs in rural areas [124], few Southern and Eastern Mediterranean countries have early warning systems, drought monitoring systems, or national drought plans. Among these countries, Morocco is an advanced country in the area of drought risk management, as it has a National Program for Drought Mitigation with two orientations: a reactive program in the short term and a drought mitigation planning in the long term. In 2001, Morocco created the National Drought Observatory, dependent on the General Secretary of Ministry of Agriculture and Rural, with the objectives of generating institutional capacity to cope with drought, and coordinating the cooperation among scientific community and the decision makers [124]. For the near-east Mediterranean countries, Jordan is a promising case, as it developed adequate policies to manage drought, although most of these policies are poorly implemented. For example, it has developed four water policies whose overall objective is to improve water resources management, including a groundwater management policy, an irrigation-water policy, a wastewater-management policy, and a water-utility policy. The irrigation-water policy considers the development of four elements directly related to drought risk management (i) drought risk management/drought monitoring and early warning systems, plans, and policies; (ii) vulnerability assessment, drought mitigation, response actions; (iii) preparedness of plans to reduce drought impacts; (iv) drought awareness, knowledge management, and education [124]. The National Centre for Agricultural Research and Technology Transfer has developed a drought-monitoring system with information from 291 weather stations and satellite information [125].

Overall, the Mediterranean region must make a transition from an emergency-management to a planned drought risk management based on the principles of risk reduction. The European countries have a clear roadmap to implement this, given the EC droughts policies. On the other hand, this transition seems to be quite difficult in Northern Africa and the Near East Mediterranean countries, as it requires substantial changes in current government policies.

9.3 Drought tools and indices

9.3.1 Methods used for drought quantification (climate, hydrological, remote sensing indices)

In spite of being a natural hazard, drought is markedly different from other natural hazards (e.g., flooding, heat waves, earthquakes) in several ways [1]. In specific, drought is often referred to as a creeping phenomenon, as the impacts of drought can be accumulated and lagged for years after the termination of a drought event. Moreover, the severity of drought depends not only on the duration, intensity, and spatial extent of a specific drought episode, but is also impacted by anthropogenic activities. With the absence of a globally recognized definition of drought and its corresponding far-reaching impacts, it is quite difficult to identify and quantify drought. All these challenges promote scientists to look for measures that allow for defining and comparing drought conditions among regions with different climatic and environmental conditions.

During the past decades, several drought indices have been developed, based on different hydroclimatic variables. These indices allow for describing and comparing droughts among different time periods and even regions with different climatic conditions. Also, they allow for assessing and monitoring drought at different time scales and delivering climatic information to different end users. The most widely recognized drought indices are the Palmer Drought Severity Index (PDSI) [126], the Standardized Precipitation Index (SPI) [127], the Standardized Precipitation Evapotranspiration Index (SPEI) [128], the Surface Water Supply Index (SWSI) [129], the Standardized Runoff Index (SRI) [130], etc. These indices were developed based on different assumptions and using various algorithms. With these indices, hydroclimatic data are often fitted to a given statistical distribution and standardized to facilitate direct comparison over both space and time. While precipitation is the common input variable in all the indices, some indices also account for relevant variables that could impact heat and energy fluxes (e.g., air temperature, wind speed, evapotranspiration, etc.). A representative example is SPI, which considers precipitation data alone, compared to other indices (e.g., SPEI) that account for several climatic information, such as precipitation and evapotranspiration. Several drought indices (e.g., SPI and SPEI) characterize drought at different time scales, allowing for defining different drought categories (e.g., meteorological, agricultural, and hydrological). Further details about the advantages and disadvantages of different climatic and hydrological drought indices are discussed in Heim [131] and Mishra and Singh [132].

In the Mediterranean, several studies have already employed some of these indices to grasp drought characteristics. The focus of these studies is placed mainly on analyzing the spatial and temporal behavior of droughts [62, 133, 134], linking it with driving forces [77], identifying drought impacts over a wide spectrum of environmental and socioeconomic sectors [12, 100, 135], and ultimately developing appropriate drought-monitoring systems. In the same context, while some studies relied on a single index, such as SPI [136, 137], other studies compared the performance of two [60, 61, 77] or multiple indices [57]. For example, Lloyd-Hughes and Saunders [138] assessed drought climatology in the whole Europe using a high spatial resolution and multitemporal data set for the 20th century. They calculated the monthly values of the SPI on a 0.5° grid for the period 1901–99. Sousa et al. [77] analyzed the spatial and temporal characteristics of drought in the Mediterranean during the 20th century using two indices: PDSI and self-calibrated PDSI (scPDSI). Garrote et al. [139] proposed a methodology to associate operational drought indicators with policy management actions for the Tagus River Basin in Spain. Also, Vasiliades et al. [68] evaluated hydrological droughts at subwatershed scale in the Pinios river basin in Greece using a water balance derived drought index within an operational context.

With the advancement in space-based sciences, many studies have employed remotely sensed data to assess and monitor droughts in the Mediterranean. Satellites can provide relatively long-term (mostly date back to the early 1980s) hydroclimatic information at fine spatial scale [140]. These data are valuable especially in data-scarce regions like the Southern and Eastern Mediterranean, as they can be used as surrogate or supplement to ground-based observation of meteorological and hydrological drought indices and provide an assessment of drought impacts as recently illustrated in a long-term (35 years) high-spatial resolution study in Spain [141]. Along the classical climatic and hydrological drought indices, new drought indices were developed using remote sensing of vegetation, such as the normalized difference vegetation index (NDVI), temperature vegetation drought index (TVDI), and vegetation condition index (VCI). Numerous studies linked drought with space-based vegetation indices in the Mediterranean [71, 142–144]. For example, Gouveia et al. [143] employed remote sensing data and a multi-scalar drought index (SPEI) to assess the impacts of drought on vegetation over the entire Mediterranean basin. More recently, Vicente-Serrano et al. [145] evaluated a detailed (1 km) spatial assessment of the links between drought and vegetation across Spain over the past four decades using SPEI and NDVI.

9.3.2 Operative tools for drought early warning and management

Drought monitoring and forecasting are essential tools for implementing appropriate mitigation measures to reduce negative impacts of droughts. In the changing climate, there is an urgent need to have effective policies and management systems to cope with the prolonged impacts of drought, which can influence both natural and human environments (e.g., water scarcity, harvest losses, desertification, forest fire, etc.). Recently, severe droughts in various Mediterranean regions have illustrated the vulnerability of the region to climate change [55, 146]. The widespread and intense droughts in the Mediterranean increased the official and public awareness of the impacts of drought on local communities. In this regard, the heat wave and drought of 2003 over Southern and Western Europe was seen as a wake-up call for Europe. This awareness has been translated into a wide range of national and multinational initiatives, which aim at mitigating the possible impacts of drought and efficiently improving current response policies and actions. These efforts are oriented primarily to advance operative tools and preparedness plans for drought forecasting and early warning.

Scientifically, many research projects have been funded to characterize and manage droughts in the Mediterranean, including—among others—Assessment of Regional

Impact of Droughts in Europe (ARIDE), Dryness and Turning into a desert in the Mediterranean basin project (SEDEMED), Proactive Management of Water Systems to face drought and water scarcity in islands and coastal areas of the Mediterranean (PRODIM), Climate Change and impact research: the Mediterranean environment (CIRCE), Mediterranean Intermittent River Management Project (MIRAGE), Drought Management Guidelines and Examples of Application (MEDROPLAN), and Water and Global Change project (WATCH). These projects employed climatological and hydrological information from different sources and products (e.g., instrumental, reanalyses, model, and remote sensing) to characterize drought episodes in the Mediterranean, develop risk indicators, and compare and prioritize different measures and indicators to alleviate the regional impacts of droughts. For instance, the MEDROPLAN developed guidelines for drought preparedness plans in the Mediterranean countries, providing a framework to minimize the impacts of droughts in the region [147]. In general, these projects have increased the knowledge on droughts in different research areas and regions in the Mediterranean, providing additional tools and experiences for policy makers. These projects enable at identifying research gaps on various drought aspects (climate, hydrology, impacts, management, and policy) in the Mediterranean and establishing the state of the art of drought-related policies.

Technically, more efforts have been made in the Mediterranean to develop operative tools to forecast and mitigate drought. Nonetheless, planning for drought mitigation needs timely and reliable climate information to help decision makers in making appropriate decisions. Unfortunately, this information is missing in many Mediterranean countries, especially in the Southern and Eastern Mediterranean, where climatic and hydrologic information is generally coarse over both space and time. In contrast, the Northern Mediterranean has more comprehensive, integrated national climate monitoring and drought early warning system. These systems were developed based on dense data, which are mostly available for public and research community. For example, the DMPs was launched in Spain in 2007 as a strategy for all river basins, with the aim of forecasting droughts and finding proper solutions that satisfy priority demands and reduce socio-economic and environmental impacts of drought [139]. The Spanish DMPs developed indicators for a quick and proper forecasting of drought. The management policies consider evaluating available resources and defining strategies to keep them sustain during anomalous dry events by developing structural and nonstructural alternatives to reduce vulnerability and impacts of drought [148]. In Turkey, the government adopted strict policy and applied several measures to diminish the impacts of frequent droughts and water stress in the country, including the renewal and modernization of traditional irrigation systems, installing water flow meters, averting unregistered wells, the construction of surface water storing structures, pricing of groundwater used in agriculture, and raising public awareness for saving water [149].

In 2007, the European Union (EU) has also developed policy tools (e.g., Water Framework Directive, WFD) to enhance communication and networking on issues related to water scarcity and droughts and to address the common challenges of droughts in the EU, including the Mediterranean. More recently, within the EU 2020 Strategy and the Resources Efficiency Roadmap, a policy review on water scarcity and droughts was implemented and integrated into the A Blueprint to Safeguard Europe's Water Resources. The technical initiatives developed by the EC to address drought issues have listed different technical measures to cope with water scarcity and droughts, highlighting the importance of adopted planned drought risk management approaches that guarantee public involvement in drought-related decisions. In addition to capacity building strategies of public audience, social groups and NGOs in several Mediterranean regions are directly involved and networked with ministries, agencies, and institutes in decision-making of drought-related issues (e.g., agriculture, water management, rural development, etc.). In many cases, there has been institutional and financial support from several cross-borders entities like the United Nations Environment Programme—Mediterranean Action Plan (UNEP/MAP), UNESCO, the World Bank, and the Global Environmental Fund (GEF) among others. In addition to the national measures to mitigate drought risk in the Mediterranean, there are also some promising local and subregional plans that adopt specific measures and tools to forecast and monitor real-time drought at a detailed spatial scale. For example, Domínguez-Castro et al. [150] have developed a detailed (1 km) probability maps for Spain that allow for detecting drought episodes of a certain severity. Recalling the multidimensional character of drought, such promising tools can contribute significantly to better sectorial management strategies. In practice, these tools can alter the mode of response of many local governments to drought from reactive to proactive. The reactive response to drought is often technically and economically inefficient, with little room for the evaluation of alternatives and the participation of stakeholders [151]. Nonetheless, albeit with these advancements, much effort is still desired to promote current policies and plans. In the Mediterranean, there is no regional directive to approach and assess drought effects over the entire region; most of current legislative processes, such as specific communication, follow-up reports and research efforts, are conducted at the national level, with few attempts to comprehensively develop long-term drought preparedness policies and plans of actions that can significantly reduce the risks and vulnerabilities to drought, especially at the local and subregional scale. As such, much effort is still desired to build prides and propose initiatives that facilitate the exchange of "efficient" practices between the Mediterranean countries and develop adaptation mainstreaming into appropriate policies and strategies. The cross-border entities (e.g., UNESCO, UNDP) can participate in initiatives, projects, and research activities to deliver knowledge, exchange experiences, and strengthen the cooperation between the Mediterranean countries.

9.4 Drought impact assessment

9.4.1 Agricultural impacts

Agricultural lands occupy c. 50% of the territory in the Mediterranean, being the main consumer of water in the region (60%–80% of water supply). As such, agricultural sector is expected to be more sensitive to any changes in drought characteristics (e.g., frequency, intensity, duration). The increase of drought frequency can threaten agricultural production and food provision in this water scarce region [152]. Crop production is very relevant for local consumption and exportation in the Mediterranean region, contributing to most of the global olive oil yield, 60% of wine yield, 20% of citrus crop, and roughly 12% of cereal crops [153]. Therefore, climate warming and water scarcity can aggravate or create conflicts among economic sectors, regions, basins, or countries since demand from agriculture is projected to rise. For instance, water scarcity problems are recurrent in the semiarid regions of south-eastern Spain (e.g., Turia and Júcar basins), where climate warming, intensive agriculture for exportation, and tourism will add further challenges to water management in this region in the future. This tension can partially be alleviated by increasing cooperation of implicated agents or by improving the efficiency of irrigation systems which, for instance, consume ca. 75% of water in Spain [154].

Models and experiments under controlled conditions suggest that increasing levels of atmospheric carbon dioxide may increase the yield of major food crops corresponding to C3 plants (e.g., wheat) [155]. This is simply due to higher rates of photosynthesis and improvements in water-use efficiency [156]. There is, however, considerable debate about whether or not these expected benefits would alleviate the negative impacts of droughts and climate warming on agriculture. These negative impacts may include a northward shift of crop growing zones. Some models that considered some adaptation of crop varieties suggest that yield losses due to increased drought (e.g., from −20% to −50% for wheat) would occur throughout Southern and Eastern Mediterranean region [157]. Nonetheless, some unmodeled variables should also be considered, particularly climatic variables related to extreme events (e.g., heat waves, frosts), climate variability affecting crop phenology, soil features, competition for water supplies, changes in the distribution, and virulence of pests and diseases, etc. [158].

Recently, agricultural trends have witnessed abrupt changes, due to variations in global food demand and production, new agricultural products (e.g., biofuels, ecological products), price instability, and political controls. However, while agriculture is subsidized in most of the Northern Mediterranean countries that correspond to the EU [159], this is not the case for the Eastern and Southern Mediterranean. In the same context, the Northern Mediterranean is a self-sufficient region in terms of food requirements, meanwhile the Eastern and Southern Mediterranean import c. 40% of their food [153]. Aridification will thus threaten food security in the less-developed Mediterranean countries, where a high population growth is projected, combined with rapid land-use changes and water use complications. However, at the national level, the possible impacts of drought on water use largely depend on economic progress and adaptive capacity. For example, specialized agri-food sectors (e.g., strawberry or truffle production) can contribute significantly to national economies, as they consume less water compared to traditional agriculture.

However, from a long-term perspective, the Mediterranean agriculture has shown a great adaptation capacity. In specific, the enhanced irrigation and fertilizers may increase agricultural yields to some degree. Nonetheless, low water and nutrient availability can constrain the production of major crops in the Mediterranean (e.g., wheat) [160], with little margin to improve its yield. In addition, the more intensive and water-demanding agriculture, as a consequence of climate warming, aridification, and the increase of human population, will negatively impact biodiversity and the functioning and resilience of terrestrial and water ecosystems. These ecosystems suffer already from rapid changes, including land-use alterations (e.g., abandonment of traditional uses of forests in European countries, forest and scrubland encroachment, etc.) [161]. There will be a higher demand for irrigated lands to safeguard food production and maintain crop stability and security [162]. Irrigation demands in the Mediterranean region are projected to increase up to 18% during the 21st century if climate warms by +5°C. Two representative examples of the possible impacts of future irrigation requirements on the Mediterranean crops include olive and grapevines. First, climate models indicate that irrigation requirements will increase for olive because it will be impacted by hotter droughts [163]. Second, warming would lead to an earlier bud bursting or flowering of grapevines. This will extend the growing season of grapevines and expose it to longer dry periods that will finally modify the quality of both grape and wine [164], even changing the French concept of terroir [165]. Although the negative impacts of the projected warming and intensified drought on the Mediterranean agriculture can vary from region to another, there will definitely be a need to increase available water resources to diminish the impacts of water shortage on food security.

9.4.2 Environmental impacts

Temperature rise and precipitation deficit accelerate aridification trends in the Mediterranean, with negative impacts on terrestrial ecosystems [143]. For example, the total area of forests in the Mediterranean is about 88 million hectares, mainly located in the Northern Mediterranean [166]. In spite of its theoretically increasing capacity to uptake and store carbon, the Mediterranean forests are frequently subjected to droughts and fires that modulate and even reverse this trend. In the Northern Mediterranean, land abandonment and forest encroachment increase wildfire risk and the competition between trees for water. On the other hand, overexploitation, overgrazing, and population pressure are the major threats to forest conservation in the Southern Mediterranean.

Warming and drier conditions will significantly alter the composition, dynamics, cover, and distribution of Mediterranean forests. These changes will also affect the biodiversity and related ecosystems services. This is basically because forests affect many protected animals, plants, and fungi species and regulate carbon and water cycles, among other functions. Droughts reduce growth and carbon uptake, thus altering hydraulic conductivity and carbon storage, which affect tree health and resistance to pests (e.g. defoliators, bark beetles). Drought has already triggered dieback episodes, affecting fir, oak, and pine forests (see Fig. 9.2) and also juniper scrublands in Spain, France, Italy, Greece, Israel, and Atlas cedar in Morocco and Algeria [14, 167–173]. Taken together, it is well evident that drought has significantly increased the vulnerability of many Mediterranean forests during the past few decades [174].

Mediterranean woodlands include many semiarid forests, which are key determinants of the interannual variability in the carbon sink capacity of terrestrial ecosystems [175]. However, droughts negatively influence the growth and health of the Mediterranean

Fig. 9.2 Image of a dead Scots pine (*Pinus sylvestris*) in a stand impacted by the 2012 drought (Corbalán, Teruel, Spain). Note how other species as junipers (*Juniperus thurifera*) survived during this drought episode.

forests forming the xeric or geographical (often southernmost) distribution limits (rear edge) of the species [14, 173]. Pine plantations are very vulnerable to drought stress, as they were often carried out during wet decades (e.g., 1970s) or in soils with maladapted trees to the frequent droughts of the late 20th century and early 21st century [176]. Although fertilization effects, due to rising CO_2 levels, are controversial in the Mediterranean forests, current paradigms consider drought as a major constrain of tree growth and forest productivity [177]. In particular, the long-term increase of water-use efficiency, caused by higher carbon availability, does not necessarily imply a growth improvement, especially with growth decline due to drought. In this context, it is still unknown how drought can interact with other major drivers of forest dynamics in the Mediterranean, including, for instance, wildfires [178] and pests [179]. Warmer and drier growing seasons could synergistically interact and amplify the negative impacts of fires and pests on the Mediterranean forests.

The negative impacts of droughts on the Mediterranean ecosystems, particularly forests, will compromise many of the services they provide. This includes natural (e.g., timber, cork, grasslands, or fungi), environmental (e.g., biodiversity, regulation of climate, soil and water cycles, carbon storage), and social services (e.g., recreational uses, traditional cultural values). The millennial relationships established between the Mediterranean societies and their environments, particularly forests, are likely to be disrupted in the future by droughts and climate warming.

9.4.3 Ecological and hydrological impacts

The main ecological impacts of droughts in stream ecosystems are the loss of habitat for aquatic organisms [180], the loss of stream connectivity, changes in the interspecific interactions, the structure of food webs, densities and size of populations, and accordingly the composition and diversity of local communities [181–184]. A representative example of the impacts of drought on food web is the decline of otter populations during drought episodes in the Mediterranean streams, which is mainly driven by the reduction of fish during these episodes [185].

Wetlands under Mediterranean climate are very fragile ecosystems. During dry conditions, the amount of water is reduced in these ecosystems, inducing changes in trophic dynamics, biodiversity loss, increase of invasive species and smoldering peat fires, and an activation of degradation processes such as eutrophication [186, 187]. Las Tablas Daimiel National Park (Central Spain) is a good example of the active degradation due to drought, enhanced by an intensive groundwater extraction. This Ramsar wetland has suffered from great deterioration and scarcity of water, which induced eutrophication, vegetation, and faunal changes [188], besides smoldering peat fires [189]. Unfortunately, this wetland is still functional only due to artificial inflows [190].

In the Mediterranean shallow lakes, the water balance is one of the key factors in the apparition of eutrophication and the associated ecological changes: a process that can be accelerated by drought [191, 192]. On the other hand, deep lakes are also vulnerable to drought impacts, although these impacts are mainly located in littoral areas. Kinneret lake (Israel) has witnessed a decrease in water table since 2013 [193], which corresponded to a reproduction of several fish species, a reduction of the littoral rock areas [194], and changes in nutrient dynamics [195].

The ground waters in the Mediterranean have also exhibited rapid changes in the last decades, mainly driven by climate variability as well as anthropogenic influences. In the Mediterranean, there is a high demand for ground water, especially in overpopulated settlements, to meet the urban, industrial, agricultural, and tourism needs. This pressure unbalanced the abstraction and natural recharge of the system. For example, in the south-eastern Spain, the abstraction (700 hm^3/year) is almost triple the estimated recharge (200 hm^3/year) [196]. With these unbalanced situations, drought periods can exacerbate the problems of the overexploitation, seawater intrusion, and water quality deterioration [197].

Alongside the direct impacts of droughts on rivers, wetlands, lakes, reservoirs, and ground waters, the decrease of water in the fresh water ecosystems can also affect other economic sectors. For example, water decrease in rivers and reservoirs can force to stop hydraulic and thermal power plant (due to lack of water for cooling). In Spain, the drought of 1990–1995 caused a loss of 210 million € in the energy sector. Also, recalling that the Mediterranean is one of the preferred tourism destinations of the world, accumulating around the 20% of the total international tourism [198], tourism sector can also be impacted by drought episodes. In specific, tourism can increase water demand and accordingly place more pressures on the limited water resources, especially during the dry season. Drought can trigger a serious water crisis in tourism destinations that can cause potable water restrictions, which may lead to an early tourist return to home, cancellations in hotel bookings, and finally harm the image of the tourist destinations: for example, Benidorm 1978 [199], Cádiz 1990s [200], and Barcelona 2007/2008 [201]. Economic impacts of droughts in the tourism sector are difficult to quantify but the drought of 2007/2008 in Barcelona had an estimated total cost of 1605 million Euros [201].

9.5 Integral assessment of the 2017 drought event over Spain: Evidence, impacts, and management strategies

Drought affected large areas of the Mediterranean during the hydrological year of 2016–17, more pronounced in summer of 2017. This mega drought trigged vast areas of the Mediterranean (e.g., Italy, France, the Iberian Peninsula, Greece, Morocco, Algeria). This drought episode was attributed to several atmospheric configurations,

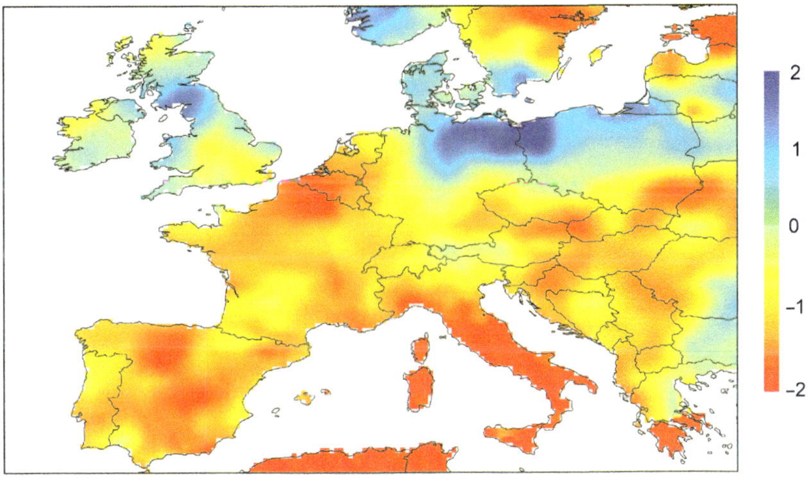

Fig. 9.3 Spatial distribution of the 6-months Standardized Precipitation Evapotranspiration Index (SPEI) in July 2017.

including circulation anomalies and a long lasting atmospheric stagnation that caused anomalous high temperatures, especially in the Iberian Peninsula [22]. Fig. 9.3 depicts the spatial distribution and the evolution of the SPEI over vast areas of the Northern Mediterranean during 2017, suggesting strong negative SPEI values in specific regions across the Mediterranean (e.g., Southern Italy, Southern Greece, mainland Spain, and northern France).

Vegetation and environment were strongly affected by this drought event. Chiogna et al. [202] and Marchina et al. [203] reported some of these impacts in Italy (e.g., altered hydro-chemical conditions in streams, a reduction of water volume). Tree physiology was also highly influenced, due to heat stress and water deficit [204]. Also, this drought episode caused drastic socioeconomic impacts in the Mediterranean. In Spain, the agricultural damages corresponding to this event were estimated at 2500 million € and the insurance companies needed to cover 725 million €. In the Duero basin (Northern Spain), several towns suffer from water restrictions. A similar situation was also reported in Italy, where cities like Rome had water supply problems, with water restrictions for almost 8 h per day during July. In France, thirty administrative departments were declared in a crisis situation, with frequent water restrictions and an increased frequency of forest fires.

Regarding drought risk management strategies and policies during this event, Spain, for example, responded to the strong decrease in water resources during the second half of 2017, as presented by several hydrological drought indicators adopted by water management agencies Spain recorded prealert and alert drought conditions in majority of the basins, and even an emergency state in specific hydrological units (e.g., Duero, Ebro, and Miño basins in the North, besides the Tagus, Guadalquivir, and the Segura basins) (Fig. 9.4). Alongside this economic cost, this dry situation had also important ecologic

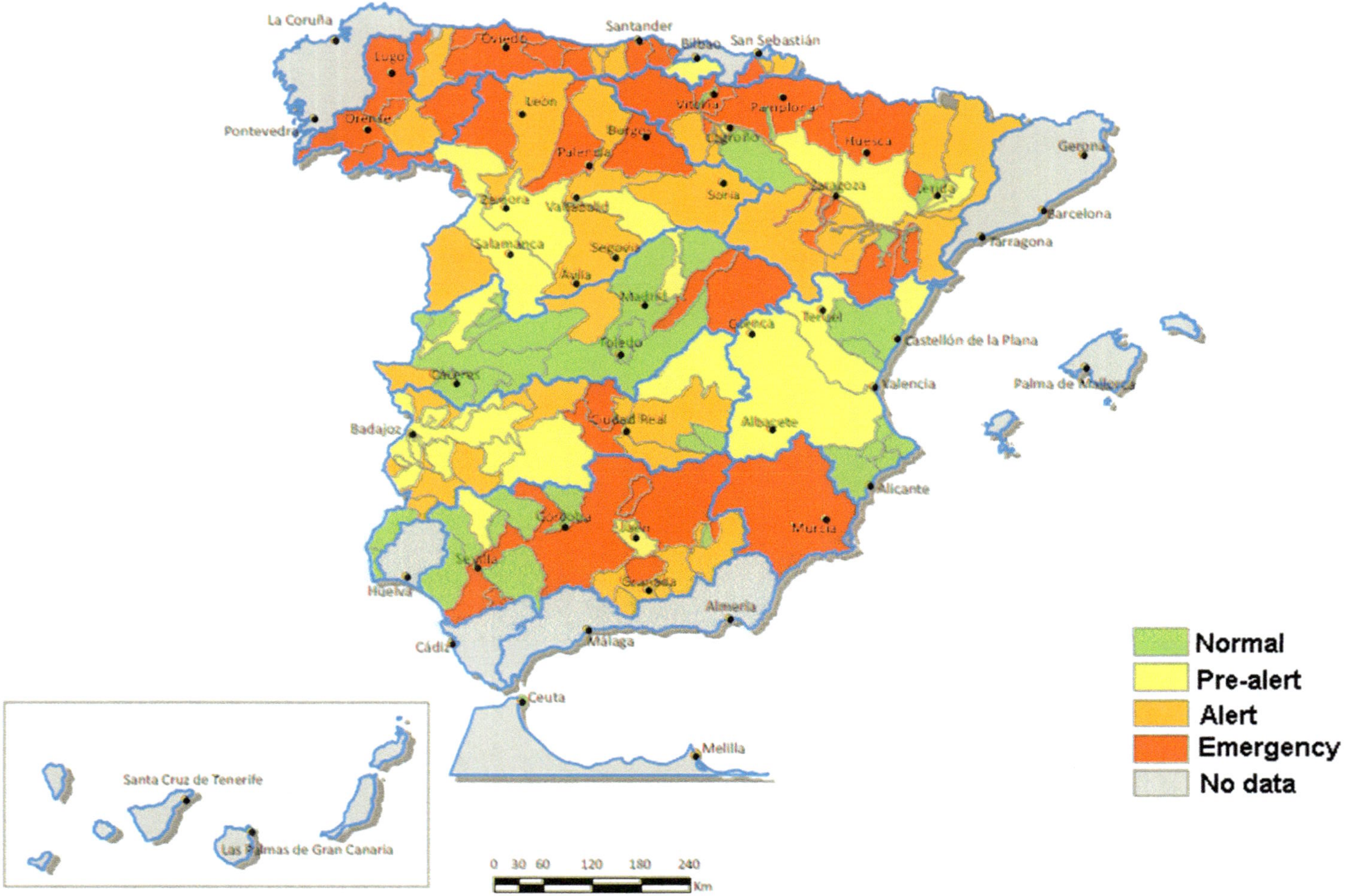

Fig. 9.4 Spatial distribution of the hydrological index in Spain in October 2017. *Courtesy of the Spanish Government: Ministerio para la transición ecológica.*

consequences in the Mediterranean. For example, the total water area of the main wetland in inner Spain ("Las Tablas de Daimiel") was reduced by almost 75%, with noticeable impacts on existing fauna. These drought conditions affected not only the arid and semiarid Spain, but extended further to the areas of Mediterranean climate as well as humid climates in the northwest.

This extreme drought situation also caused several political problems and conflicts between politicians, farmers, and environmental associations of different regions, highlighting the need to promote different assistance measures by the national and regional governments across Spain. Also, this episode illustrated the lack of prevision and the number of pressures presented to different sectors and territories, providing lessons on how to develop better preparedness and mitigation plans in order to face the probable intensification of the projected droughts in the region, as suggested by the state-of-the-art climate models [118].

In reality, the management of this drought episode was a perfect example of what is not recommended in order to manage droughts [205, 206]. The majority of available measures were mostly reactive, with less consideration of how to predict or promote the preparedness to drought. Although Spain has well-established drought plans, including drought monitoring and early warning protocols [26, 190, 191], besides a developed agricultural insurance system [24], all these measures combined were insufficient to cope with the 2017 extraordinary drought episode, providing less skill in improving the adaptation or mitigating the negative effects of this event.

These measures were mostly characterized by subsides or credits. For example, the national government published different laws in 2017 in order to establish subsides and other types of assistance to sectors largely affected by drought, mostly livestock and agriculture. While the law 10/2017 declared on 9th June 2017 stressed that agricultural insurance only covers nonirrigated lands, exceptions were also made to irrigated lands (especially wood crops) that were impacted by the extraordinary drought event in 2017. Following this action, the national government considered some measures in two of the most affected regions: Castilla y Leon and the Segura basin (Southeast Spain). In Castilla y Leon, precipitation during the hydrologic year was 55% lower than the average conditions, combined with only 34.2% of the normal water capacity of reservation during the early summer. In the Segura basin, the water reserves were only 32%, promoting an excessive use of high-cost desalinized water. In these regions, the payment of the taxes by farmers and the payments to the social security system were postponed. Alternatively, farmers received more subsidies following the European Agricultural Policy. At the regional scale, some governments (e.g., Castilla Leon) regulated credits of 920 million €, with no interest, to the farmers to cope with this situation.

The situation was very critical in the headwaters of the Tagus basin in which two large reservoirs (Entrepeñas and Buendía) were at 11.8% of their normal capacity. These two reservoirs are the origin of water transfer to the Segura basin for intensive agriculture, urban, and tourism uses. This situation induced important social and regional conflicts.

This kind of drought crisis is very difficult to manage, albeit with the presence of drought plans and strict protocols to manage these hazards once emergency is declared [26]. For example, in the case of the Tagus-Segura water transfer, the Minister council has the competence to decide the amount of water transferred to the Segura basin under emergency situations. In response to political and territorial pressures, the minister council decided to transfer 7.5 hm^3 in May 2017, which did not meet the requirements for this critical situation. This action caused a political conflict between the regional governments of the regions that transfer and receive water, besides conflicts with the central government and the water management authorities of the basin. The conflicts continued during the summer, with several agents involved (e.g., ecologists associations, irrigation exploitations, civil society, etc.), stressing the vulnerability of socioeconomic systems that depend on external water resources affected by overexploitation and strong interannual variability. In November 2017, the Ministry of Agriculture announced an alternative plan, in which the desalinization capacity of the plants located in the southeast Mediterranean coastland will increase from 40 to 120 hm^3/year to deal with the dramatic water deficits in the southeast.

While the Tagus-Segura water transfer was the most critical system that required a rapid action from the Spanish authorities in 2017, other regions were also impacted by the 2017 extreme hydrological drought, with several negative impacts. A number of initiatives were necessary to develop in order to deal with the 2017 drought from a hydrological point of view, activating the drought plans and establishing water restrictions for irrigation and water supply in these regions.

9.6 Concluding remarks

In this chapter, we have stressed the large complexity of drought processes in the Mediterranean region. This is one of the most frequent meteorological hazards that affect the region, with a wide spectrum of agricultural, hydrological, environmental, and socioeconomic implications. While all these drought types have a climatological background, other anthropogenic factors can accelerate the impacts of these drought events on the hydrologic system. Historically, the Mediterranean had witnessed several catastrophic droughts, which forced local communities to look for proper strategies and plans to cope with these events. In the instrumental era, the availability of dense and continuous meteorological records has provided evidence on the strong variability of droughts in the Mediterranean, particularly in the recent decades, as forced by temperature rise, precipitation deficit, and the increase of atmospheric evaporative demand. However, the atmospheric mechanisms that trigger droughts in the Mediterranean region are complex, which makes it difficult to propose a unique scenario for drought development in the region.

Recalling the difficulties of quantifying and monitoring droughts in general and in the Mediterranean in particular, different drought indices and metrics have been proposed—some of them based on climatic information while others included hydrological information and/or remote sensing data. Here, it should be stressed that—under the strong increase of the atmospheric evaporative demand (AED) in the region—it is highly recommended to characterize and quantify drought using drought indices that consider not only precipitation data but also the AED. Several recent drought episodes in the Mediterranean highlighted the relative importance of AED, compared to precipitation deficit, in drought evolution.

Hydrological droughts show clear combination of climatic and human drivers, including land cover changes, water demands, river regulation, and water management strategies. While climate variability is a dominant factor controlling hydrological drought variability in the Mediterranean over the past decades, human drivers are also of particular importance, especially with the increased green water use by natural vegetation and crops. Water transfers and large basins have also contributed to the increase in the frequency and severity of hydrological droughts, especially downstream the main reservoirs.

To sum up, this chapter stressed the strong complexity of managing anomalous dry events like that recorded in 2017. In a country with well-developed drought plans, monitoring systems, and adaptation strategies (e.g., a large reservoir network and an advanced agrarian insurance system) like Spain, it was extremely difficult to manage this extraordinary drought event. Under these conditions, sectorial and territorial conflicts emerged, illustrating how regional authorities needed to apply nonadaptive measures (e.g., subsides) to contribute to mitigating this phenomenon. Under future climate scenarios, the extreme hydrological droughts like the event of 2017 can be intensified and even become normal in the future, which stresses the need to develop more appropriate management strategies and adaptation plans in the Mediterranean.

References

[1] D.A. Wilhite, Drought as a natural hazard: concepts and definitions, in: D.A. Wilhite (Ed.), Drought A Glob. Assessment, vol. 1, Drought Mitigation Center Faculty Publications, 2000, pp. 3–18.

[2] D.A. Wilhite, R.S. Pulwarty, Drought as hazard: understanding the natural and social context, in: Drought Water Crises: Integrating Science, Management, and Policy, Taylor & Francis Group, CRC Press, Boca Raton, 2017, pp. 3–22.

[3] D.A. Wilhite, M.D. Svoboda, M.J. Hayes, Understanding the complex impacts of drought: a key to enhancing drought mitigation and preparedness, Water Resour. Manag. 21 (2007) 763–774, https://doi.org/10.1007/s11269-006-9076-5.

[4] S.M. Vicente-Serrano, Foreword: drought complexity and assessment under climate change conditions, Cuad. Investig. Geogr. 42 (2016), https://doi.org/10.18172/cig.2961.

[5] P. Páscoa, C.M. Gouveia, A. Russo, R.M. Trigo, The role of drought on wheat yield interannual variability in the Iberian Peninsula from 1929 to 2012, Int. J. Biometeorol. (2016) 1–13, https://doi.org/10.1007/s00484-016-1224-x.

[6] A. Scaini, N. Sánchez, S.M. Vicente-Serrano, J. Martínez-Fernández, SMOS-derived soil moisture anomalies and drought indices: a comparative analysis using in situ measurements, Hydrol. Process. 29 (2015), https://doi.org/10.1002/hyp.10150.
[7] I. Giuntoli, B. Renard, J.-P. Vidal, A. Bard, Low flows in France and their relationship to large-scale climate indices, J. Hydrol. 482 (2013) 105–118, https://doi.org/10.1016/j.jhydrol.2012.12.038.
[8] L. Caillouet, J.-P. Vidal, E. Sauquet, A. Devers, B. Graff, Ensemble reconstruction of spatio-temporal extreme low-flow events in France since 1871, Hydrol. Earth Syst. Sci. 21 (2017) 2923–2951, https://doi.org/10.5194/hess-21-2923-2017.
[9] J. Lorenzo-Lacruz, S.M. Vicente-Serrano, J.C. González-Hidalgo, J.I. López-Moreno, N. Cortesi, Hydrological drought response to meteorological drought in the Iberian Peninsula, Clim. Res. 58 (2013), https://doi.org/10.3354/cr01177.
[10] J. Lorenzo-Lacruz, C. Garcia, E. Morán-Tejeda, Groundwater level responses to precipitation variability in Mediterranean insular aquifers, J. Hydrol. 552 (2017) 516–531, https://doi.org/10.1016/J.JHYDROL.2017.07.011.
[11] J. Carnicer, M. Coll, M. Ninyerola, X. Pons, G. Sánchez, J. Peñuelas, Widespread crown condition decline, food web disruption, and amplified tree mortality with increased climate change-type drought, Proc. Natl. Acad. Sci. U. S. A. 108 (2011) 1474–1478, https://doi.org/10.1073/pnas.1010070108.
[12] Á. Rubio-Cuadrado, J.J. Camarero, M. del Río, M. Sánchez-González, R. Ruiz-Peinado, A. Bravo-Oviedo, L. Gil, F. Montes, Drought modifies tree competitiveness in an oak-beech temperate forest, For. Ecol. Manag. 429 (2018) 7–17, https://doi.org/10.1016/j.foreco.2018.06.035.
[13] E. Pasho, J.J. Camarero, M. de Luis, S.M. Vicente-Serrano, Impacts of drought at different time scales on forest growth across a wide climatic gradient in north-eastern Spain, Agric. For. Meteorol. 151 (2011), https://doi.org/10.1016/j.agrformet.2011.07.018.
[14] J.J. Camarero, A. Gazol, G. Sangüesa-Barreda, J. Oliva, S.M. Vicente-Serrano, To die or not to die: early warnings of tree dieback in response to a severe drought, J. Ecol. 103 (2015), https://doi.org/10.1111/1365-2745.12295.
[15] J.G. Pausas, S. Fernández-Muñoz, Fire regime changes in the Western Mediterranean Basin: from fuel-limited to drought-driven fire regime, Clim. Chang. 110 (2012) 215–226, https://doi.org/10.1007/s10584-011-0060-6.
[16] J. Ruffault, T. Curt, N.K. Martin-Stpaul, V. Moron, R.M. Trigo, Extreme wildfire events are linked to global-change-type droughts in the northern Mediterranean, Nat. Hazards Earth Syst. Sci. 18 (2018) 847–856, https://doi.org/10.5194/nhess-18-847-2018.
[17] F. Domínguez-Castro, P. Ribera, R. García-Herrera, J.M. Vaquero, M. Barriendos, J.M. Cuadrat, J.M. Moreno, Assessing extreme droughts in Spain during 1750–1850 from rogation ceremonies, Clim. Past 8 (2012) 705–722, https://doi.org/10.5194/cp-8-705-2012.
[18] S.M. Vicente-Serrano, J.M. Cuadrat, North Atlantic oscillation control of droughts in north-east Spain: Evaluation since 1600 A.D, Clim. Chang. 85 (2007), https://doi.org/10.1007/s10584-007-9285-9.
[19] P. Lionello, P. Malanotte-Rizzoli, R. Boscolo, P. Alpert, V. Artale, L. Li, J. Luterbacher, W. May, R. Trigo, M. Tsimplis, U. Ulbrich, E. Xoplaki, The Mediterranean climate: an overview of the main characteristics and issues, Dev. Earth Environ. Sci. 4 (2006) 1–26, https://doi.org/10.1016/S1571-9197(06)80003-0.
[20] F. Raymond, A. Ullmann, P. Camberlin, B. Oueslati, P. Drobinski, Atmospheric conditions and weather regimes associated with extreme winter dry spells over the Mediterranean basin, Clim. Dyn. 50 (2018) 4437–4453, https://doi.org/10.1007/s00382-017-3884-6.
[21] R.M. Trigo, J.A. Añel, D. Barriopedro, R. García-Herrera, L. Gimeno, R. Nieto, R. Castillo, M.R. Allen, N. Massey, The record winter drought of 2011–12 in the Iberian Peninsula, Bull. Am. Meteorol. Soc. 94 (2013) 41–45.
[22] R. García-Herrera, D. Garrido-Pérez, J.M. Barriopedro, C. Ordóñez, R. Vicente-Serrano, S.M. Nieto, L. Gimeno, R. Sorí, P. Yiou, The European 2016/2017 drought, J. Clim. (2019) https://journals.ametsoc.org/doi/full/10.1175/JCLI-D-18-0331.1.
[23] J.M. García-Ruiz, J.I. López-Moreno, S.M. Vicente-Serrano, T. Lasanta-Martínez, S. Beguería, Mediterranean water resources in a global change scenario, Earth-Sci. Rev. 105 (2011), https://doi.org/10.1016/j.earscirev.2011.01.006.

[24] J.C.C. García, Drought and agricultural insurance in Spain, in: Drought: Research and Science-Policy Interfacing—Proceedings of the International Conference on Drought: Research and Science-Policy Interfacing, 2015, pp. 371–375. https://www.scopus.com/inward/record.uri?eid=2-s2.0-84941560626&partnerID=40&md5=704397c30aca40902946dc51e8e1cb06.
[25] J. Vogt, The European drought observatory, in: 2011 GEOSS Workshop XL Managing Drought-Through Earth Observation. GEOSS 2011, 2011. https://doi.org/10.1109/GEOSS.2011.5948940.
[26] R. Maia, S.M. Vicente-Serrano, Drought planning and management in the Iberian Peninsula, in: D. Wilhite, R. Pulwarty (Eds.), Drought Water Crises. Sci. Technol. Manag. Issues, CRC, 2017, pp. 481–506.
[27] L.W. Mays, Ancient Water Technologies, Springer, Netherlands, 2010.
[28] A.W. Schneider, S.F. Adalı, "No harvest was reaped": demographic and climatic factors in the decline of the neo-Assyrian empire, Clim. Chang. 127 (2014) 435–446, https://doi.org/10.1007/s10584-014-1269-y.
[29] J.G. Manning, F. Ludlow, A.R. Stine, W.R. Boos, M. Sigl, J.R. Marlon, Volcanic suppression of Nile summer flooding triggers revolt and constrains interstate conflict in ancient Egypt, Nat. Commun. 8 (2017) 900, https://doi.org/10.1038/s41467-017-00957-y.
[30] R. Brázdil, A. Kiss, J. Luterbache, D.J. Nash, L. Řezníčková, Documentary data and the study of past droughts: a global state of the art, Clim. Past 14 (2018) 1915–1960, https://doi.org/10.5194/cp-14-1915-2018.
[31] R. Brázdil, P. Dobrovolný, M. Trnka, U. Büntgen, L. Řezníčková, O. Kotyza, H. Valášek, P. Štěpánek, Documentary and instrumental-based drought indices for the Czech lands back to AD 1501, Clim. Res. 70 (2016) 103–117, https://doi.org/10.3354/cr01380.
[32] F. Domínguez-Castro, R. García-Herrera, Documentary sources to investigate multidecadal variability of droughts | Fuentes documentales para investigar la variabilidad multidecadal de las sequías, Cuad. Investig. Geogr. 42 (2016) 13–27, https://doi.org/10.18172/cig.2936.
[33] I. Telelis, Climatic fluctuations in the eastern Mediterranean and the Middle East AD 300–1500 from byzantine documentary and proxy physical Paleoclimatic evidence—a comparison, Jahrb. Der Österreichischen Byzantinistik. 58 (2008) 167–208, https://doi.org/10.1553/joeb58s167.
[34] F. Domínguez-Castro, J.C. de Miguel, J.M. Vaquero, M.C. Gallego, R. García-Herrera, Climatic potential of Islamic chronicles in Iberia: extreme droughts (ad 711–1010), The Holocene 24 (2014) 370–374, https://doi.org/10.1177/0959683613518591.
[35] J. Martin-Vide, M.B. Vallvé, The use of rogation ceremony records in climatic reconstruction: a case study from Catalonia (Spain), Clim. Chang. 30 (1995) 201–221, https://doi.org/10.1007/BF01091842.
[36] M. Barriendos, Climatic variations in the Iberian Peninsula during the late maunder minimum (AD 1675–1715): an analysis of data from rogation ceremonies, The Holocene 7 (1997) 105–111, https://doi.org/10.1177/095968369700700110.
[37] F. Domínguez-Castro, J.I. Santisteban, M. Barriendos, R. Mediavilla, Reconstruction of drought episodes for Central Spain from rogation ceremonies recorded at the Toledo cathedral from 1506 to 1900: a methodological approach, Global Planet. Change 63 (2008) 230–242, https://doi.org/10.1016/j.gloplacha.2008.06.002.
[38] F.S. Rodrigo, M. Barriendos, Reconstruction of seasonal and annual rainfall variability in the Iberian peninsula (16th–20th centuries) from documentary data, Global Planet. Change 63 (2008) 243–257, https://doi.org/10.1016/j.gloplacha.2007.09.004.
[39] M.I. Fernández-Fernández, M.C. Gallego, F. Domínguez-Castro, R.M. Trigo, J.M. Vaquero, The climate in Zafra from 1750 to 1840: temperature indexes from documentary sources, Clim. Chang. 141 (2017) 671–684, https://doi.org/10.1007/s10584-017-1910-7.
[40] F. Domínguez-Castro, R. García-Herrera, P. Ribera, M. Barriendos, A shift in the spatial pattern of Iberian droughts during the 17th century, Clim. Past 6 (2010) 553–563, https://doi.org/10.5194/cp-6-553-2010.
[41] E. Piervitali, M. Colacino, Evidence of drought in western Sicily during the period 1565–1915 from liturgical offices, Clim. Chang. 49 (2001) 225–238, https://doi.org/10.1023/A:1010746612289.
[42] E. Xoplaki, P. Maheras, J. Luterbacher, Variability of climate in meridional Balkans during the periods 1675–1715 and 1780–1830 and its impact on human life, Clim. Chang. 48 (2001) 581–615, https://doi.org/10.1023/A:1005616424463.

[43] N. Chbouki, C.W. Stockton, D.E. Myers, Spatio-temporal patterns of drought in Morocco, Int. J. Climatol. 15 (1995) 187–205, https://doi.org/10.1002/joc.3370150205.
[44] J. Esper, D. Frank, U. Büntgen, A. Verstege, J. Luterbacher, E. Xoplaki, Long-term drought severity variations in Morocco, Geophys. Res. Lett. (2007), https://doi.org/10.1029/2007GL030844.
[45] E. Tejedor, M.A. Saz, J. Esper, J.M. Cuadrat, M. de Luis, Summer drought reconstruction in northeastern Spain inferred from a tree ring latewood network since 1734, Geophys. Res. Lett. 44 (2017) 8492–8500, https://doi.org/10.1002/2017GL074748.
[46] R. Touchan, G.M. Garfin, D.M. Meko, G. Funkhouser, N. Erkan, M.K. Hughes, B. S. Wallin, Preliminary reconstructions of spring precipitation in southwestern Turkey from tree-ring width, Int. J. Climatol. 23 (2003) 157–171, https://doi.org/10.1002/joc.850.
[47] Ü. Akkemik, A. Aras, Reconstruction (1689–1994 AD) of April-August precipitation in the southern part of Central Turkey, Int. J. Climatol. 25 (2005) 537–548, https://doi.org/10.1002/joc.1145.
[48] R. Touchan, K.J. Anchukaitis, D.M. Meko, S. Attalah, C. Baisan, A. Aloui, Long term context for recent drought in northwestern Africa, Geophys. Res. Lett. 35 (2008)L13705. https://doi.org/10.1029/2008GL034264.
[49] R. Touchan, K.J. Anchukaitis, D.M. Meko, M. Sabir, S. Attalah, A. Aloui, Spatiotemporal drought variability in northwestern Africa over the last nine centuries, Clim. Dyn. 37 (2011) 237–252, https://doi.org/10.1007/s00382-010-0804-4.
[50] S. Brewer, S. Alleaume, J. Guiot, A. Nicault, Historical droughts in Mediterranean regions during the last 500 years: a data/model approach, Clim. Past 3 (2007) 355–366, https://doi.org/10.5194/cp-3-355-2007.
[51] A. Nicault, S. Alleaume, S. Brewer, M. Carrer, P. Nola, J. Guiot, Mediterranean drought fluctuation during the last 500 years based on tree-ring data, Clim. Dyn. 31 (2008) 227–245, https://doi.org/10.1007/s00382-007-0349-3.
[52] B.I. Cook, K.J. Anchukaitis, R. Touchan, D.M. Meko, E.R. Cook, Spatiotemporal drought variability in the Mediterranean over the last 900 years, J. Geophys. Res. Atmos. 121 (2016) 2060–2074, https://doi.org/10.1002/2015JD023929.
[53] F. Giorgi, Climate change hot-spots, Geophys. Res. Lett. 33 (2006), https://doi.org/10.1029/2006GL025734.
[54] T. Caloiero, S. Veltri, P. Caloiero, F. Frustaci, Drought analysis in Europe and in the Mediterranean Basin using the standardized precipitation index, Water 10 (2018) 1043, https://doi.org/10.3390/w10081043.
[55] M. Hoerling, J. Eischeid, J. Perlwitz, X. Quan, T. Zhang, P. Pegion, M. Hoerling, J. Eischeid, J. Perlwitz, X. Quan, T. Zhang, P. Pegion, On the increased frequency of Mediterranean drought, J. Clim. 25 (2012) 2146–2161, https://doi.org/10.1175/JCLI-D-11-00296.1.
[56] J. Spinoni, J.V. Vogt, G. Naumann, P. Barbosa, A. Dosio, Will drought events become more frequent and severe in Europe? Int. J. Climatol. 38 (2018) 1718–1736, https://doi.org/10.1002/joc.5291.
[57] J. Spinoni, G. Naumann, J.V. Vogt, P. Barbosa, The biggest drought events in Europe from 1950 to 2012, J. Hydrol. Reg. Stud. 3 (2015) 509–524, https://doi.org/10.1016/j.ejrh.2015.01.001.
[58] M. Dubrovský, M. Hayes, P. Duce, M. Trnka, M. Svoboda, P. Zara, Multi-GCM projections of future drought and climate variability indicators for the Mediterranean region, Reg. Environ. Chang. 14 (2014) 1907–1919, https://doi.org/10.1007/s10113-013-0562-z.
[59] J.C. González-Hidalgo, S.M. Vicente-Serrano, D. Peña-Angulo, C. Salinas, M. Tomas-Burguera, S. Beguería, High-resolution spatio-temporal analyses of drought episodes in the western Mediterranean basin (Spanish mainland, Iberian Peninsula), Acta Geophys. 66 (2018), https://doi.org/10.1007/s11600-018-0138-x.
[60] S.M. Vicente-Serrano, J.-I. Lopez-Moreno, S. Beguería, J. Lorenzo-Lacruz, A. Sanchez-Lorenzo, J.M. García-Ruiz, C. Azorin-Molina, E. Morán-Tejeda, J. Revuelto, R. Trigo, F. Coelho, F. Espejo, Evidence of increasing drought severity caused by temperature rise in southern Europe, Environ. Res. Lett. (2014), https://doi.org/10.1088/1748-9326/9/4/044001.
[61] P. Páscoa, C.M. Gouveia, A. Russo, R.M. Trigo, Drought trends in the Iberian Peninsula over the last 112 years, Adv. Meteorol. 2017 (2017) 1–13, https://doi.org/10.1155/2017/4653126.
[62] P.T. Nastos, C.S. Zerefos, Spatial and temporal variability of consecutive dry and wet days in Greece, Atmos. Res. 94 (2009) 616–628, https://doi.org/10.1016/j.atmosres.2009.03.009.

[63] J.-P. Vidal, E. Martin, L. Franchistéguy, F. Habets, J.-M. Soubeyroux, M. Blanchard, M. Baillon, Multilevel and multiscale drought reanalysis over France with the Safran-Isba-Modcou hydrometeorological suite, Hydrol. Earth Syst. Sci. 14 (2010) 459–478, https://doi.org/10.5194/hess-14-459-2010.
[64] İ. Dabanlı, A.K. Mishra, Z. Şen, Long-term spatio-temporal drought variability in Turkey, J. Hydrol. 552 (2017) 779–792, https://doi.org/10.1016/j.jhydrol.2017.07.038.
[65] S.M. Vicente-Serrano, J.M. Cuadrat-Prats, Trends in drought intensity and variability in the middle Ebro valley (NE of the Iberian peninsula) during the second half of the twentieth century, Theor. Appl. Climatol. 88 (2007) 247–258, https://doi.org/10.1007/s00704-006-0236-6.
[66] T. Caloiero, R. Coscarelli, E. Ferrari, B. Sirangelo, Analysis of dry spells in southern Italy (Calabria), Water 7 (2015) 3009–3023, https://doi.org/10.3390/w7063009.
[67] G. Marini, N. Fontana, A.K. Mishra, Investigating drought in Apulia region, Italy using SPI and RDI, Theor. Appl. Climatol. 137 (2018) 383–397, https://doi.org/10.1007/s00704-018-2604-4.
[68] L. Vasiliades, A. Loukas, N. Liberis, A water balance derived drought index for Pinios River basin, Greece, Water Resour. Manag. 25 (2011) 1087–1101, https://doi.org/10.1007/s11269-010-9665-1.
[69] P. Greve, B. Orlowsky, B. Mueller, J. Sheffield, M. Reichstein, S.I. Seneviratne, Global assessment of trends in wetting and drying over land, Nat. Geosci. 7 (2014) 716–721, https://doi.org/10.1038/NGEO2247.
[70] A.J. Teuling, A.F. Van Loon, S.I. Seneviratne, I. Lehner, M. Aubinet, B. Heinesch, C. Bernhofer, T. Grünwald, H. Prasse, U. Spank, Evapotranspiration amplifies European summer drought, Geophys. Res. Lett. 40 (2013) 2071–2075, https://doi.org/10.1002/grl.50495.
[71] M. Turco, J. von Hardenberg, A. AghaKouchak, M.C. Llasat, A. Provenzale, R.M. Trigo, On the key role of droughts in the dynamics of summer fires in Mediterranean Europe, Sci. Rep. 7 (2017) 81, https://doi.org/10.1038/s41598-017-00116-9.
[72] H.K. Kafle, H.J. Bruins, Climatic trends in Israel 1970–2002: warmer and increasing aridity inland, Clim. Chang. 96 (2009) 63–77, https://doi.org/10.1007/s10584-009-9578-2.
[73] P.H. Gleick, Water, drought, climate change, and conflict in Syria, Weather Clim. Soc. 6 (2014) 331–340, https://doi.org/10.1175/WCAS-D-13-00059.1.
[74] C.P. Kelley, S. Mohtadi, M.A. Cane, R. Seager, Y. Kushnir, Climate change in the Fertile Crescent and implications of the recent Syrian drought, Proc. Natl. Acad. Sci. U. S. A. 112 (2015) 3241–3246, https://doi.org/10.1073/pnas.1421533112.
[75] B.I. Cook, J.S. Mankin, K.J. Anchukaitis, Climate change and drought: from past to future, Curr. Clim. Chang. Rep. 4 (2018) 164–179, https://doi.org/10.1007/s40641-018-0093-2.
[76] F. Giorgi, P. Lionello, Climate change projections for the Mediterranean region, Glob. Planet. Change 63 (2008) 90–104, https://doi.org/10.1016/j.gloplacha.2007.09.005.
[77] P.M. Sousa, R.M. Trigo, P. Aizpurua, R. Nieto, L. Gimeno, R. Garcia-Herrera, Trends and extremes of drought indices throughout the 20th century in the Mediterranean, Nat. Hazards Earth Syst. Sci. 11 (2011) 33–51, https://doi.org/10.5194/nhess-11-33-2011.
[78] A. Mariotti, M.V. Struglia, N. Zeng, K.-M. Lau, The hydrological cycle in the Mediterranean region and implications for the water budget of the Mediterranean Sea, J. Clim. 15 (2002) 1674–1690, https://doi.org/10.1175/1520-0442(2002)015<1674:THCITM>2.0.CO;2.
[79] D. Shohami, U. Dayan, E. Morin, Warming and drying of the eastern Mediterranean: additional evidence from trend analysis, J. Geophys. Res. Atmos. 116 (2011), https://doi.org/10.1029/2011JD016004.
[80] A. Giannini, Oceanic forcing of sahel rainfall on interannual to interdecadal time scales, Science 302 (2003) 1027–1030, https://doi.org/10.1126/science.1089357.
[81] R. Seager, N. Naik, G.A. Vecchi, Thermodynamic and dynamic mechanisms for large-scale changes in the hydrological cycle in response to global warming, J. Clim. 23 (2010) 4651–4668, https://doi.org/10.1175/2010JCLI3655.1.
[82] A.F. Van Loon, K. Stahl, G. Di Baldassarre, J. Clark, S. Rangecroft, N. Wanders, T. Gleeson, A.I.J.M. Van Dijk, L.M. Tallaksen, J. Hannaford, R. Uijlenhoet, A.J. Teuling, D.M. Hannah, J. Sheffield, M. Svoboda, B. Verbeiren, T. Wagener, H.A.J. Van Lanen, Drought in a human-modified world: reframing drought definitions, understanding, and analysis approaches, Hydrol. Earth Syst. Sci. 20 (2016) 3631–3650, https://doi.org/10.5194/hess-20-3631-2016.

[83] S.M. Vicente-Serrano, J.I. López-Moreno, S. Beguería, J. Lorenzo-Lacruz, C. Azorin-Molina, E. Morán-Tejeda, Accurate computation of a streamflow drought index, J. Hydrol. Eng. 17 (2012), https://doi.org/10.1061/(ASCE)HE.1943-5584.0000433.
[84] S. Rangecroft, A.F. Van Loon, G. Coxon, J.A. Breña-Naranjo, F. Van Ogtrop, H.A.J. Van Lanen, Using paired catchments to quantify the human influence on hydrological droughts, Hydrol. Earth Syst. Sci. Discuss. (2018) 1–23, https://doi.org/10.5194/hess-2018-215.
[85] J.P. Bloomfield, B.P. Marchant, Analysis of groundwater drought building on the standardised precipitation index approach, Hydrol. Earth Syst. Sci. 17 (2013) 4769–4787, https://doi.org/10.5194/hess-17-4769-2013.
[86] H.A.J. Van Lanen, N. Wanders, L.M. Tallaksen, A.F. Van Loon, Hydrological drought across the world: impact of climate and physical catchment structure, Hydrol. Earth Syst. Sci. 17 (2013) 1715–1732, https://doi.org/10.5194/hess-17-1715-2013.
[87] A.F. Van Loon, Hydrological drought explained, Wiley Interdiscip. Rev. Water 2 (2015) 359–392, https://doi.org/10.1002/wat2.1085.
[88] M.G. Floriancic, I. van Meerveld, M. Smoorenburg, M. Margreth, F. Naef, J.W. Kirchner, P. Molnar, Spatio-temporal variability in contributions to low flows in the high Alpine Poschiavino catchment, Hydrol. Process. 32 (2018) 3938–3953, https://doi.org/10.1002/hyp.13302.
[89] J.I. López-Moreno, S.M. Vicente-Serrano, J. Zabalza, S. Beguería, J. Lorenzo-Lacruz, C. Azorin-Molina, E. Morán-Tejeda, Hydrological response to climate variability at different time scales: a study in the Ebro basin, J. Hydrol. 477 (2013) 175–188, https://doi.org/10.1016/j.jhydrol.2012.11.028.
[90] Y. Yang, S. Zhang, T.R. McVicar, H.E. Beck, Y. Zhang, B. Liu, Disconnection between trends of atmospheric drying and continental runoff, Water Resour. Res. 54 (2018) 4700–4713, https://doi.org/10.1029/2018WR022593.
[91] S.M. Vicente-Serrano, C. Azorin-Molina, A. Sanchez-Lorenzo, J. Revuelto, J.I. López-Moreno, J.C. González-Hidalgo, E. Moran-Tejeda, F. Espejo, Reference evapotranspiration variability and trends in Spain, 1961–2011, Glob. Planet. Change 121 (2014), https://doi.org/10.1016/j.gloplacha.2014.06.005.
[92] R. Prăvălie, A. Piticar, B. Roşca, L. Sfîcă, G. Bandoc, A. Tiscovschi, C. Patriche, Spatio-temporal changes of the climatic water balance in Romania as a response to precipitation and reference evapotranspiration trends during 1961–2013, Catena 172 (2019) 295–312, https://doi.org/10.1016/j.catena.2018.08.028.
[93] U. Maček, N. Bezak, M. Šraj, Reference evapotranspiration changes in Slovenia, Europe, Agric. For. Meteorol. 260–261 (2018) 183–192, https://doi.org/10.1016/j.agrformet.2018.06.014.
[94] S.M. Vicente-Serrano, C. Azorin-Molina, A. Sanchez-Lorenzo, E. Morán-Tejeda, J. Lorenzo-Lacruz, J. Revuelto, J.I. López-Moreno, F. Espejo, Temporal evolution of surface humidity in Spain: recent trends and possible physical mechanisms, Clim. Dyn. 42 (2014), https://doi.org/10.1007/s00382-013-1885-7.
[95] J.H. Stagge, D.G. Kingston, L.M. Tallaksen, D.M. Hannah, Observed drought indices show increasing divergence across Europe, Sci. Rep. 7 (2017), https://doi.org/10.1038/s41598-017-14283-2.
[96] L. Gudmundsson, S.I. Seneviratne, X. Zhang, Anthropogenic climate change detected in European renewable freshwater resources, Nat. Clim. Chang. 7 (2017) 813–816, https://doi.org/10.1038/nclimate3416.
[97] J.-P. Vidal, E. Martin, N. Kitova, J. Najac, J.-M. Soubeyroux, Evolution of spatio-temporal drought characteristics: validation, projections and effect of adaptation scenarios, Hydrol. Earth Syst. Sci. 16 (2012) 2935–2955, https://doi.org/10.5194/hess-16-2935-2012.
[98] S. Beguería, J.I. López-Moreno, A. Lorente, M. Seeger, J.M. García-Ruiz, Assessing the effect of climate oscillations and land-use changes on streamflow in the Central Spanish Pyrenees, Ambio 32 (2003) 283–286, https://doi.org/10.1579/0044-7447-32.4.283.
[99] S.M. Vicente-Serrano, J. Zabalza-Martínez, G. Borràs, J.I. López-Moreno, E. Pla, D. Pascual, R. Savé, C. Biel, I. Funes, N. Martín-Hernández, M. Peña-Gallardo, S. Beguería, M. Tomas-Burguera, Effect of reservoirs on streamflow and river regimes in a heavily regulated river basin of Northeast Spain, Catena 149 (2017) 727–741, https://doi.org/10.1016/j.catena.2016.03.042.

[100] S.M. Vicente-Serrano, J. Zabalza-Martínez, G. Borràs, J.I. López-Moreno, E. Pla, D. Pascual, R. Savé, C. Biel, I. Funes, C. Azorin-Molina, A. Sanchez-Lorenzo, N. Martín-Hernández, M. Peña-Gallardo, E. Alonso-González, M. Tomas-Burguera, A. El Kenawy, Extreme hydrological events and the influence of reservoirs in a highly regulated river basin of northeastern Spain, J. Hydrol. Reg. Stud. 12 (2017), https://doi.org/10.1016/j.ejrh.2017.01.004.

[101] P. Döll, K. Fiedler, J. Zhang, Global-scale analysis of river flow alterations due to water withdrawals and reservoirs, Hydrol. Earth Syst. Sci. 13 (2009) 2413–2432, https://doi.org/10.5194/hess-13-2413-2009.

[102] E. Morán-Tejeda, J. Lorenzo-Lacruz, J.I. López-Moreno, A. Ceballos-Barbancho, J. Zabalza, S. M. Vicente-Serrano, Reservoir management in the duero basin (Spain): impact on river regimes and the response to environmental change, Water Resour. Manag. 26 (2012), https://doi.org/10.1007/s11269-012-0004-6.

[103] J. Lorenzo-Lacruz, S.M. Vicente-Serrano, J.I. López-Moreno, E. Morán-Tejeda, J. Zabalza, Recent trends in Iberian streamflows (1945–2005), J. Hydrol. 414–415 (2012), https://doi.org/10.1016/j.jhydrol.2011.11.023.

[104] D. Martínez-Granados, J.F. Maestre-Valero, J. Calatrava, V. Martínez-Alvarez, The economic impact of water evaporation losses from water reservoirs in the Segura Basin, SE Spain, Water Resour. Manag. 25 (2011) 3153–3175, https://doi.org/10.1007/s11269-011-9850-x.

[105] J. Lorenzo-Lacruz, S.M. Vicente-Serrano, J.I. López-Moreno, S. Beguería, J.M. García-Ruiz, J.M. Cuadrat-Prats, The impact of droughts and water management on various hydrological systems in the headwaters of the Tagus River (Central Spain), J. Hydrol. 386 (2010) 13–26, https://doi.org/10.1016/j.jhydrol.2010.01.001.

[106] J.I. López-Moreno, S.M. Vicente-Serrano, S. Begueria, J.M. Garcia-Ruiz, M.M. Portela, A.B. Almeida, Dam effects on droughts magnitude and duration in a transboundary basin: the lower river tagus, pain and Portugal, Water Resour. Res. 45 (2009)W02405. https://doi.org/10.1029/2008WR007198.

[107] A. Aouad-Rizk, J.-O. Job, S. Khalil, T. Touma, C. Bitar, C. Bocquillon, W. Najem, Snow in Lebanon: a preliminary study of snow cover over Mount Lebanon and a simple snowmelt model, Hydrol. Sci. J. 50 (2005) 555–569, https://doi.org/10.1623/hysj.50.3.555.65023.

[108] A. Fayad, S. Gascoin, G. Faour, P. Fanise, L. Drapeau, J. Somma, A. Fadel, A. Al Bitar, R. Escadafal, Snow observations in Mount Lebanon (2011–2016), Earth Syst. Sci. Data 9 (2017) 573–587, https://doi.org/10.5194/essd-9-573-2017.

[109] J.I. López-Moreno, J.M. García-Ruiz, Influence of snow accumulation and snowmelt on streamflow in the central Spanish Pyrenees | influence de l'accumulation et de la fonte de la neige Sur les écoulements dans les Pyrénées centrales espagnoles, Hydrol. Sci. J. 49 (2004) 787–802.

[110] J.I. López-Moreno, Recent variations of snowpack depth in the central Spanish Pyrenees, Arctic, Antarct. Alp. Res. 37 (2005) 253–260, https://doi.org/10.1657/1523-0430(2005)037[0253:RVOSDI]2.0.CO;2.

[111] E. Morán-Tejeda, A. Ceballos-Barbancho, J.M. Llorente-Pinto, Hydrological response of Mediterranean headwaters to climate oscillations and land-cover changes: the mountains of Duero River basin (Central Spain), Glob. Planet. Change 72 (2010) 39–49, https://doi.org/10.1016/j.gloplacha.2010.03.003.

[112] J.I. López-Moreno, S.M. Vicente-Serrano, Atmospheric circulation influence on the interannual variability of snow pack in the Spanish Pyrenees during the second half of the 20th century, Nord. Hydrol. 38 (2007) 33–44, https://doi.org/10.2166/nh.2007.030.

[113] E. Morán-Tejeda, J. Lorenzo-Lacruz, J.I. López-Moreno, K. Rahman, M. Beniston, Streamflow timing of mountain rivers in Spain: recent changes and future projections, J. Hydrol. 517 (2014) 1114–1127, https://doi.org/10.1016/j.jhydrol.2014.06.053.

[114] M. Zampieri, E. Scoccimarro, S. Gualdi, A. Navarra, Observed shift towards earlier spring discharge in the main Alpine rivers, Sci. Total Environ. 503–504 (2015) 222–232, https://doi.org/10.1016/j.scitotenv.2014.06.036.

[115] A. Bard, B. Renard, M. Lang, Observed trends in the hydrologic regime of alpine catchments [Tendances observées Sur les régimes hydrologiques de l'Arc Alpin], Houille Blanche (2012) 38–43, https://doi.org/10.1051/lhb/2012006.

[116] A. Dai, T. Zhao, J. Chen, Climate change and drought: a precipitation and evaporation perspective, Curr. Clim. Chang. Rep. 4 (2018) 301–312, https://doi.org/10.1007/s40641-018-0101-6.

[117] G. Naumann, L. Alfieri, K. Wyser, L. Mentaschi, R.A. Betts, H. Carrao, J. Spinoni, J. Vogt, L. Feyen, Global changes in drought conditions under different levels of warming, Geophys. Res. Lett. 45 (2018) 3285–3296, https://doi.org/10.1002/2017GL076521.

[118] G. Forzieri, L. Feyen, R. Rojas, M. Flörke, F. Wimmer, A. Bianchi, Ensemble projections of future streamflow droughts in Europe, Hydrol. Earth Syst. Sci. 18 (2014) 85–108, https://doi.org/10.5194/hess-18-85-2014.

[119] W. DePau, Drought Early Warning Systems in West Asia and North Africa, Syrian Arab Republic, 2004.

[120] D.A. Wilhite, Drought monitoring as a component of drought preparedness planning, in: A. Iglesias, A. Cancelliere, D.A. Wilhite, L. Garrote, F. Cubillo (Eds.), Coping with Drought Risk in Agriculture and Water Supply Systems: Drought Management and Policy Development in the Mediterranean, Springer Netherlands, Dordrecht, 2009, pp. 3–19, https://doi.org/10.1007/978-1-4020-9045-5_1.

[121] T. Estrela, E. Vargas, Drought management plans in the European Union. The case of Spain, Water Resour. Manag. 26 (2012) 1537–1553, https://doi.org/10.1007/s11269-011-9971-2.

[122] CHG, Plan Especíal de Sequía de la Confederación Hidrográfica del Guadiana (Drought Management Plan for Guadiana River Basin), https://www.chguadiana.es/sites/default/files/2018-12/PESCHGn.pdf, 2018.

[123] R. Maia, Drought management in the Iberian Peninsula shared river basins: developments and perspectives, in: E. 7th I.C. on W.R.C. and R.R.U.C. Instability, Cyprus, 2009, http://www.nireas-iwrc.org/index.php?link=publications_details.php&serial=83.

[124] FAO, Drought characteristics and management in North Africa and the Near East, in: FAO Water Reports 45, Rome, 2018.

[125] M. Semawi, A. Alokour, R. Hani, Drought conditions and management strategies in Jordan, in: WMO, UNCCD, FAO, CBA and UNW-DPC (Ed.), Capacity Development to Support National Drought Management Policies, Cairo, Egypt, 2014.

[126] W.C. Palmer, Meteorological Drought. Research Paper No. 45, 1965, 58 p, US Department of Commerce Weather Bureau, Washington, DC, Research Paper, 1965.

[127] T.B. McKee, N.J. Doesken, J. Kleist, The relationship of drought frequency and duration to time scales, in: Eighth Conference on Applied Climatology, 1993, pp. 179–184.

[128] S.M. Vicente-Serrano, S. Beguería, J.I. López-Moreno, S.M. Vicente-Serrano, S. Beguería, J.I. López-Moreno, A multiscalar drought index sensitive to global warming: the standardized precipitation evapotranspiration index, J. Clim. 23 (2010) 1696–1718, https://doi.org/10.1175/2009JCLI2909.1.

[129] B.A. Shafer, L.E. Dezman, Development of a surface water supply index (SWSI) to assess the severity of drought conditions in snowpack runoff areas, in: Proceedings Western Snow Conference, 1982, pp. 164–175. https://www.scopus.com/inward/record.uri?eid=2-s2.0-0020244777&partnerID=40&md5=a1b32a3ed0a368ac271a95ae267a8bd7.

[130] S. Shukla, A.W. Wood, Use of a standardized runoff index for characterizing hydrologic drought, Geophys. Res. Lett. 35 (2008), https://doi.org/10.1029/2007GL032487.

[131] R.R. Heim, A review of twentieth-century drought indices used in the United States, Bull. Am. Meteorol. Soc. 83 (2002) 1149–1165, https://doi.org/10.1175/1520-0477(2002)083<1149:AROTDI>2.3.CO;2.

[132] A.K. Mishra, V.P. Singh, A review of drought concepts, J. Hydrol. 391 (2010) 202–216, https://doi.org/10.1016/j.jhydrol.2010.07.012.

[133] J. Lorenzo-Lacruz, E. Morán-Tejeda, S.M. Vicente-Serrano, J.I. López-Moreno, Streamflow droughts in the Iberian Peninsula between 1945 and 2005: spatial and temporal patterns, Hydrol. Earth Syst. Sci. 17 (2013) 119–134, https://doi.org/10.5194/hess-17-119-2013.

[134] J.R. Coll, E. Aguilar, L. Ashcroft, Drought variability and change across the Iberian Peninsula, Theor. Appl. Climatol. 130 (2017) 901–916, https://doi.org/10.1007/s00704-016-1926-3.

[135] P. Páscoa, C.M. Gouveia, A. Russo, R.M. Trigo, The role of drought on wheat yield interannual variability in the Iberian Peninsula from 1929 to 2012, Int. J. Biometeorol. 61 (2017) 439–451, https://doi.org/10.1007/s00484-016-1224-x.
[136] S.M. Vicente-Serrano, Differences in spatial patterns of drought on different time scales: an analysis of the Iberian Peninsula, Water Resour. Manag. 20 (2006), https://doi.org/10.1007/s11269-006-2974-8.
[137] I. Livada, V.D. Assimakopoulos, Spatial and temporal analysis of drought in Greece using the standardized precipitation index (SPI), Theor. Appl. Climatol. 89 (2007) 143–153, https://doi.org/10.1007/s00704-005-0227-z.
[138] B. Lloyd-Hughes, M.A. Saunders, A drought climatology for Europe, Int. J. Climatol. 22 (2002) 1571–1592, https://doi.org/10.1002/joc.846.
[139] L. Garrote, A. Iglesias, M. Moneo, A. Garrido, A. Gómez, A. Lapeña, S. Benbeniste, F. Cubillo, J. Ibáñez, Application of the drought management guidelines in Spain [part 2. Examples of application], in: A. Iglesias, M. Moneo, A. López-Francos (Eds.), Drought Manag. Guidel. Tech. Annex. (Options Méditerranéennes Série B. Etudes Rech. n. 58), CIHEAM/EC MEDA Water, Zaragoza, 2007, pp. 373–406.
[140] F.N. Kogan, Global drought watch from space, Bull. Am. Meteorol. Soc. 78 (1997) 621–636, https://doi.org/10.1175/1520-0477(1997)078<0621:GDWFS>2.0.CO;2.
[141] S.M. Vicente-Serrano, C. Azorin-Molina, M. Peña-Gallardo, M. Tomas-Burguera, F. Dominguez-Castro, N. Martin-Hernández, S. Begueria, A. El Kenawy, I. Noguera, M. Garcia, A high-resolution spatial assessment of the impacts of drought variability on vegetation activity in Spain from 1981 to 2015, Nat. Hazards Earth Syst. Sci. 19 (2019) 1189–1213, https://doi.org/10.5194/nhess-19-1189-2019.
[142] M. Zribi, G. Dridi, R. Amri, Z. Lili-Chabaane, Analysis of the effects of drought on vegetation cover in a Mediterranean region through the use of SPOT-VGT and TERRA-MODIS long time series, Remote Sens. 8 (2016) 992, https://doi.org/10.3390/rs8120992.
[143] C.M. Gouveia, R.M. Trigo, S. Beguería, S.M. Vicente-Serrano, Drought impacts on vegetation activity in the Mediterranean region: an assessment using remote sensing data and multi-scale drought indicators, Glob. Planet. Chang. 151 (2017), https://doi.org/10.1016/j.gloplacha.2016.06.011.
[144] E. Loulli, D.G. Hadjimitsis, Remote sensing based indices for drought assessment in the east Mediterranean region, in: C.M. Neale, A. Maltese (Eds.), Remote Sensing for Agriculture, Ecosystems, and Hydrology, vol. XX, SPIE, 2018, p. 40, https://doi.org/10.1117/12.2325331.
[145] S.M. Vicente-Serrano, C. Azorin-Molina, M. Peña-Gallardo, M. Tomas-Burguera, F. Dominguez-Castro, N. Martin-Hernández, S. Begueria, A. El Kenawy, I. Noguera, M. Garcia, A high-resolution spatial assessment of the impacts of drought variability on vegetation activity in Spain from 1981 to 2015, Nat. Hazards Earth Syst. Sci. (Discussion) 2018 (2018) 1–52, https://doi.org/10.5194/nhess-2018-356.
[146] J. Spinoni, P. Barbosa, A. De Jager, N. McCormick, G. Naumann, J.V. Vogt, D. Magni, D. Masante, M. Mazzeschi, A new global database of meteorological drought events from 1951 to 2016, J. Hydrol. Reg. Stud. 22 (2019) 100593, https://doi.org/10.1016/J.EJRH.2019.100593.
[147] A. Ouassou, T. Ameziane, A. Ziyad, M. Belghiti, Application of the drought management guidelines in Morocco, in: MEDROPLAN Mediterranean Drought Preparedness and Mitigation PLAN Med Planning, 2005, http://agris.fao.org/agris-search/search.do?recordID=QC2007600102.
[148] P. Paneque, Drought management strategies in Spain, Water 7 (2015) 6689–6701, https://doi.org/10.3390/w7126655.
[149] M. Dogdu, C. Sagnak, Climate change, drought and over pumping impacts on groundwaters: two examples from Turkey, in: Third Int. BALWOIS Conf. Balk. Water Obs. Inf. Syst., Ohrid, the Former Yugoslav Republic of Macedonia, 2008.
[150] F. Domínguez-Castro, S.M. Vicente-Serrano, M. Tomás-Burguera, M. Peña-Gallardo, S. Beguería, A. El Kenawy, Y. Luna, A. Morata, High-spatial-resolution probability maps of drought duration and magnitude across Spain, Nat. Hazards Earth Syst. Sci. 19 (2019) 611–628, https://doi.org/10.5194/nhess-19-611-2019.
[151] A. Iglesias, A. Cancelliere, D. Gabiña, A. López-Francos, M. Moneo, G. Rossi, Drought Management Guidelines, Brussels, Belgium, (2007).

[152] G. Nelson, M.W. Rosegrant, J. Koo, R. Robertson, T. Sulser, T. Zhu, C. Ringler, S. Msangi, A. Palazzo, M. Batka, M. Magalhães, R.A. Santos, M. Ewing, D. Lee, Climate Change: Impact on Agriculture and Costs of Adaptation, (2009) https://doi.org/10.2499/0896295354.

[153] C. Rosenzweig, F.N. Tubiello, Impacts of global climate change on Mediterranean agriculture: current methodologies and future directions. An introductory essay, Mitig. Adapt. Strateg. Glob. Chang. 1 (1997) 219–232, https://doi.org/10.1023/B:MITI.0000018269.58736.82.

[154] I.N. de Estadística, Estadísticas e indicadores del agua 2008, in: Colección Cifras, Instituto Nacional de Estadística, 2008. http://www.ine.es/revistas/cifraine/0108.pdf.

[155] G.J. Fitzgerald, M. Tausz, G. O'Leary, M.R. Mollah, S. Tausz-Posch, S. Seneweera, I. Mock, M. Löw, D.L. Partington, D. Mcneil, R.M. Norton, Elevated atmospheric [CO_2] can dramatically increase wheat yields in semi-arid environments and buffer against heat waves, Glob. Chang. Biol. 22 (2016) 2269–2284, https://doi.org/10.1111/gcb.13263.

[156] D. Deryng, J. Elliott, C. Folberth, C. Müller, T.A.M. Pugh, K.J. Boote, D. Conway, A.C. Ruane, D. Gerten, J.W. Jones, N. Khabarov, S. Olin, S. Schaphoff, E. Schmid, H. Yang, C. Rosenzweig, Regional disparities in the beneficial effects of rising CO_2 concentrations on crop water productivity, Nat. Clim. Chang. 6 (2016) 786–790, https://doi.org/10.1038/nclimate2995.

[157] H.M. El-Shaer, C. Rosenzweig, A. Iglesias, M.H. Eid, D. Hillel, Impact of climate change on possible scenarios for Egyptian agriculture in the future, Mitig. Adapt. Strateg. Glob. Chang. 1 (1997) 233–250, https://doi.org/10.1007/BF00517805.

[158] M. Miraglia, H.J.P. Marvin, G.A. Kleter, P. Battilani, C. Brera, E. Coni, F. Cubadda, L. Croci, B. De Santis, S. Dekkers, L. Filippi, R.W.A. Hutjes, M.Y. Noordam, M. Pisante, G. Piva, A. Prandini, L. Toti, G.J. van den Born, A. Vespermann, Climate change and food safety: an emerging issue with special focus on Europe, Food Chem. Toxicol. 47 (2009) 1009–1021, https://doi.org/10.1016/j.fct.2009.02.005.

[159] E. Ferragina, G. Canitano, Geopolitical Implications of Water and Food Security in Southern and Eastern Mediterranean Countries, (2015) pp. 33–59.https://www.researchgate.net/publication/279186882_Geopolitical_Implications_of_Water_and_Food_Security_in_Southern_and_Eastern_Mediterranean_Countries.

[160] N.D. Mueller, J.S. Gerber, M. Johnston, D.K. Ray, N. Ramankutty, J.A. Foley, Closing yield gaps through nutrient and water management, Nature 490 (2012) 254–257, https://doi.org/10.1038/nature11420.

[161] W. Cramer, J. Guiot, M. Fader, J. Garrabou, J.-P. Gattuso, A. Iglesias, M.A. Lange, P. Lionello, M.C. Llasat, S. Paz, J. Peñuelas, M. Snoussi, A. Toreti, M.N. Tsimplis, E. Xoplaki, Climate change and interconnected risks to sustainable development in the Mediterranean, Nat. Clim. Chang. 8 (2018) 972–980, https://doi.org/10.1038/s41558-018-0299-2.

[162] A. Iglesias, S. Quiroga, M. Moneo, L. Garrote, From climate change impacts to the development of adaptation strategies: challenges for agriculture in Europe, Clim. Chang. 112 (2012) 143–168, https://doi.org/10.1007/s10584-011-0344-x.

[163] L. Tanasijevic, M. Todorovic, L.S. Pereira, C. Pizzigalli, P. Lionello, Impacts of climate change on olive crop evapotranspiration and irrigation requirements in the Mediterranean region, Agric. Water Manag. 144 (2014) 54–68, https://doi.org/10.1016/j.agwat.2014.05.019.

[164] H. Fraga, I. de Cortázar Atauri, A.C. Malheiro, J.A. Santos, Modelling climate change impacts on viticultural yield, phenology and stress conditions in Europe, Glob. Chang. Biol. 22 (2016) 3774–3788, https://doi.org/10.1111/gcb.13382.

[165] C. van Leeuwen, P. Darriet, The impact of climate change on viticulture and wine quality, J. Wine Econ. 11 (2016) 150–167, https://doi.org/10.1017/jwe.2015.21.

[166] FAO, Plan Bleu, State of Mediterranean Forests 2018, Food and Agricultural Organization, Rome, 2018.

[167] Y. Preisler, F. Tatarinov, J.M. Grünzweig, D. Bert, J. Ogée, L. Wingate, E. Rotenberg, S. Rohatyn, N. Her, I. Moshe, T. Klein, D. Yakir, Mortality versus survival in drought-affected Aleppo pine forest depends on the extent of rock cover and soil stoniness, Funct. Ecol. 33 (2019) 901–912, https://doi.org/10.1111/1365-2435.13302.

[168] R.M. Navarro-Cerrillo, M. Sarmoum, A. Gazol, F. Abdoun, J.J. Camarero, The decline of Algerian Cedrus atlantica forests is driven by a climate shift towards drier conditions, Dendrochronologia 55 (2019) 60–70, https://doi.org/10.1016/j.dendro.2019.04.003.
[169] A. Gazol, G. Sangüesa-Barreda, E. Granda, J.J. Camarero, Tracking the impact of drought on functionally different woody plants in a Mediterranean scrubland ecosystem, Plant Ecol. 218 (2017) 1009–1020, https://doi.org/10.1007/s11258-017-0749-3.
[170] M. Colangelo, J.J. Camarero, G. Battipaglia, M. Borghetti, V. De Micco, T. Gentilesca, F. Ripullone, A multi-proxy assessment of dieback causes in a Mediterranean oak species, Tree Physiol. 37 (2017) 617–631, https://doi.org/10.1093/treephys/tpx002.
[171] M. Cailleret, M. Nourtier, A. Amm, M. Durand-Gillmann, H. Davi, Drought-induced decline and mortality of silver fir differ among three sites in southern France, Ann. For. Sci. 71 (2014) 643–657, https://doi.org/10.1007/s13595-013-0265-0.
[172] M. Vennetier, B. Vila, E. Liang, F. Guibal, A. Taahbet, C. Gadbin-Henry, Impact of climate change on pine forest productivity and on the shift of a bioclimatic limit in a Mediterranean area. Option Méditerranéennes, séries A n°75 p. 189–197, Options Méditerr. Serie A (2007) 189–197.
[173] D. Sarris, D. Christodoulakis, C. Körner, Recent decline in precipitation and tree growth in the eastern Mediterranean, Glob. Chang. Biol. 13 (2007) 1187–1200, https://doi.org/10.1111/j.1365-2486.2007.01348.x.
[174] R. Sánchez-Salguero, J.J. Camarero, M. Carrer, E. Gutiérrez, A.Q. Alla, L. Andreu-Hayles, A. Hevia, A. Koutavas, E. Martínez-Sancho, P. Nola, A. Papadopoulos, E. Pasho, E. Toromani, J. A. Carreira, J.C. Linares, Climate extremes and predicted warming threaten Mediterranean Holocene firs forests refugia, Proc. Natl. Acad. Sci. U. S. A. 114 (2017) E10142–E10150, https://doi.org/10.1073/pnas.1708109114.
[175] B. Poulter, D. Frank, P. Ciais, R.B. Myneni, N. Andela, J. Bi, G. Broquet, J.G. Canadell, F. Chevallier, Y.Y. Liu, S.W. Running, S. Sitch, G.R. Van Der Werf, Contribution of semi-arid ecosystems to interannual variability of the global carbon cycle, Nature 509 (2014) 600–603, https://doi.org/10.1038/nature13376.
[176] R. Sánchez-Salguero, R.M. Navarro-Cerrillo, J.J. Camarero, Á. Fernández-Cancio, Selective drought-induced decline of pine species in southeastern Spain, Clim. Chang. 113 (2012) 767–785, https://doi.org/10.1007/s10584-011-0372-6.
[177] J.J. Camarero, A. Gazol, J.C. Tardif, F. Conciatori, Attributing forest responses to global-change drivers: limited evidence of a CO_2-fertilization effect in Iberian pine growth, J. Biogeogr. 42 (2015) 2220–2233, https://doi.org/10.1111/jbi.12590.
[178] R. Alfaro-Sánchez, J.J. Camarero, R. Sánchez-Salguero, V. Trouet, J.D.L. Heras, How do droughts and wildfires alter seasonal radial growth in mediterranean Aleppo pine forests? Tree-Ring Res. 74 (2018) 1–14, https://doi.org/10.3959/1536-1098-74.1.1.
[179] G. Sangüesa-Barreda, J.J. Camarero, A. García-Martín, R. Hernández, J. la Riva, Remote-sensing and tree-ring based characterization of forest defoliation and growth loss due to the Mediterranean pine processionary moth, For. Ecol. Manag. 320 (2014) 171–181, https://doi.org/10.1016/j.foreco.2014.03.008.
[180] E.H. Stanley, S.G. Fisher, N.B. Grimm, Ecosystem expansion and contraction in streams, Bioscience 47 (1997) 427–435, https://doi.org/10.2307/1313058.
[181] P.S. Lake, Ecological effects of perturbation by drought in flowing waters, Freshw. Biol. 48 (2003) 1161–1172, https://doi.org/10.1046/j.1365-2427.2003.01086.x.
[182] A.J. Boulton, Parallels and contrasts in the effects of drought on stream macroinvertebrate assemblages, Freshw. Biol. 48 (2003) 1173–1185, https://doi.org/10.1046/j.1365-2427.2003.01084.x.
[183] E. Kalogianni, A. Vourka, I. Karaouzas, L. Vardakas, S. Laschou, N.T. Skoulikidis, Combined effects of water stress and pollution on macroinvertebrate and fish assemblages in a Mediterranean intermittent river, Sci. Total Environ. 603–604 (2017) 639–650, https://doi.org/10.1016/j.scitotenv.2017.06.078.
[184] M.E. Ledger, L.E. Brown, F.K. Edwards, A.M. Milner, G. Woodward, Drought alters the structure and functioning of complex food webs, Nat. Clim. Chang. 3 (2013) 223–227, https://doi.org/10.1038/nclimate1684.

[185] J. Ruiz-Olmo, J.M. López-Martín, S. Palazón, The influence of fish abundance on the otter (*Lutra lutra*) populations in Iberian Mediterranean habitats, J. Zool. 254 (2001)S0952836901000838. https://doi.org/10.1017/S0952836901000838.

[186] S. Gascón, D. Boix, J. Sala, X.D. Quintana, Changes on macrobenthic fauna of a Mediterranean salt marsh (Empordà wetlands, NE Iberian Peninsula) after a severe drought, with special emphasis on the Corophium orientale population, Vie Milieu 57 (2007) 3–12. https://www.scopus.com/inward/record.uri?eid=2-s2.0-34547917187&partnerID=40&md5=94d421c300447e3b4e95d4daaf10b479.

[187] H. Aguilera, L. Moreno, J.G. Wesseling, M.E. Jiménez-Hernández, S. Castaño, Soil moisture prediction to support management in semiarid wetlands during drying episodes, Catena 147 (2016) 709–724, https://doi.org/10.1016/j.catena.2016.08.007.

[188] M. Alvarez-Cobelas, S. Cirujano, S. Sánchez-Carrillo, Hydrological and botanical man-made changes in the Spanish wetland of Las Tablas de Daimiel, Biol. Conserv. 97 (2001) 89–98, https://doi.org/10.1016/S0006-3207(00)00102-6.

[189] M.V. Moreno, M. Conedera, E. Chuvieco, G.B. Pezzatti, Fire regime changes and major driving forces in Spain from 1968 to 2010, Environ. Sci. Pol. 37 (2014) 11–22, https://doi.org/10.1016/j.envsci.2013.08.005.

[190] S. Castaño-Castaño, P. Martínez-Santos, P.E. Martínez-Alfaro, Evaluating infiltration losses in a Mediterranean wetland: Las Tablas de Daimiel National Park, Spain, Hydrol. Process. 22 (2008) 5048–5053, https://doi.org/10.1002/hyp.7124.

[191] A. Özen, B. Karapınar, İ. Kucuk, E. Jeppesen, M. Beklioglu, Drought-induced changes in nutrient concentrations and retention in two shallow Mediterranean lakes subjected to different degrees of management, Hydrobiologia 646 (2010) 61–72, https://doi.org/10.1007/s10750-010-0179-x.

[192] J. Coppens, A. Özen, Ü.N. Tavşanoğlu, Ş. Erdoğan, E.E. Levi, C. Yozgatlıgil, E. Jeppesen, M. Beklioğlu, Impact of alternating wet and dry periods on long-term seasonal phosphorus and nitrogen budgets of two shallow Mediterranean lakes, Sci. Total Environ. 563–564 (2016) 456–467, https://doi.org/10.1016/j.scitotenv.2016.04.028.

[193] A. Tal, The implications of climate change driven depletion of Lake Kinneret water levels: the compelling case for climate change-triggered precipitation impact on Lake Kinneret's low water levels, Sci. Total Environ. 664 (2019) 1045–1051, https://doi.org/10.1016/j.scitotenv.2019.02.106.

[194] A. Gasith, S. Gafny, M. Goren, Response of the fish assemblage of rocky habitats to lake level fluctuations: possible effect of varying habitat choice, Adv. Limnol. 55 (2000) 317–331.

[195] M. Gophen, The impact of available nitrogen deficiency on long-term changes in the Lake Kinneret ecosystem, Open J. Ecol. 05 (2015) 147–157, https://doi.org/10.4236/oje.2015.54013.

[196] E. Custodio, Aquifer overexploitation: what does it mean? Hydrogeol. J. 10 (2002) 254–277, https://doi.org/10.1007/s10040-002-0188-6.

[197] K. Mazi, A.D. Koussis, G. Destouni, Intensively exploited Mediterranean aquifers: resilience to seawater intrusion and proximity to critical thresholds, Hydrol. Earth Syst. Sci. 18 (2014) 1663–1677, https://doi.org/10.5194/hess-18-1663-2014.

[198] UNWTO, U.N.W.T. Organization, World Tourism Barometer, Madrid, World Tourism Organization, 2009.

[199] E. Martínez-Ibarra, Climate, water and tourism: causes and effects of droughts associated with urban development and tourism in Benidorm (Spain), Int. J. Biometeorol. 59 (2015) 487–501, https://doi.org/10.1007/s00484-014-0851-3.

[200] M.R. Llamas, E. Custodio, A. de la Hera, J.M. Fornés, Groundwater in Spain: increasing role, evolution, present and future, Environ. Earth Sci. 73 (2015) 2567–2578, https://doi.org/10.1007/s12665-014-4004-0.

[201] J. Martin-Ortega, M. González-Eguino, A. Markandya, The costs of drought: the 2007/2008 case of Barcelona, Water Policy 14 (2012) 539–560, https://doi.org/10.2166/wp.2011.121.

[202] G. Chiogna, P. Skrobanek, T.S. Narany, R. Ludwig, C. Stumpp, Effects of the 2017 drought on isotopic and geochemical gradients in the Adige catchment, Italy, Sci. Total Environ. 645 (2018) 924–936, https://doi.org/10.1016/j.scitotenv.2018.07.176.

[203] C. Marchina, C. Natali, G. Bianchini, The Po River water isotopes during the drought condition of the year 2017, Water (Switzerland) 11 (2019), https://doi.org/10.3390/w11010150.

[204] M. Haworth, G. Marino, C. Brunetti, D. Killi, A. De Carlo, M. Centritto, The impact of heat stress and water deficit on the photosynthetic and stomatal physiology of olive (*Olea europaea* L.)—a case study of the 2017 heat wave, Plants 7 (2018), https://doi.org/10.3390/plants7040076.

[205] D.A. Wilhite, Combating drought through preparedness, Nat. Resour. Forum 26 (2002) 275–285, https://doi.org/10.1111/1477-8947.00030.

[206] D.A. Wilhite, A methodology for drought preparedness, Nat. Hazards 13 (1996) 229–252, https://doi.org/10.1007/BF00215817.

Further reading

J. Andreu, J. Ferrer-Polo, M.A. Pérez, A. Solera, Decision support system for drought planning and management in the Jucar River Basin, Spain, in: 18th World IMACS Congr. MODSIM09 Int. Congr. Model. Simul. Interfacing Model. Simul. with Math. Comput. Sci. Proc., 2009, pp. 3223–3229. https://www.scopus.com/inward/record.uri?eid=2-s2.0-80053024716&partnerID=40&md5=0dc2f67c466727c3b1106500bf715d03.

J. Andreu, J. Ferrer-Polo, M.A. Pérez, A. Solera, J. Paredes-Arquiola, Drought Planning and Management in the Júcar River Basin, Spain, (2013).

CHAPTER 10

Land-use change and impacts

Noemí Lana-Renault[a], Enrique Morán-Tejeda[b], Mariano Moreno de las Heras[c,d], Jorge Lorenzo-Lacruz[a], Nacho López-Moreno[e]
[a]University of La Rioja, Logroño, Spain
[b]Department of Geography, University of the Balearic Islands, Palma de Mallorca, Spain
[c]Institute of Environmental Assessment and Water Research, CSIC, Barcelona, Spain
[d]Desertification Research Centre, CSIC-UV-GV, Valencia, Spain
[e]Pyrenean Institute of Ecology, CSIC, Zaragoza, Spain

10.1 Introduction

Land-use change is considered one of the main driving forces of Global Change [1–3], a term that refers to planetary-scale recent anthropogenic transformation of the environment [4]. Human activities have left a strong imprint on the Earth's land surface for thousands of years, and even more so in recent centuries, with a significant impact on natural resources. One of the major environmental effects of land-cover changes worldwide is the alteration of the hydrological cycle, which directly affects the quantity and quality of freshwater supplies [5].

The Mediterranean basin is one of the main *hotspots* of Global Change [6, 7]. First, it is one of the regions most severely affected by climate change [8–10]. Many studies predict that these climate alterations will continue in the future, inducing general increases in temperatures for all seasons, decreases in precipitation (particularly during the summer season), longer and more frequent droughts, and higher compression of precipitation in extreme events [8, 11, 12]. In addition, the Mediterranean basin is characterized by a long history of human presence and landscape transformation. Paleoenvironmental and historical records in Mediterranean landscapes reveal that land-use changes of varying magnitude have affected the region continuously for the past 10,000 years [13, 14], causing intense deforestation due to the expansion of livestock pasture and cultivation in large areas. At present, the Mediterranean region is a complex mosaic-structured landscape, determined by the heterogeneity of Mediterranean topography, soils, and vegetation as well as by the long-term legacy of past human activities [15].

Mediterranean landscapes show at present highly dynamic behavior, changing rapidly according to demographic evolution and market forces. A clear distinction can be made between major land-use trends affecting the northern rim of the Mediterranean basin and the southern- and eastern-rim Mediterranean regions. The northern rim is characterized by the stabilization of population and the coexistence of contrasting patterns of land use, with land abandonment in remote and low-productive mountain areas and concentration

Water Resources in the Mediterranean Region
https://doi.org/10.1016/B978-0-12-818086-0.00010-8

of economic activity in lowlands and coastal areas. On the contrary, for the southern- and eastern-rim regions, rapid population growth involves increasing demand for agricultural production, associated with deforestation, cropland expansion, and agriculture intensification [16, 17].

Water resources, usually scarce and unevenly distributed throughout the Mediterranean region, are highly vulnerable to Global Change [13, 17, 18]. Water demand for human uses is expected to increase in the future, while both climate and land-use changes are expected to decrease freshwater supplies, especially for the southern- and eastern-rim Mediterranean regions [19]. Water resources are particularly relevant for these regions, where the economic development and well-being of a rapidly growing population will depend to a great extent on the accessibility of water. Hence, innovative mitigation strategies are required to optimize the management of water resources in these areas and thus reduce the very real threat that one of the primary world's water crisis *hotspots* will worsen [19, 20].

10.2 State of the art

10.2.1 Recent land-use changes in the Mediterranean region

The Mediterranean basin has sustained important human population settlements since Classical Antiquity [14]. Today the region contains about 560 million inhabitants, unequally distributed between northern (225 million inhabitants) and southern and eastern countries (335 million inhabitants) [21]. This disparity can be explained by their contrasting demographic dynamics (Fig. 10.1). Whereas population growth has stabilized in recent decades in the northern countries (0.08% annual growth rate), growth rates are still high (1.72%) in the southern and eastern countries, where population is expected to increase to 450 million by 2050 [21, 22]. These increases in population are associated

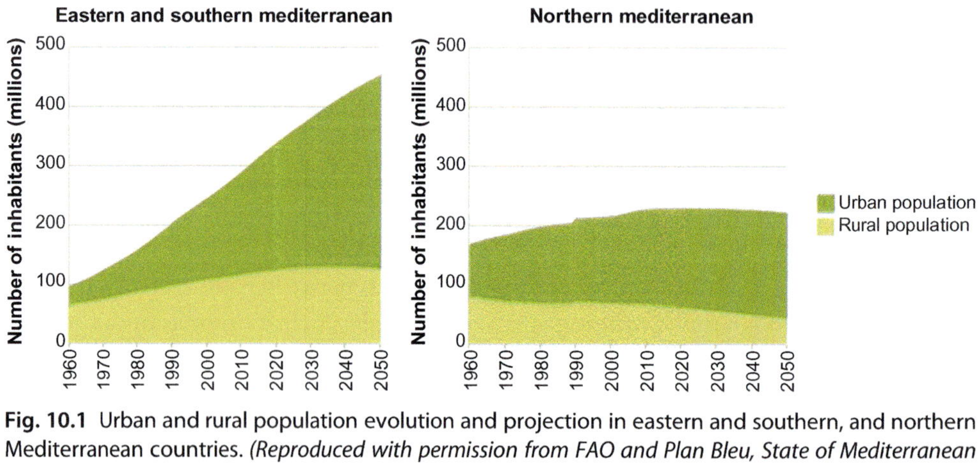

Fig. 10.1 Urban and rural population evolution and projection in eastern and southern, and northern Mediterranean countries. *(Reproduced with permission from FAO and Plan Bleu, State of Mediterranean Forests 2018, 2018. http://www.fao.org/3/CA2081EN/ca2081en.PDF (Accessed 12 November 2019).)*

with strong urban growth (i.e., urban population increased from 48% in 1960 to 68% in 2015; [21]), which occurs predominantly on the coast. This tendency is expected to continue so that urban population in the southern and eastern countries will represent more than 75% by 2050 (Fig. 10.1). Rural population has also increased in these countries since 1960, though at a much lower rate, and is expected to stabilize by 2030. In contrast, rural population in the northern countries has declined considerably since 1960, while urban population has slightly increased, although it is expected to stabilize by 2020.

Population decline in rural areas of the northern countries has resulted in abandonment of farmland and reduction of livestock grazing, with the loss of traditional agricultural practices (e.g., use of cultivation terraces and shifting agriculture in marginal areas) and subsequent general regeneration of forests and shrubs [23–25]. Vegetation recovery in former agricultural land has been accompanied by extensive afforestation programs by national forest services to improve the environmental value of abandoned lands and to control hydrological and soil erosion processes [26, 27]. Revegetation in abandoned lands has important environmental consequences, including changes in the water cycle (see Section 10.2.2), soil erosion and sediment transfer alterations, stream morphology modifications, and both landscape heterogeneity and biodiversity changes, with further implications for wildfire risk due to increased fuel load and biomass [24, 28]. However, lowlands and coastal areas have increased both population density and intensive agriculture aimed at national and international markets [29, 30]. These changes have led to the expansion of irrigated areas and a general increase in agricultural productivity, favoring new crops (e.g., maize, alfalfa, rice, new citrus and exotic fruit trees) with higher water requirements than traditional Mediterranean crops (e.g., cereals, olives, almonds, grapes) [31].

In the southern and eastern countries, high demographic pressure has resulted in significant deforestation due to land clearance for cultivation, livestock breeding, and wood gathering for heating [32, 33]. In some of these countries, for example Morocco and Algeria, large areas of forests disappeared during the colonial period due to logging or were severely damaged by war and extreme drought [34]. By 1990, wooded lands occupied 1.7%, 4.2%, and 13% of the total surface areas of Algeria, Tunisia, and Morocco, respectively. Actual forest distribution is markedly smaller than the ecological potential of these countries [35]. Since the 1990s, forest decline has slowed down, assisted by forest protection and restoration by public bodies. In the Rif area, for instance, the implementation of UN Development Programs since 2000 has stimulated woodland preservation by the local population [36]. Several protected areas, such as the Toubkal National Park in Morocco, have also been created in order to preserve the remaining forest ecosystems [36]. In some other areas, major forest plantation programs, such as the *Green Barrage* in Algeria, have been developed to fight desertification [35]. However, the existence of large rural populations, particularly in mountain areas, explains the persistence of significant human pressure on forest ecosystems. Cultivation of marginal lands, extraction of wood for fuel and overgrazing by small to medium-size households result in significant woodland degradation [37–39]. Intense urbanization and the associated expansion of agricultural production to feed the rapidly increasing population are at present among the main causes of deforestation in these regions [36, 40]. The expansion of cultivated land has been frequently accompanied by agriculture intensification and transformation to irrigation crops, in many cases stimulated by national governments (e.g., the *Plan Maroc Vert* of the Moroccan government) to reduce dependency on food imports and increase food production competitiveness in international agricultural markets [41, 42].

As well as this demographic evolution of the Mediterranean countries, tourism development is another key factor explaining land-use trends. The Mediterranean region has become a top tourist destination, with 30% of all global tourist arrivals [43]. The number of international tourists grew from 58 million in 1970 to more than 349 million in 2015, with a projected 500 million visitors by 2030 [43]. Tourism is concentrated in the coastal areas and results in strong urbanization processes in the entire Mediterranean basin [30, 44, 45], including forest regression and fragmentation from urban sprawl [46, 47], and a notable increase in water demand (see Section 10.2.2).

Fig. 10.2 summarizes the changes in vegetation that have taken place in the Mediterranean basin during the last 35 years. The analysis comprises remote sensing data derived from the Global Inventory and Mapping Studies-Normalized Difference Vegetation Index (GIMMS-NDVI) dataset (https://glam1.gsfc.nasa.gov/), obtained from NOAA-AVHRR satellite images between 1982 and 2015. Fig. 10.2A shows the average annual NDVI for the study period: large contrasts in vegetation activity among regions can be observed. There is a clear contrast in vegetation activity between the northern and the southern and eastern rims of the basin, with annual NDVI reaching values greater than 6 (high vegetation activity) for large extensions in Europe and parts of Anatolia, particularly in France, the northern and mountain areas of the Iberian Peninsula, most of Italy, the Balkan Peninsula, and northern Turkey. Annual NDVI on the southern and eastern rims rarely exceed a value of 2 (low vegetation activity), even though particular areas show higher vegetation activity (e.g., coastal sectors in Morocco and Algeria). Fig. 10.2B shows the magnitude of the change in NDVI values per year between 1982 and 2015, based on linear regression analysis, while Fig. 10.2C shows those regions that have experienced significant changes (either positive or negative) in NDVI during the study period, based on the nonparametric Spearman-Rho test ($\rho < 0.05$). Despite the evident contrast in average vegetation activity between the northern and southern rims (Fig 10.2A), greening up dominates throughout the Mediterranean basin, favored by woodland recovery [49, 50], lower pressures on forest resources, and the expansion of irrigated lands. Recent studies have also identified global warming and the rising of atmospheric CO_2 concentrations as additional factors contributing to increased vegetation growth and expansion (e.g., [51]). Significant positive trends in annual NDVI were observed in mountain areas of Spain, mainland France, the Italian, Balkan and Anatolian

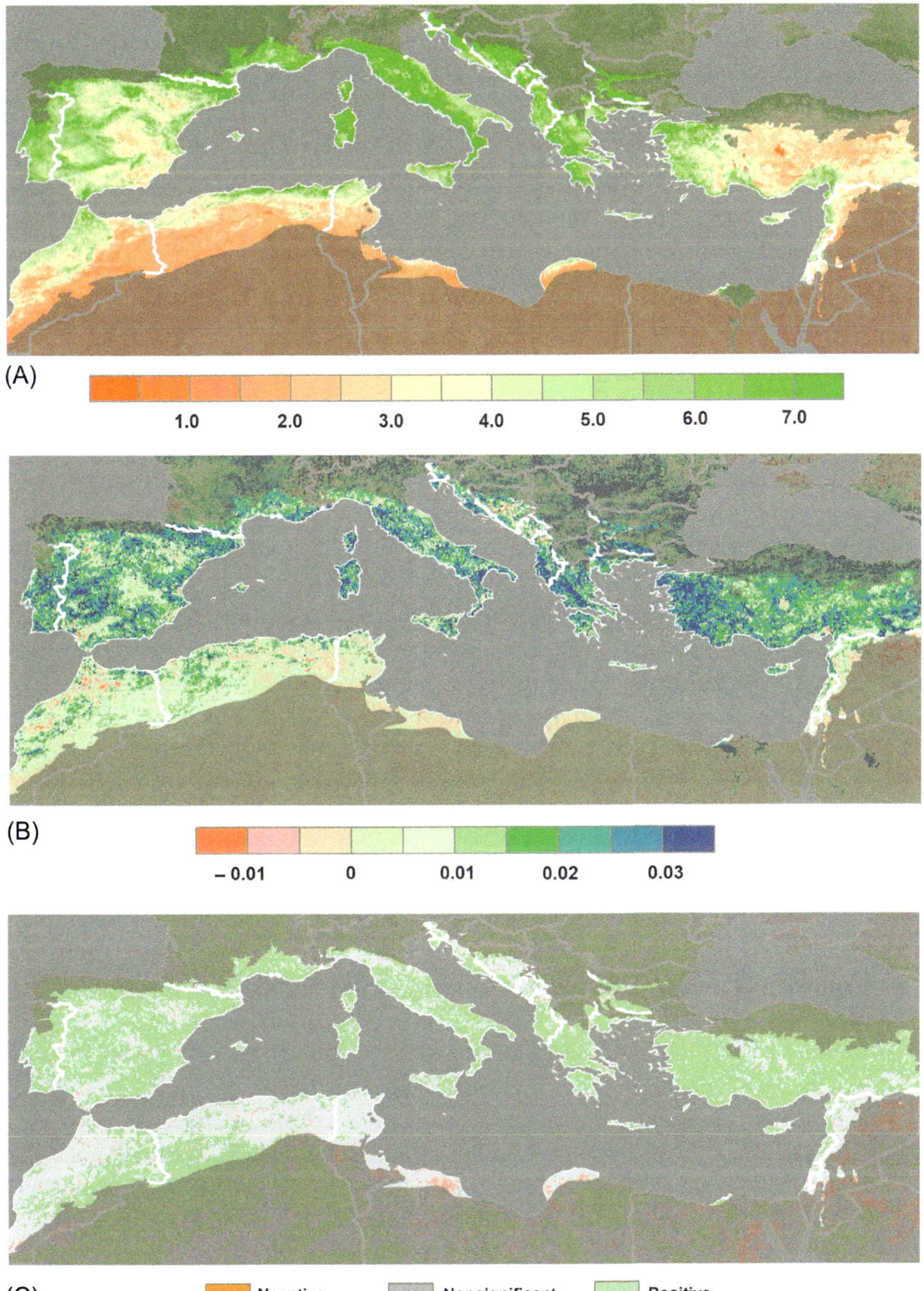

Fig. 10.2 NDVI (Normalized Difference Vegetation Index) evolution in the Mediterranean region for the period 1982–2015: (A) mean annual NDVI sums; (B) annual change in the NDVI (slope of the linear regression); (C) significant changes in the NDVI values according to the Spearman-Rho test ($\rho < 0.05$). The limits of the Mediterranean region are based on [48].

Peninsulas, and the Atlas mountain range. However, some exceptions to this general recovery pattern can be found, illustrated by significant negative trends in various areas of Morocco, Algeria, Tunisia, and Libya.

10.2.2 Hydrological impacts of land-use changes

The direct effect of land-use changes is the transformation of land cover, which further alters the water balance and partitioning of precipitation between evapotranspiration, runoff, and groundwater flow [5]. Given that water resources in the Mediterranean region are particularly threatened by Global Change, the study of the impact of land-use changes on the different components of the hydrological cycle is central to hydrological research [52].

10.2.2.1 Hydrological impact of revegetation on formerly cultivated lands

Vegetation regulates the transfer of water between the land surface and the atmosphere, through rainfall interception and evapotranspiration processes. Trees intercept and lose (through evaporation) large quantities of rainfall. A comprehensive review carried out by Llorens and Domingo [53] in the Mediterranean area of Europe showed that mean rainfall interception ranges from 18% to 32% for scrub and forest vegetation, which illustrates the major repercussion of vegetation cover for the availability of water resources. Transpiration demand is thought to be lower in forests than in short (grasslands) vegetation because of their high surface (or stomatal) resistance [54]. However, forest trees can mobilize more soil water than grasses because they have deeper root systems [55]. Additionally, their larger evaporative potential (due to the lower aerodynamic resistance of forest cover) guarantees substantial and continuous water transfer to the atmosphere. Thus, it is broadly accepted that evapotranspiration is higher under forest cover than under grassland vegetation.

As a result of the effects of forest vegetation on rainfall interception and increased evapotranspiration, water yield usually increases when forests are transformed into systems with lower biomass and decreases after revegetation processes [56, 57]. However, the hydrological impact of these processes varies greatly and depends on several factors, mainly:

- *Climate conditions.* The impact of forest clearance on water yield is higher in humid than in drier climates [56, 58]. The main reason for this is that, under wet climate, differences in evapotranspiration rates between land covers are larger. In an extensive modeling and metaanalysis study using worldwide hydrological data from a large number of experimental catchments, Zhang et al. [55] showed that, for annual precipitation below 500 mm, the differences in water loss (including both transpiration and interception processes) between forest cover and short vegetation were very small. It is worth noting that for the Mediterranean basin, large areas on the southern rim are classified as arid or semiarid, with annual precipitation below 300 mm. In these areas, characterized by very high evapotranspirative demand and low precipitation, water loss to the atmosphere may be similar for all types of natural vegetation cover. In addition, wet climates frequently contain a period of water surplus, which allows water recharge in deep soil layers that may favor subsequent water release, particularly in open areas and cleared forests [57]. However, the effects of clearance are shorter in wet climates because vegetation recovers strongly and more rapidly [56, 58].

- *The extent and spatial distribution of the land-cover disturbance.* The review carried out by Stednick [58] in the United States reported a minimal threshold of forest reduction at the catchment scale between 20% and 50% before significant changes in water yield are detected. In oak forests under Mediterranean conditions of the western United States (California), Lewis et al. [59] showed that a 14% reduction in tree cover had no significant effect on streamflow. A similar result was found by Gökbulak et al. [60] for an oak forest in Turkey, where an 18% reduction of forest cover caused no significant increases in river discharge.
- *Soil characteristics and depth.* In his review of forests and water, Andréassian [57] indicated that the effects of a change in vegetation cover are particularly noticeable where the soils are deep enough, as the volume of water extracted by forest trees (with deep roots) is higher than that extracted by shorter vegetation. In addition, both Cosandey [61], in the French Massif Central, and Stednick [58], in paired catchments distributed all over the United States, suggested that if the (high) permeability that characterizes forest soils is maintained after forest clearance, tree removal will only have a very small impact on water yield.
- *Characteristics of the vegetation.* In their review, Bosch and Hewlett [56] indicated that the impact of forest clearance was greater for conifers, followed by deciduous hardwood forest and scrubs. These results were confirmed by Sahin and Hall [62], who showed that, for the same level of reduction in tree canopy cover, the resulting increases in annual water yield were higher for conifers (20–25 mm/year) and deciduous hardwood forest (17–19 mm/year) than for scrubs or eucalyptus cover (5–6 mm). Andréassian [57] argued that the effects of forest thinning on water yield may also depend on the age and physiology of vegetation, as tree transpiration is sometimes much higher in young individuals than in older trees. He illustrated these effects with an example of eucalyptus plantations, where water yield decreased rapidly after 2 years of tree clearance due to strong transpiration by young eucalyptus trees. These are important considerations when assessing the hydrological impact of revegetation on abandoned farmlands, as the new forest may comprise pioneer and/or introduced species with different physiological characteristics from the original tree vegetation [63].

In arid and semiarid areas, the impact of a change in vegetation cover on water yield should be small, given the negligible differences in evapotranspiration between different types of vegetation [55]. However, some authors [64, 65] have indicated that the water yield effects of land-cover management can be reinforced for semiarid woodlands located in (i) areas where the water table is close to the ground surface, as tree and shrub vegetation can transpire important amounts of water in these locations; (ii) areas with a mismatch of rainfall and evapotranspirative demand between cold and warm periods (as in the Mediterranean region), because this favors water recharge to the deep soil layers during the cold/wet period, and (iii) areas where soil/bedrock properties allow a rapid vertical transmission of water to deep layers and stores. The complexity of the soil-vegetation-water relationships in drylands is illustrated by Wilcox et al. [66] in a semiarid (annual rainfall 400–500 mm/year) catchment in Texas (Concho River watershed). They observed a notable decrease in streamflow during the second half of the 20th century, due to a strong process of vegetation recovery. Interestingly, whereas the effect was clear for the stormflows, the baseflows did not change or slightly increased. The authors explained this phenomenon by the higher infiltration rates of revegetated soils after landscape recovery.

An extensive body of literature has examined, at different spatial scales, the hydrological impact of vegetation recovery throughout the Mediterranean region. Whereas great effort has been devoted to the identification of changes in streamflow data series due to Global Change (including both climate and land-use change) at the regional scale, fewer studies have focused on detecting the effect of land-cover changes on observed flows, regardless of climate trends. Gallart and Llorens [67] reported a decrease in streamflow discharge in several Spanish rivers over the period 1921–96 and suggested that this negative trend should be explained, not only by climate variability and increased water consumption due to irrigation but also by increased evapotranspiration from vegetation recovery in the headwaters. This was confirmed in the Spanish Pyrenees by Beguería et al. [68], who observed a decline in river discharges over the previous 50 years and demonstrated that this decline was due to an expansion of shrubs and forests in formerly cultivated areas. Similar results were reported by Morán-Tejeda et al. [69] in the Duero basin, Gallart et al. [70] in the eastern Pyrenees, and López-Moreno et al. [71] in the Ebro basin. Additionally, López-Moreno et al. [72] indicated that vegetation recovery in the Pyrenees explains to a great extent the negative trends in high flow occurrence observed over the last 50 years.

More studies have assessed the impact of land-use changes on river water yields by using modeling approaches. Literature reviews by García-Ruiz et al. [13] and Merheb et al. [52] included several examples of studies that used different models for streamflow simulation under land-use change scenarios. For instance, in a large agricultural catchment in northwestern Anatolia, Albek et al. [73] applied the HSPF hydrological model to show that the existence of deep-rooted vegetation cover over the catchment may decrease streamflow by 37%. Based on the Hydrological Land Use Change Model, Delgado et al. [74] found a significant decrease in summer low flows following farmland abandonment and forest recovery in a river basin of northeastern Spain. Ranzi et al. [75]

found that the impact of increasing vegetation cover in the headwaters of a catchment in the Italian pre-Alps was balanced by an increase in urban areas in the lowlands so that only a slight decrease in flood peaks and volumes was detected. On the contrary, Antonelli et al. [76], using in northern Morocco a model to forecast the hydrological impact of expected increases in dryland farming (and associated decreases in natural areas, including forest), showed that such land-use changes may cause about 10% increase in runoff.

The hydrological impact of land-use changes has also been examined at the small catchment scale. This scale permits examination in greater detail of how a change in land cover affects the hydrological processes that control river discharge (see Section 10.3). In the Mediterranean region, the seminal work of Cosandey et al. [77] in southern France demonstrated that forests noticeably reduce runoff coefficients and slightly decrease annual flooding. This is usually explained by the effects of rainfall interception and forest water consumption that reduce soil water content and limit catchment hydrological responses, as shown by Gallart et al. [78] and Llorens et al. [79] in small catchments of northeastern Spain. Similar results were obtained by García-Ruiz et al. [80] in several catchments in the Pyrenees with different land cover, including natural forests, vegetation recovery on abandoned farmlands, and bare-soil badlands. Lana-Renault et al. [81] showed that these catchments responded very differently to an exceptional rainstorm event, with longer runoff response times in the more densely vegetated catchments, which demonstrated that vegetation cover was a major factor affecting the control of floods, at least at this spatial scale and under dry preceding conditions. Research in these catchments also revealed that vegetation cover, associated with deep and highly permeable soils, affected flood hydrograph characteristics. Thus, the forest catchment usually had lower peakflows, slower response times, and very slow recession limbs, suggesting a large contribution of subsurface flows [80, 82], which is consistent with the general perception that stormflow in such forest environments is dominated by subsurface processes [77, 83]. Interestingly, recent research in a neighboring afforested catchment [84] showed, at least for wet conditions, different behavior, characterized by rapid responses and sharp hydrographs. These faster and more acute hydrological responses of the afforested catchment suggest the occurrence of excess saturation overland flow, probably favored by soils with low storage capacity (e.g., shallow soils) that saturate easily [85]. These findings corroborate results obtained by Cosandey and Didon-Lescot [86] in southern France and Robinson et al. [87] in Germany, who showed that the impact of afforestation was small or nonexistent during extreme rainfall. In arid and semiarid areas, revegetation on formerly cultivated areas is limited by scarce and irregular rainfall and shows a very complex pattern [88]. Long-term research in southeastern Spain has shown that the number of runoff events occurring in an abandoned farmland catchment was lower than in a nearby badland catchment, especially for low-magnitude rainfall, and that the hydrological response was somehow delayed [89].

The key role of land cover on water yield has also been confirmed at the plot scale. In the Pyrenees, Nadal-Romero et al. [90] analyzed the hydrological response of various traditional and current land uses in Mediterranean mountain areas, showing that both cereal crops and fallow lands had higher runoff coefficients than shrub cover. Using rainfall simulations on present and former agricultural land in central Portugal, Nunes et al. [91] demonstrated the positive effects of plant colonization on controlling runoff production, obtaining the most rapid and greatest response to precipitation in cereal crops, and very low or no overland flow on abandoned land, especially on surfaces covered by forest revegetation patches. In contrast, several studies of semiarid areas [92–94] have shown that crust formation after land abandonment is a common process that reduces infiltration capacity and increases overland flow. However, in such environments, the soil surface is extremely heterogeneous and combines patches of bare surfaces, crusts, and vegetation that act as sources and sinks for surface water resources, which explains these environments' complex hydrological response [89, 95, 96].

Revegetation also affects soil moisture dynamics and the recharge of aquifers. Research in forest plantations (see for example Callegari et al. [97] in Italy or Maestre and Cortina [98] for a review of *Pinus halepensis* plantations in the Mediterranean semiarid region) has demonstrated that there is lower soil water content under trees, due to rainfall interception and transpiration processes. These effects were also illustrated by plot measurements in Israel, indicating that, under semiarid conditions, total water uptake by a *P. halepensis* stand can reach up to 93% of the effective rainfall [99]. Long-term research in the more humid Vallcebre catchments in the eastern Pyrenees [79] has shown that soil water content is usually lower under forest than in areas covered by grass. However, under wet conditions, catchment topography was the factor that controlled soil water distribution, limiting the effect of vegetation [78]. Research at this site also showed that high transpiration by Scots pines (*P. sylvestris*) under dry conditions greatly reduces deep-water stores at the catchment scale [100]. Similarly, Bellot et al. [101], using field measurements and a hydrological modeling approach in a semiarid landscape in Spain, found that pine plantations reduce deep drainage (i.e., aquifer recharge) much more than shrubs do.

The influence of forest on snow accumulation and distribution is another significant factor to take into account when assessing the impact of land-use change on runoff generation in Mediterranean mountain environments. Forest reduces beneath-canopy snow accumulation due to interception and subsequent sublimation. It may also alter (either accelerate or reduce) snow-melting rates, depending on the climate and where the forest stand is located [102]. For example, López-Moreno and Latron [103] and Revuelto et al. [104], in studies of two forest stands in the central Pyrenees, suggested that forests reduce the annual peak of snow-water equivalent and lead to earlier snowmelt. The magnitude of these changes was proportional to forest density: they were more intense during warm and dry years. Revuelto et al. [105] analyzed the effect of harvesting the lowest branches of pine trees (a common practice in forestry to reduce fire risk). This forest treatment resulted in an increase in snow accumulation close to 10%, which demonstrated that simple forest management actions can have a positive impact on the water balance of mountainous catchments. Other results have also pointed to the potential impact of expected shift-up changes of the tree line in Mediterranean mountains, as a consequence of reduced livestock grazing and generalized increases in temperatures [106].

10.2.2.2 Hydrological impact of the expansion of irrigated land, tourism and urban development

Irrigation agriculture is the most water-demanding sector in the Mediterranean region, accounting for 65% of the total water demand, with a clear contrast between the northern (48%) and the southern and eastern rims (82%) of the Mediterranean basin [107]. These figures reflect the significant role that irrigation plays in agricultural production to overcome the deficit in precipitation in the region, especially in those countries that are most affected by climate aridity. Conversion of rain-fed to irrigation agriculture usually involves an increase in cropping intensity and a shift to higher market value crops with greater water requirements. In the Mediterranean region of Turkey, Tanrivermis [31], analyzing land-use changes between 1950 and 1997, found that modernization of agriculture and expansion of irrigated land were accompanied by a large increase in agricultural land devoted to vegetables and orchards, from 646,000 to 1,524,000 ha. In the large-scale irrigation scheme of Tadla, 170 km southeast of Casablanca (Morocco), Bekkar et al. [108] reported that the conversion to micro-irrigation involved a change to higher value crops such as vegetables (melon, tomatoes, etc.) or citrus fruits, which implied a likely increase in water consumption at the plot level. Such an increase was also illustrated by Rodríguez-Díaz et al. [109] in southern Spain, where the introduction of citrus plantations in newly irrigated land resulted in a 20% increase of crop evapotranspiration. Additionally, some studies have reported that the greater and more frequent water supply in irrigated lands may increase crop transpiration and subsequent water consumption (see [110] for examples).

Urbanization and tourism are also important water-demanding sectors. Both the increase of low-density urban sprawl [111] and the development of new types of tourism (i.e., "quality" or "residential" tourism) in the northern countries have boosted water consumption. March and Saurí [112], for instance, showed that the water usage of single residential households in the metropolitan area of Barcelona was about 100 L per day and capita higher than for more compact areas (with multiresidential households). Similarly, Rico-Amorós et al. [113] estimated that, on the western coast of Spain, water consumption in single detached houses could be up to five times higher than in apartments, especially where single houses have swimming pools and/or gardens, two characteristic features of residential tourism. On the island of Mallorca, Hof and Schmitt [114] showed that the higher water consumption of quality tourism was mainly caused by garden irrigation, which accounted for more than 70% of total annual water use. A more recent study in the Balearic Islands [115] showed that evaporation from swimming pools added almost 10 L of water per tourist and guest night, which represented 5% of the total urban water consumption (i.e., 4.8 hm^3 of water lost by evaporation from pools in 2015). In the southern and eastern countries, urbanization and tourism growth are also expected to increase water demand. For instance, in the Souss-Massa Basin in western Morocco, water demand to satisfy the requirements of the increasingly golf courses and tourist areas near Agadir is projected to double by 2030 [116], although it will remain small in absolute terms compared to other water uses (i.e., agricultural, urban, and industrial).

To satisfy the water needs of the growing irrigation agriculture sector and expanding urbanization, complex infrastructures have been constructed in recent decades, such as reservoirs, canals, and inland waterways to store and transfer water to areas where the demanding sectors are concentrated, and pumping systems to abstract groundwater from aquifers. Reservoir construction and management directly affect river regimes. A decline in downstream river discharge due to flow diversion to irrigation channels has been observed in the whole Mediterranean region (e.g., [117–120]). This decrease can be so dramatic that it sometimes severely threatens the minimum ecological requirements for maintaining the biological functionality of the flows [121]. Another reason that contributes to reducing downstream river discharge is water evaporation from the reservoirs, as shown by Leduc et al. [122] in Tunisia and Warner [123] in the Rhône basin in France. Reservoirs also affect flood intensity and distribution, increasing peakflow return periods [117, 124]. General overexploitation of aquifers has also been reported over the entire Mediterranean region. For instance, overpumping of the alluvial aquifer for intensive agriculture in the Souss-Massa basin has resulted in water level decline of between 0.5 and 2.5 m/year in the past three decades [116]. In Lebanon, Shaban [125] observed a clear regression of the water levels in several aquifers over the last 40 years due to human activities, as well as a 50% decrease in the number of springs. Groundwater abstraction in southeastern Spain, mainly for irrigation but also for tourism, was determined as 700 hm^3/year at the beginning of the 21th century for an estimated recharge of 200 hm^3/year [126, 127], resulting in a very large structural water deficit. On the island of Mallorca, García and Servera [128] showed that the rapid growth of tourism between 1970 and 2000 caused a steep decline in available groundwater resources, leading to the construction of desalination plants. The consequences of intensive aquifer exploitation are the exhaustion of groundwater resources, but also the loss of freshwater springs and wetlands, with the associated loss of biodiversity [129]. Other problems from groundwater overexploitation detected in the Mediterranean area include extensive soil subsidence [130] and seawater intrusion into aquifers near coastal areas [131, 132].

Both agriculture and urbanization can deteriorate the quality of freshwater resources. Intensive agriculture usually causes water degradation due to nutrient and chemical leaching to groundwater and streams. Hydrochemical and isotopic studies in the

Souss-Massa irrigation scheme showed a deterioration of the overall quality of groundwater, with a markedly high concentration of nitrates and almost 40% of the wells exceeding the regulatory quality thresholds for drinking water [116]. In the Ebro basin (Spain), Lassaletta et al. [133] observed an increase of 46% in nitrate concentrations in stream waters over the period 1981–2005 due to the growing use of fertilizers, associated with the expansion of irrigated areas. However, a more recent study by Lutz et al. [134] in the same region showed a recent decrease in nitrate pollution in streamflows, largely attributable to improvements in agricultural practice after the implementation of UE and Spanish regulations. Urbanization may also be the main cause of freshwater pollution due to untreated or insufficiently treated wastewater [135, 136]. The establishment of efficient urban sewage treatment facilitates recovery, to a large extent, of water quality, as shown by Aguilera et al. [137] in the Ebro basin. These authors observed an overall decrease in phosphate water concentrations since the 1990s, coinciding with the improvement of wastewater treatment in the most important cities of the area.

10.2.3 Land-use change and water management

Water demand in the Mediterranean basin has doubled since the late 1950s [138] and it is expected that this upward trend will be maintained in the future. Land use-change trends on the one hand indicate that cropland expansion and intensification will continue. Valleys and coastal areas will face strong urbanization due to population growth and the development of tourism, while vegetation will still expand in low-productive marginal areas [42, 139, 140]. On the other hand, climate change scenarios point to future decreases in the availability of freshwater resources [13, 18], especially on the southern rim of the Mediterranean region [19]. The future of water resources on this southern side of the Mediterranean basin is of the utmost importance, as this area will see the highest rates of population growth in the coming decades. Less water available will impose major constraints on its future socio-economic development. These worrying trends urge improvement of land and water management, which requires both good understanding of the impact of land-use changes on water resources and the application of sustainable land and water management strategies.

10.2.3.1 Forest management at the watershed scale

We have shown earlier in this chapter that vegetation recovery in formerly cultivated areas reduces runoff production and deep-water percolation, with important consequences for streamflow discharge. Decades of forest hydrology research have provided sufficient knowledge on the impact of forest management strategies on water yields. From these studies (see Section 10.2.1), we can draw some interesting management recommendations for the Mediterranean region. First, if the main landscape management objective is to increase water yield, forest management interventions (i.e., forest thinning) should focus more on areas with humid and subhumid conditions, rather than a dryer climate. In fact, forest thinning experiments and modeling in semiarid Mediterranean landscapes of eastern Spain dominated by marginal holm oak (*Quercus ilex*) formations and mature Aleppo pine (*P. halepensis*) afforestation indicate that tree clearance has a negligible impact on water yield for areas with under 350 mm annual precipitation [141, 142]. Positively, the most important *hotspots* of forest expansion for the Mediterranean region are located at the headwaters of the river basins in the mountain ranges of the northern-rim Mediterranean countries, under more humid situations that provide optimal conditions for water yield-focused forest management applications. Though producing negligible impact on water yield, forest thinning applications in more arid areas usually result in increased water availability for tree transpiration, so providing better conditions for the growth of the forest stands, improving their resilience to natural disturbances (droughts, plagues), and providing a practical way to prevent wildfire risk and propagation [141, 143]. Second, to obtain positive effects in terms of water yield and/or forest stand improvements, forest clearance should be sufficiently large, i.e., at least 20% reduction of forest cover and up to 50% reduction under drier conditions [58, 144, 145]. Finally, it is important to keep in mind that the effects of forest treatments are temporary. As they are particularly short under humid conditions with strong vegetation recovery, these treatments need to be regularly applied. At present, however, there is an intense debate among scientists and land managers regarding forest intervention (i.e., forest clearing) versus leaving areas to undergo secondary succession (also known as "rewilding" or "woody encroachment"). There is clear evidence that vegetation expansion reduces runoff production and that the presence of dense, continuous forests favors the occurrence of large wildfires due to increased fuel load [146]. Moreover, for many authors, rewilding may constitute a significant loss of traditional, cultural landscapes and development opportunities for the local population [147, 148]. On the other hand, vegetation recovery is generally thought to provide important environmental benefits, such as increases in carbon sequestration [149, 150], and to be the best alternative for nature conservation and biodiversity [151]. Nevertheless, some authors differ, arguing that revegetation may cause in some circumstances landscape homogenization and species simplification [152]. In any case, land and forest managers should be aware of these conflicting positions. Alternative land management strategies have been proposed, such as shrub clearing [153]. However, no clear consensus about the optimal land management strategy for these abandoned areas has yet been reached.

10.2.3.2 Water use and management

Urbanization, tourism development, and the expansion of irrigated land are expected to increase water demand in the medium term over the entire Mediterranean region, jeopardizing the availability of freshwater resources. Under this projected scenario of increased water use, many authors argue for mitigation measures based on the improvement of water efficiency in both the agricultural and domestic sectors (e.g., Mediterranean Strategy for

Sustainable Development [153a]). Numerous studies (e.g., [19, 20, 140, 154]) have investigated the patterns and trends of water resources under sustainable development scenarios that follow different assumptions, including improvements in water consumption, in water supply infrastructures, and in the use of land. These studies demonstrate that such mitigation measures reduce water withdrawal and subsequent water stress in the region. However, some of the proposed water-saving measures, particularly in the agriculture sector (e.g., micro-irrigation), need to be treated with caution as they can cause, as shown by Molle and Tanouti [110], farming practices (overirrigation, changes in cropping patterns, or crop expansion) that may result in higher water consumption. Other authors have considered sustainable land management strategies. Malek et al. [140], for instance, explored the impact of alternative land cover and land management options (including crop production, crop type, irrigation systems, population growth, demand for built-up areas, etc.) on freshwater resources. Their results indicate that increasing the efficiency of rain-fed systems and enhancing traditional mosaic-like landscapes improve the status of water resources. In woody crops, there is an increasing tendency toward the introduction of alternative cultivation practices with protective covers (e.g., grass cover, mulching, or geotextiles) that show positive effects for water conservation: for instance, their application can decrease runoff and provides favorable soil moisture conditions [155–157].

10.3 Description of the main methodologies used to assess the hydrological impact of land-use changes

10.3.1 Field observations, measurements, and experiments at different spatial scales

Fundamental advance in hydrological research has come from field observation and measurement [158, 159]. As Burt and McDonnell [158] maintain, most hypotheses have emerged from curiosity in the field and from asking oneself how the natural system works. Although a lot of knowledge has been accumulated on water flux mechanisms (e.g., runoff generation processes, flow paths, water delivery to the streams, etc.), these processes need to be understood better in a changing environment characterized by rapid transitions in land use, uncertainties in climate change projections, and increasing water demand. For instance, some authors claim that there is still a lot to explain about how managed landscapes (e.g., forest plantations, eco-restoration projects) affect important water fluxes and processes such as water storage or water yield, especially in the long term (see [158], for examples).

Controlled field experiments have given first-hand experience of the impact of land-cover changes on water yield. One of the most widely used experimental approaches is that of paired watersheds. This approach was originally developed at the beginning of the 20th century by the US Forest Service, to answer the practical question of what effects forest management (mainly forest thinning, but also afforestation programs) has on streamflow discharge. The first paired-watershed experiment was established in the Wagon Wheel Gap site in Colorado, United States, [160] and provided sound evidence that forest cutting increases water yield and that forest regrowth reduces it. This approach rapidly gained recognition among foresters and hydrologists and became accepted worldwide during the rest of the 20th century. The approach [57] is based on monitoring two similar catchments jointly during a given time period (calibration period). Then, the land use of one of the catchments (treated catchment) is modified, while the other one remains untouched (control catchment). The relationship between the streamflows of the two catchments established during the calibration period can be used to reconstitute the streamflow of the treated catchment and thus assess the impact of the land-use change (Fig. 10.3). The knowledge gained using this approach has provided an essential aid to decision-making in watershed management.

Alternative approaches to assessing the hydrological impact of land-cover changes have been used at the catchment scale. One of these is based on single catchment experiments in which land cover is modified during the study period (e.g., [161]). The main drawback of this approach is that studies need long periods, so as to avoid the influence of

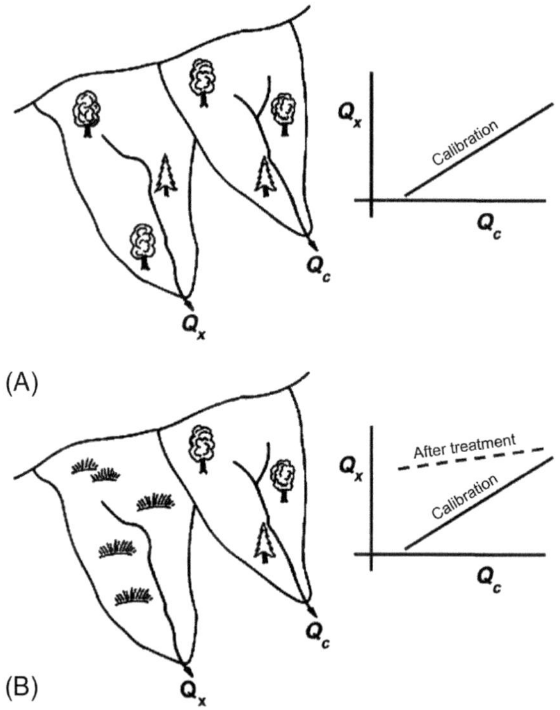

Fig. 10.3 Sketch of a watershed experiment, from Hewlett [160a]. (A) Calibration period and (B) treatment period. *(From V. Andréassian, Waters and forests: from historical controversy to scientific debate, J. Hydrol. 291 (2004) 1–27. https://doi.org/10.1016/j.jhydrol.2003.12.015.)*

climate variability. Another approach is based on the comparison of the hydrological responses of neighboring catchments with different land covers (e.g., [61]). However, this is only valid if all other attributes of the studied catchments (e.g., climate, topography, lithology) are equivalent. These approaches are usually applied in small ($<10\,km^2$) catchments, as this spatial scale has several advantages: (i) the small size enables high density and quality measurements that facilitate the understanding of the complexities of the hydrological processes; (ii) the intensively collected datasets provide essential information for model development and testing [162], especially for physically based approaches that require large parameterization efforts; and (iii) the catchment scale allows the study of the interactions between slopes and channels.

Studies at the forest-stand and plot-scale levels (up to hundreds or few thousands of square meters) are also useful for assessing the hydrological impact of different land covers and land-cover managements. However, the results must be interpreted with caution, given the size of the plots/stands, which only enable the study of processes operating at smaller scales (e.g., rainfall partitioning, water infiltration, or runoff generation processes). Given its limitations, this spatial scale should be only used to study the hydrological response of specific land covers. It is strongly recommended when its results are used for comparative purposes within the same or equivalent sites [145, 163]. Eventually, the data obtained facilitate the interpretation and modeling of hydrological changes at larger scales [142, 164].

Finally, the effects of land-use change can also be assessed by hydrological modeling (see Section 10.3.2). However, field observations are still needed to calibrate and validate the model outputs. As noted by Hewlett et al. [165] long ago, modeling studies require high-quality input data for providing significant advances in process understanding and useful outputs for land management. This is especially true for the Mediterranean region, where the inherent complexity of its hydrology, which is affected by a very strong seasonal and interannual variability, compromises the understanding of the rainfall-runoff relationships [166]. Thus, as experimental process observations, field measurements, and data modeling are all indispensable and complementary, they should be jointly conducted.

Field data acquisition is essential if the hydrological impact of land-use changes is to be properly understood and forecast. In particular, long-term data series are extremely valuable, as they reflect the actual evolution of the landscapes in their particular context of changes in land use and climate. Additionally, they facilitate the analysis of low-frequency events, such as extreme floods and droughts. These low-frequency hydrological events are a principal concern for land managers and policy makers, as they cause the largest and most significant environmental impact [162]. Unfortunately, data scarcity is still a great handicap for many Mediterranean regions [52, 167]. We believe that long-term data-acquisition approaches should be promoted more, despite their requiring high economic investment and commitment of human resources.

10.3.2 Hydrological modeling

Recent changes in climate and land cover are unquestionable, as is their impact on the water cycle. Industrialization and intense use of land resources by humans seem unstoppable. The most feasible way to measure the possible impact of future changes in climate and land cover is hydrological modeling [168]. Hydrological models simplify the hydrological cycle in natural systems. They usually represent the conversion of rainfall into runoff by means of empirical (statistical), conceptual (which involve empirical observations with calibrated parameters), or physically based relationships (by means of differential equations) between the different components of the system (see types of hydrological models in [169]). Both statistical and conceptual models have been used to model the hydrological impact of land-cover changes (e.g., [68, 170]), taking advantage of the relatively low quantity of data required to run the models. However, because of the spatially varying nature of land cover and the complex physical and biological mechanisms involved in vegetation-soil-atmosphere interactions, the most suitable types of model for calculating the hydrological impact of land-cover changes are what are known as physical or process-based distributed models (e.g., [74, 171]). These models require representation of the watershed's entire geography in order to simulate the physical processes for water routing across the catchment: climate (precipitation and temperature at least), topography, soil characteristics (e.g., granulometry, organic matter, preceding soil moisture), land-cover types, river/channel network distribution, etc. These features are entered in the model as data series or spatial gridded layers. The parameters that define such physical characteristics are included in the equations that simulate water movement through the system. Parameterization of process-based modeling approaches requires extensive calibration of parameters. Then, the calibrated model needs to be further validated against observed data (typically by comparing estimated vs. observed streamflow) for a (boundary condition) time period with valid climate and land-cover observations. Physical models may be powerful tools for understanding the various hydrological processes that modulate the responses of the catchments, the driving forces of major hydrological changes, and the interactions between the several components of the water cycle. For example, these approaches enable calculation of the effect of a broad variety of changes in climate, vegetation cover, and soil properties on water flow and storage, include the importance of soil depth on transpiration rates, and take into account the control that soil properties and preceding soil moisture exert on the timing and magnitude of surface flows [172, 173].

One of the most extended applications of hydrological models involves running experiments in which climate and land cover are forced to change over a period of time (most commonly by including projections of the likely evolution of the climate and the landscape) to simulate runoff generation under such environmental changes and to evaluate water availability in the medium term. The Mediterranean basin is ideal for such experiments for several reasons: (i) the scarce and highly variable water availability in the region, with climate projections predicting even dryer conditions in the coming decades, (ii) frequent flooding due to the particular nature of a harsh climate and rugged topography, and (iii) intense soil occupation, with rapid changes in land uses and land cover for socio-economic and demographic reasons.

Future changes in climate variables in hydrological models are usually derived from atmospheric simulations with Global Circulation Models that take into account future socio-economic (e.g., SRES scenarios, [174]) and/or greenhouse gas emission (e.g., RPCs scenarios, [175]) pathways for the 21st century. Depending on the quantity of greenhouse gas (mainly CO_2) emissions in the scenarios, different changes in future temperature and precipitation can be expected. To adapt the spatial resolution of the climate simulations to that of the hydrological simulation, downscaling is recommended [176, 177], in such a way that the climate inputs for the hydrological model (either from meteorological stations or from gridded datasets) are forced to change in line with the expected climate evolution for the particular region of analysis. For the Mediterranean region, a range of temperature and precipitation change values can be expected for the rest of the present century [8]. In the case of land cover, the most common procedure consists of three phases of analysis: (i) running the hydrological models for both current and past land-use conditions [74], (ii) implementing the expected changes in land cover (e.g., revegetation by secondary succession, transformation of agricultural lands, deforestation/afforestation) in the model simulations, and (iii) comparing the different hydrological simulations in a diachronic fashion (see [121]). Land-use scenarios can be generated by means of land-use models, examining plausible story lines for the socio-economic evolution of the planet in a way that is consistent with the climate scenarios used [178] so that hydrological simulations for future climate and land-cover conditions will be robust [176].

Results from experiments carried out in Mediterranean catchments that include the likely evolution of climate and land cover generally show overall decreases in streamflow and aquifer recharge over the coming decades. However, there is discrepancy about the role of land cover versus climate variations in the projected decreases in water yield. While some studies indicate that land-cover change (i.e., increased forest cover) played a dominant role in runoff reduction for a Pyrenean catchment [179], other studies of the region suggest that the largest part of runoff decrease was attributable to changes in climate (i.e., reduced precipitation and increased temperature) [121, 176, 177]. Differences can be explained by the degree of future afforestation considered in the hydrological simulation or by the current coverage of forest in the catchments. Delgado et al. [74] discussed how the great variability of climate in the Mediterranean region, together with the spatial scale of the modeling experiments, may overshadow the hydrological consequences of changes in land cover. Further, Morán-Tejeda et al. [177] observed that the choice of model for hydrological simulation may be a source of uncertainty, since some models are more sensitive to changes in land cover than others, depending on the complexity of the modeling routines and parameterization of vegetation-related processes. This is just to indicate that there is still room for improvement. The experiments cited above simulated the hydrological consequences of simultaneous changes in climate and land cover. However, interactions and feedback between the two changing processes remain unexplored. Will afforestation rates in the Mediterranean region remain the same as now under a warmer and dryer climate? What will the hydrological responses to land-cover change be like under warmer and dryer conditions? Scenarios and model set-ups for future experiments need to be able to answer such questions if they are to provide more reliable projections for future water resources in the Mediterranean basin.

10.4 Applications

10.4.1 Case 1: Small scale. Field observation in small catchments: Comparing flood events under different farmland-abandonment scenarios

This case study recently published by Lana-Renault et al. [85] analyzed the hydrological response of two small catchments that are representative of different revegetation scenarios following farmland abandonment. Their behavior was compared to that of a natural forest catchment. The three catchments ($<3\,km^2$) are located in the Spanish Pyrenees and share similar lithology, topography, and climate conditions, which enables the effects of soil and land cover on their hydrological responses to be separated. One of the catchments, which ceased active agricultural management in the 1950s, was left to undergo a natural process of revegetation and contains at present a mosaic of shrubs (64%) and forest (27%) at different stages of succession (from now on, the shrub catchment). The second catchment, afforested in the 1960s, has most of its surface currently covered by forest, mainly pines (*P. nigra* and *P. sylvestris*) of moderate size. The third catchment is covered by a dense natural forest of *P. sylvestris*. All three catchments were equipped in the same fashion, including (i) tipping buckets to collect rainfall; (ii) a complete weather station; (iii) a gauging station at the outlet, at which the water level for the main stream was recorded by an ultrasound and/or pressure sensor; and (iv) several piezometers, installed at various distances from the main channel and at different topographical locations to record the depth of the water table. In the natural forest catchment, throughfall and stemflow were also measured in three forest stand plots, each equipped with 25 pluviometers. Further details about data acquisition and treatment can be found in Lana-Renault et al. [85].

Analysis of the catchments' responses to similar rainfall events under contrasting preceding conditions showed significant hydrological differences between the land covers studied. A large (30–40 mm) and intense rainfall event that took place under dry preceding conditions (Fig. 10.4) produced the highest response in the shrub catchment, reaching a runoff coefficient (i.e., runoff to rainfall ratio) of 0.18 and a specific

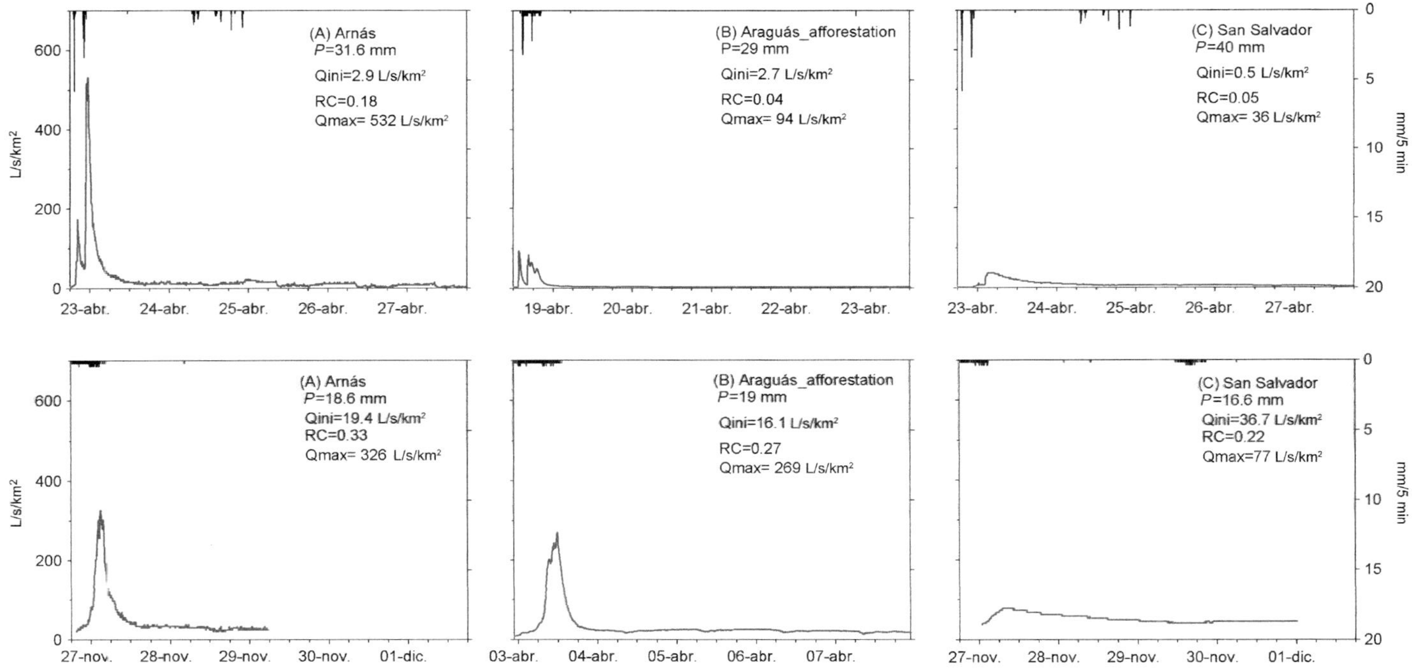

Fig. 10.4 Flood hydrographs in the (A) shrub (Arnás), (B) afforested (Araguás_afforestation), and (C) natural forest (San Salvador) catchments for similar rainfall (*P*) characteristics under dry (upper panels) and wet (lower panels) preceding conditions. Qini: baseflow discharge at the start of the event; RC: runoff coefficient; Qmax: peakflow. *(Modified from N. Lana-Renault, M. López-Vicente, E. Nadal-Romero, R. Ojanguren, J.A. Llorente, P. Errea, D. Regüés, P. Ruiz-Flaño, M. Khorchani, J. Arnáez, N. Pascual, Catchment based hydrology under post farmland abandonment scenarios, Geophys. Res. Lett. 44 (2018) 503–534. https://doi.org/10.18172/cig.3475.)*

peakflow discharge of 532 L/s/km^2. The responses in the afforested and natural forest catchments were much lower, accounting for runoff coefficients of 0.04 and 0.05, respectively. Interestingly, peakflow-specific discharge in the afforested catchment was more than double that in the natural forest (94 vs 36 L/s/km^2). In both the abandoned catchments (shrub and afforested), the hydrological responses were rapid (<65 min) and the recessions were short (55–85 min), whereas for the natural forest, the responses were slower (about 4 h 30 min) and the recession limb much longer (about 14 h). A moderate (16–19 mm), low-intensity rainfall event caused, under wet preceding conditions, similar streamflow responses in both the shrub and the afforested catchments, with runoff coefficients of 0.33 and 0.27, respectively, and peakflow-specific discharges of 326 and 269 L/s/km^2, respectively. The response times in the two abandoned-farmland scenarios were 3 h and almost 4 h, respectively, while the recessions lasted slightly less than 3 h. The natural forest catchment was characterized by a moderate response (runoff coefficient 0.22), a much gentler peakflow-specific discharge (77 L/s/km^2), a much slower response (7 h), and a very long recession (>5 days).

The observed differences between the flood hydrographs suggest the dominance of distinct runoff generation processes under the three land cover types considered in the analysis. Runoff production in the afforested catchment is similar to that in the natural forest catchment, reaching very low values under dry preceding conditions and moderate-to-high values under wet conditions. These findings demonstrate the effect of trees on rainfall interception and evapotranspiration processes. However, previous research [84] showed that very intense rainfall events caused higher runoff coefficients in the afforested catchment than in the natural forest. Interestingly, these two catchments showed completely different hydrograph characteristics. Higher peakflows were observed in the afforested catchment, which, under wet conditions, were comparable to the peakflow generated in the shrub catchment. Under these conditions, the hydrographs are characterized by rapid responses and short recession limbs in both catchments, findings that are usually associated with excess-saturation overland flow, more likely in soils with low storage capacity (e.g., shallow soils) that saturate easily [180]. The occurrence of these processes is supported by significant correlation between storm-flow discharge and rainfall in these two catchments. Thus, although the catchments differ significantly in vegetation cover, the soil properties of the shrubs and afforested catchments are probably similar, as they result from intense agricultural practices in the past, a finding that may explain their similar hydrological responses under wet conditions. Under dry conditions, the rapid hydrological responses of the abandoned catchments suggest the occurrence of fast overland flow processes, probably resulting from excess infiltration runoff in areas devoid of vegetation that are well connected to the main channel. The larger hydrological response in the shrub catchment indicates that such processes may involve a larger area than in the afforested catchment. Previous research [181] had already demonstrated that under such dry preceding conditions (and especially in response to high-intensity rainfall), infiltration excess runoff over areas with sparse vegetation and low-permeability (degraded) soils occurred in the shrub catchment. The slower response and longer recession limb in the natural forest catchment suggest that a large contribution of a slow flow component (i.e., subsurface flow within the soil matrix) dominates runoff in this catchment, under both dry and wet preceding conditions. This is supported by previous studies that showed strong correlation between streamflow response, preceding rainfall and water table levels [80, 182]. This study illustrates how a detailed analysis of flood characteristics and related hydrological variables at the small catchment scale can help understanding of the influence of vegetation cover and soils on runoff generation processes and subsequent catchment streamflow responses.

10.4.2 Case 2: Medium scale. Physical modeling: Using SWAT to assess the effect of land cover on the provision of hydrological services

The following example was developed by Carvalho-Santos et al. [183] with the aim of assessing the effect of land cover and climate change on the provision of hydrological services of a medium-scale watershed. Here, we focus only on the effect of land-cover change on water supply services, including water quantity, timing, and quality. The study was carried out in the Vez catchment (252 km^2) in northern Portugal, with a Mediterranean-type rainfall regime in the lowlands and land cover dominated by shrubs. This study was chosen to show an application of the well-known Soil and Water Assessment Tool (SWAT; [184]) to evaluate the hydrological impact of land-cover change trends and management options that typically occur in the Mediterranean region: (i) forest conservation with increases in common oak (*Quercus robur*) density and/or dominance; (ii) Eucalyptus/pine plantations; (iii) agricultural uses with expansion of corn crops and vineyards (agri/vine); and (iv) degraded low vegetation resulting from grazing and/or wildfires. SWAT was originally developed in the early 1990s by the USDA Agricultural Research Service to predict the impact of land management on water, sediment and agricultural chemical yields in medium-to-large ungauged catchments. It is a semi-distributed, physical model that operates on a daily time step and can simulate surface flow, subsurface flow, soil erosion, sediment deposition, and the movement of nutrients through watersheds [184, 185]. In this study, the model was set up with the ArcSWAT interface for ArcGis. Table 10.1 lists the input data used for the model set-up. Observed daily stream discharge was used to calibrate and validate the model; for nitrates, observed data consisted of one measurement per month. The model performed well, especially for monthly stream discharges: nitrate simulations were considered valid. A detailed explanation of the model parameterization and calibration can be found in Carvalho-Santos et al. [183].

Overall, the results of this study showed that the effects on the provision of water services of the alternative land-cover scenarios used in SWAT testing were, in general, moderate, compared to the current scenario that is dominated by shrubland (Table 10.2).

Table 10.1 SWAT data variables for model set-up and calibration/validation.

Variables	Description
DEM	20 m resolution grid derived from contour line map 1:25,000
Land cover	Land-cover map 2006, 1:25,000. Classes were aggregated into seven main cover classes Validation period—Land-cover map 1990, 1:25,000
Soil	Soil map 1:100,000, with five main soil types (Humic Regosols, Humic Leptosols, Dystric Fluvisols, Antrosols, and Urban)
Rainfall	Inside the basin, hourly rainfall (mm) data were collected for five stations and turned into daily values. Calibration period: 1999–2008 (4-year initialization period) Validation: 1984–89
Climate (other variables)	Hourly values were converted into daily temperature (min., max. °C), solar radiation, relative humidity, and wind speed (m/s) from climate station Ponte da Barca
River discharge	Hourly observed river stages were converted into daily river discharge values (m^3/s) and averaged for the basin using stage-discharge relationships Calibration period: 2003–08 Validation: 1984–89
Nitrates (NO_3)	Daily observations (one per month) were taken from two quality stations (period 2003–08)

Adapted from C. Carvalho-Santos, J.P. Nunes, A.T. Monteiro, L. Hein, J.P. Honrado, Assessing the effects of land cover and future climate conditions on the provision of hydrological services in a medium-sized watershed of Portugal, Hydrol. Process. 30 (2016) 720–738. https://doi.org/10.1002/hyp.10621.

Table 10.2 Land-cover effects on water services provision in the Vez watershed.

		Water supply quantity		Water timing		Water quality		
Land-cover scenario	**ET (mm)**	**Total annual water yield (mm/year)**	**% change**	**Q5 (m^3/s)**	**Q95 (m^3/s)**	**NO_3–N (mg/L)**	**No. of days with NO_3–N exceeding 5.6 mg/L**	**N exports (kg/ha year)**
Current cover	590	1025		29.3	0.16	1.9	74	1.04
Eucalyptus/pine	660	955	−7	28.2	0.15	2.1	104	1.11
Oak	597	1017	−1	30	0.16	1.9	64	1.36
Agri/vine	532	1083	+6	31	0.17	2.2	68	1.44
Low vegetation	557	1058	+3	30.3	0.18	1.7	52	1.32

Q5, high flows; *Q95*, low flows.

Adapted from C. Carvalho-Santos, J.P. Nunes, A.T. Monteiro, L. Hein, J.P. Honrado, Assessing the effects of land cover and future climate conditions on the provision of hydrological services in a medium-sized watershed of Portugal, Hydrol. Process. 30 (2016) 720–738. https://doi.org/10.1002/hyp.10621.

A small reduction in total water yield was found when forest area increased (−1% and −7% for the forest conservation and plantation scenarios, respectively). In contrast, total water yield increased by 6% and 3% for the scenarios of agriculture expansion and low vegetation, respectively. Results for water timing indicated that low-flow values were similar between all land covers. High flows were affected slightly more by land cover, with the plantation and the agricultural scenarios showing the lowest and highest peakflow values, respectively. When water quality was assessed, agri/vine was the land-use scenario causing the most important nitrogen (N) exports and major NO_3–N concentrations at the river outlet. The plantation scenario showed daily-simulated NO_3–N concentrations frequently exceeding 5.6 mg/L (established as the guideline value for standard surface water quality in Portugal) in the river, which the authors associated with a smaller river discharge and increased number of days with low flows. This study provides a good example of SWAT application and shows the usefulness of complex, spatially explicit, physical hydrological models that provide a more complete picture of the water cycle by evaluating accurately the hydrological impact of land-use changes. In particular, the results of the study suggest that selecting one particular land-cover scenario over other options does not compromise the overall provision of hydrological services for the watershed analyzed, as each land-use scenario performs a particular hydrological service provision better than the others (e.g., low vegetation for water supply versus forest plantation for flood regulation).

10.4.3 Case 3: Large scale. Statistical modeling: Assessing land-cover changes and recent hydrological evolution at the headwaters of the Duero river basin

The following case, published in Morán-Tejeda et al. [186], describes the hydrological impact of land-use changes in the Duero river basin, the second largest river basin in Spain, with an extension of 97,300 km^2. Large climatic and hydrological datasets, together with a thorough digitalization of land-use maps from 1957 and 2003, allowed empirical (statistical) modeling of streamflows from the river network and inferences on the impact of land-cover changes on the observed decline in water resources. The headwaters—Cantabrian, Iberian, and Central System mountains—of the Duero basin were intensely occupied by agriculture for the first half of the 20th century, but the massive migration from rural to urban areas that took place during the 1960s led to the abandonment of fields, pastures, terraces, and traditional land-use activities. Revegetation from secondary succession and afforestation policies increased overall forest cover by about 15% in the region, which was even larger (30% and 18% forest expansion) when the north and south headwaters, respectively, were included (Table 10.3). Replacement of pastures and crops by shrubland was also observed, leading to an overall increase of woody canopy in the basin. Hydro-climatic analysis focused on gauged streams at the headwaters of the basin, with catchment areas generally smaller than 1000 km^2 and mean

Table 10.3 Changes in land cover in the Duero basin between 1966 and 2003.

	Total Duero basin			North fringe			South fringe		
Land use	**1966**	**2003**	**Δ %**	**1966**	**2003**	**Δ %**	**1966**	**2003**	**Δ %**
Water bodies	313	422	25.89	55	94	41.33	19	44	56.45
Urban-bare rock	30	1024	97.04	5	179	97.46	5	116	95.94
Grassland/ shrubland	11,956	14,050	14.91	5475	4018	−36.25	2251	3460	34.94
Forest	20,127	23,452	14.18	4866	7062	31.1	2388	2922	18.26
Cropland	46,482	39,932	−16.4	5962	5041	−18.26	5106	3235	−57.83

From E. Morán-Tejeda, A. Ceballos-Barbancho, J.M. Llorente-Pinto, J.I. López-Moreno, Land-cover changes and recent hydrological evolution in the Duero Basin (Spain), Reg. Environ. Change 12 (2012) 17–33. https://doi.org/10.1007/s10113-011-0236-7.

elevations above 1000 m.a.s.l. As the streams studied had natural flow regimes, it was assumed that any changes in streamflows over time were due to natural causes.

Of the 21 rivers studied, 19 had negative Mann-Kendall trend slopes in annual discharges for the period 1961–2006, 11 of which were statistically significant at the 95% confidence level (example in Fig. 10.5). However, when looking at the precipitation series recorded at stations located within or near the river catchments, none of them showed a significant negative trend for the studied period. Temperature measured in the catchments showed an overall increase during the study period, but this alone cannot explain such a reduction in river discharges. Decreasing streamflow under stationary precipitation is a first ready indicator for the hydrological impact of the observed land-cover increases.

To further quantify the relative roles of precipitation and temperature in the evolution of river discharge, linear models were constructed for each individual catchment. In these statistical models, precipitation explained a significant amount of variance of river discharges in 20 out of the 21 cases, whereas temperature helped to improve the amount of variance explained in 7 cases. In other words, climate could significantly explain the variability of streamflow, but not its linear, negative trend (i.e., overall decrease) during the study period. Such reductions should be attributed mainly to the increased forest cover observed in the headwaters. However, the forest cover increase could not be included in the tested statistical models, since it was computed from two static pictures of land cover for 1957 and 2003. To verify such a hypothesis, the residuals from the models (i.e., the departures of estimated streamflow from observed streamflow) were considered as a proxy of the effect of forest increase in the models. Under forest cover increase scenarios, the model-estimated streamflow would tend to become larger over time than the observed streamflow values (because the model does not consider forest cover increase as a variable) and, therefore, the residuals would also tend to depict

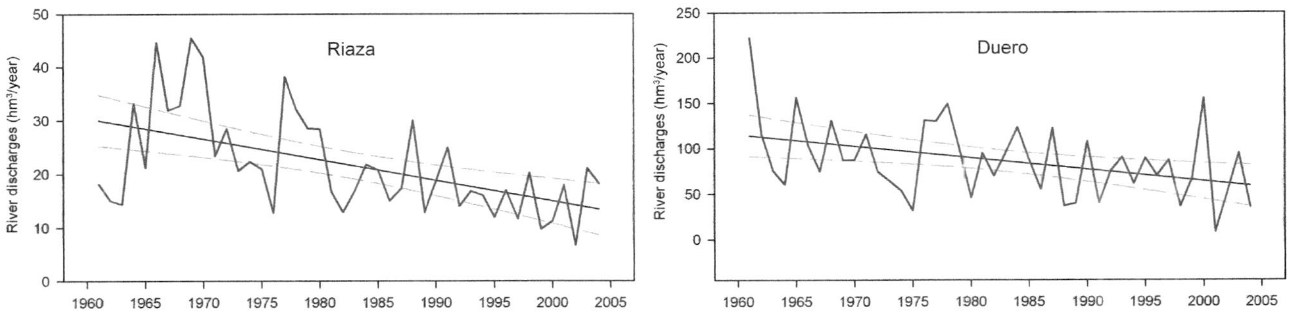

Fig. 10.5 Decreases of streamflow in two headwater streams from the Duero basin. *(From E. Morán-Tejeda, A. Ceballos-Barbancho, J.M. Llorente-Pinto, J.I. López-Moreno, Land-cover changes and recent hydrological evolution in the Duero Basin (Spain), Reg. Environ. Chang. 12 (2012) 17–33. https://doi.org/10.1007/s10113-011-0236-7.)*

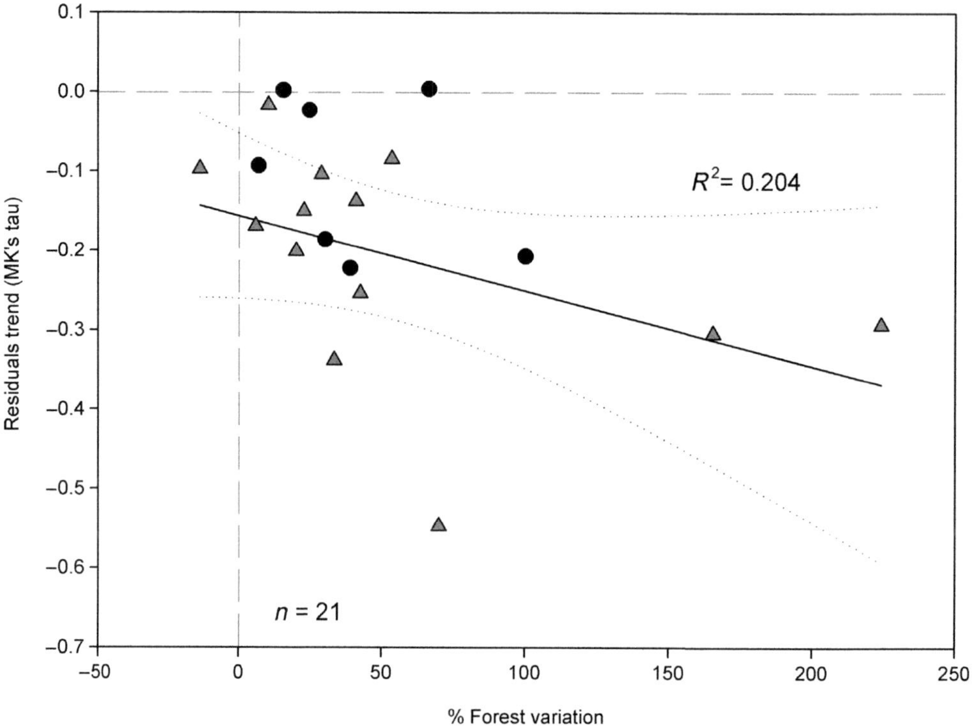

Fig. 10.6 Linear regression between forest variations and trends in the residuals of streamflow ~climate linear models. *(From E. Morán-Tejeda, A. Ceballos-Barbancho, J.M. Llorente-Pinto, J.I. López-Moreno, Land-cover changes and recent hydrological evolution in the Duero Basin (Spain), Reg. Environ. Change 12 (2012) 17–33. https://doi.org/10.1007/s10113-011-0236-7.)*

negative trends when plotted against time. Results derived from calculating trend slopes to the residuals indicated that for 18 out of 21 statistical models, the residuals showed a negative trend coefficient (residuals showing larger negative values), with 6 of them significant at the 95% confidence level. When plotting the trend coefficients of the residuals versus the percentage of forest variation within the catchments (Fig. 10.6), a negative linear trend is observed, indicating that the residuals show more negative values (i.e., a larger streamflow overprediction in the statistical models) where forest increase was larger. In addition, when applying regionally aggregated streamflow and climate variables, the resulting models also showed a larger decrease of residuals in the northern sector of the basin, where overall land-cover increase was larger than in the southern sector, with less evidence of land-cover increases. Although these results are not based on direct observation of the hydrological impact of forest cover increase, they provide a good example of how statistical modeling can be used to infer the impact of land-cover change on water resources in large river basins for which large hydro-climatic and land-cover datasets are available.

10.5 Main conclusions

1. The Mediterranean basin has had for centuries a history of human pressure and landscape transformation. Recent land-use changes show contrasting patterns between the northern and southern rims of the basin, caused by different demographic trends. In the northern countries, population decline in rural areas has resulted in farmland and grazing abandonment and subsequent vegetation recovery of former agricultural land, whereas the lowlands and coastal areas have experienced concentration of population and economic activities, including intensive agriculture and tourism. In the southern and eastern countries, strong population growth explains the persistence of significant pressures on forest ecosystems in rural areas as well as the present pulses in both increased urbanization and the associated expansion of cultivated land, mostly devoted to irrigated crops. However, the general trend in the entire Mediterranean basin is a greening up over the last 35 years, favored by woodland recovery, relatively lower pressures on forest resources, and the expansion of irrigated lands.
2. Vegetation expansion on former agricultural land usually causes a decrease in water yields due to increased rainfall interception and transpiration by vegetation. However, the impact of these may vary greatly, as it depends on several factors, including climate conditions (humid or drier climate); the extent and spatial distribution of land-cover disturbance; the characteristics and depth of the soils; and the type, age, or physiology of the vegetation. Thus, in arid and semiarid areas, the hydrological impact of a change in vegetation cover is usually smaller, although it may be much greater under specific conditions. Revegetation reduces soil water content and deep-water percolation; additionally, it lessens beneath-canopy snow accumulation and alters snow-melting rates.
3. The expansion of new land uses such as irrigated land and urbanization increases water withdrawals. Conversion of rain-fed to irrigation agriculture usually involves more intense cropping and a shift to more water-demanding crops, which increase water consumption. The recent development of low-density urban sprawl and "residential" tourism, which usually includes single houses with garden and swimming pool, has boosted water consumption. To satisfy this growing water demand, reservoirs have been constructed, very often with negative impact on downstream river discharges. Alternatively, groundwater has been extracted, resulting in most cases in overexploitation of the aquifers.
4. Demographics, land-use, and climate scenarios forecast a decline in freshwater supplies and an increase in water demand, especially in the southern and eastern rims of the Mediterranean basin. Changing water availability and demand will necessitate mitigation strategies to improve the management of water resources. This requires a good understanding of the effects of land-use changes on water resources and the application of innovative land- and water-management strategies. Such strategies

include forest management at the watershed scale, and water use and management based not only on efficient water consumption and water supply infrastructure but also on more sustainable use of the land.

5. Field data acquisition (especially long-term data series) at different spatial scales is essential to examine the hydrological impact of land-use changes. Hydrological modeling (including empirical, conceptual, and physically based models) is a powerful tool for understanding the interactions between different components of the water cycle, grasping the driving forces of major hydrological changes, and predicting hydrological changes under different land-use (and climate) scenarios. These two indispensable and complementary approaches should be jointly conducted. A selection of three study cases illustrates the application of different approaches at different spatial scales (field measurements at the small catchment scale, physical modeling at the medium scale, and statistical modeling at the large scale) for assessing the hydrological impact of land-use changes.

Acknowledgments

Support for this research was provided by the projects MASCC-DYNAMITE (PCIN-2017-061/AEI), TransHyMed (CGL2016-75957-R, AEI/FEDER, EU), and ESPAS (CGL2015-65569-R), funded by the State Research Agency of the Spanish Ministry of Science, Innovation and Universities and the Spanish Ministry of Economy and Competitiveness. M. Moreno-de-las-Heras is beneficiary of a Juan de la Cierva Incorporación fellowship (IJCI-2015-26463) funded by funded by the State Research Agency of the Spanish Ministry of Science, Innovation and Universities.

References

[1] P.M. Vitousek, Beyond global warming: ecology and global Change, Ecology 75 (1994) 1861–1876, https://doi.org/10.2307/1941591.

[2] W. Steffen, Global Change and the Earth System: A Planet under Pressure, Springer, 2005.

[3] E. Nadal-Romero, E. Cammeraat, Geo-ecology in the anthropocene, Geogr. Res. Lett. 45 (2019) 5, https://doi.org/10.18172/cig.3876.

[4] D.A. DellaSala, Global change, in: D.A. DellaSala, M.I. Goldstein (Eds.), Encyclopedia of the Anthropocene, vol. 2, Elsevier, 2018, pp. 347–349. https://doi.org/10.1016/B978-0-12-809665-9.05355-6.

[5] J.A. Foley, R. DeFries, G.P. Asner, C. Barford, G. Bonan, S.R. Carpenter, F.S. Chapin, M. T. Coe, G.C. Daily, H.K. Gibbs, J.H. Helkowski, T. Holloway, E.A. Howard, C.J. Kucharik, C. Monfreda, J.A. Patz, I.C. Prentice, N. Ramankutty, P.K. Snyder, Global consequences of land use, Science 309 (2005) 570–574, https://doi.org/10.1126/science.1111772.

[6] D. Schröter, W. Cramer, R. Leemans, I.C. Prentice, M.B. Araújo, N.W. Arnell, A. Bondeau, H. Bugmann, T.R. Carter, C.A. Gracia, A.C. De La Vega-Leinert, M. Erhard, F. Ewert, M. Glendining, J.I. House, S. Kankaanpää, R.J.T. Klein, S. Lavorel, M. Lindner, M.J. Metzger, J. Meyer, T.D. Mitchell, I. Reginster, M. Rounsevell, S. Sabaté, S. Sitch, B. Smith, J. Smith, P. Smith, M.T. Sykes, K. Thonicke, W. Thuiller, G. Tuck, S. Zaehle, B. Zierl, Ecology: ecosystem service supply and vulnerability to global change in Europe, Science 310 (2005) 1333–1337, https://doi.org/10.1126/science.1115233.

[7] F. Giorgi, Climate change hot-spots, Geophys. Res. Lett. 33 (2006) L08707, https://doi.org/10.1029/2006GL025734.
[8] F. Giorgi, P. Lionello, Climate change projections for the Mediterranean region, Glob. Planet. Change 63 (2008) 90–104, https://doi.org/10.1016/j.gloplacha.2007.09.005.
[9] IPCC, Climate change 2014: synthesis report, in: Contribution of Working Groups I, II and III to the Fifth Assessment Report of the Intergovernmental Panel on Climate Change, Geneva, 2014.
[10] J. Guiot, W. Cramer, Climate change: the 2015 Paris agreement thresholds and Mediterranean basin ecosystems, Science 354 (2016) 465–468, https://doi.org/10.1126/science.aah5015.
[11] N.S. Diffenbaugh, J.S. Pal, F. Giorgi, X. Gao, Heat stress intensification in the Mediterranean climate change hotspot, Geophys. Res. Lett. 34 (2007) L11706, https://doi.org/10.1029/2007GL030000.
[12] M. Dubrovský, M. Hayes, P. Duce, M. Trnka, M. Svoboda, P. Zara, Multi-GCM projections of future drought and climate variability indicators for the Mediterranean region, Reg. Environ. Chang. 14 (2014) 1907–1919, https://doi.org/10.1007/s10113-013-0562-z.
[13] J.M. García-Ruiz, J.I. López-Moreno, S.M. Vicente-Serrano, T. Lasanta-Martínez, S. Beguería, Mediterranean water resources in a global change scenario, Earth Sci. Rev. 105 (2011) 121–139, https://doi.org/10.1016/j.earscirev.2011.01.006.
[14] A. Grove, O. Rackham, The Nature of Mediterranean Europe : An Ecological History, Yale University Press, 2003.
[15] J. Blondel, J. Aronson, J.-Y. Bodiou, G. Boeuf, The Mediterranean Region : Biological Diversity in Space and Time, Oxford University Press, 2010.
[16] M.C. Hansen, R.S. DeFries, Detecting long-term global forest change using continuous fields of tree-cover maps from 8-km advanced very high resolution radiometer (AVHRR) data for the years 1982–99, Ecosystems 7 (2004) 695–716. https://doi.org/10.2307/3659068.
[17] Y. Shen, T. Oki, N. Utsumi, S. Kanae, N. Hanasaki, Projection of future world water resources under SRES scenarios: water withdrawal, Hydrol. Sci. J. 53 (2008) 11–33, https://doi.org/10.1623/hysj.53.1.11.
[18] P.C.D. Milly, K.A. Dunne, A.V. Vecchia, Global pattern of trends in streamflow and water availability in a changing climate, Nature 438 (2005) 347–350, https://doi.org/10.1038/nature04312.
[19] M. Milano, D. Ruelland, S. Fernandez, A. Dezetter, J. Fabre, E. Servat, J.-M. Fritsch, S. Ardoin-Bardin, G. Thivet, Current state of Mediterranean water resources and future trends under climatic and anthropogenic changes, Hydrol. Sci. J. 58 (2013) 498–518, https://doi.org/10.1080/02626667.2013.774458.
[20] D. Jorda-Capdevila, D. Gampe, V. Huber García, R. Ludwig, S. Sabater, L. Vergoñós, V. Acuña, Impact and mitigation of global change on freshwater-related ecosystem services in southern Europe, Sci. Total Environ. 651 (2019) 895–908, https://doi.org/10.1016/j.scitotenv.2018.09.228.
[21] FAO and Plan Bleu, State of Mediterranean Forests 2018, (2018).
[22] PRB, 2019 World Population Data Sheet–Population Reference Bureau, (2019).
[23] M. Debussche, J. Lepart, A. Dervieux, Mediterranean landscape changes: evidence from old postcards, Glob. Ecol. Biogeogr. 8 (1999) 3–15, https://doi.org/10.1046/j.1365-2699.1999.00316.x.
[24] J.M. García-Ruiz, N. Lana-Renault, Hydrological and erosive consequences of farmland abandonment in Europe, with special reference to the Mediterranean region—a review, Agric. Ecosyst. Environ. 140 (2011) 317–338, https://doi.org/10.1016/j.agee.2011.01.003.
[25] M. Moreno-de-las-Heras, F. Lindenberger, J. Latron, N. Lana-Renault, P. Llorens, J. Arnáez, A. Romero-Díaz, F. Gallart, Hydro-geomorphological consequences of the abandonment of agricultural terraces in the Mediterranean region: key controlling factors and landscape stability patterns, Geomorphology 333 (2019) 73–91, https://doi.org/10.1016/j.geomorph.2019.02.014.
[26] L.M. Ortigosa, J.M. Garcia-Ruiz, E. Gil-Pelegrin, Land reclamation by reforestation in the Central Pyrenees, Mt. Res. Dev. 10 (1990) 281–288, https://doi.org/10.2307/3673607.
[27] D.R. Vallauri, J. Aronson, M. Barbero, An analysis of Forest restoration 120 years after reforestation on Badlands in the southwestern Alps, Restor. Ecol. 10 (2002) 16–26, https://doi.org/10.1046/j.1526-100X.2002.10102.x.
[28] J.G. Pausas, S. Fernández-Muñoz, Fire regime changes in the Western Mediterranean basin: from fuel-limited to drought-driven fire regime, Clim. Chang. 110 (2012) 215–226, https://doi.org/10.1007/s10584-011-0060-6.

[29] P. Serra, X. Pons, D. Saurí, Land-cover and land-use change in a Mediterranean landscape: a spatial analysis of driving forces integrating biophysical and human factors, Appl. Geogr. 28 (2008) 189–209, https://doi.org/10.1016/j.apgeog.2008.02.001.
[30] J. Bellot, A. Bonet, J. Peña, J.R. Sánchez, Human impacts on land cover and water balances in a coastal Mediterranean County, Environ. Manag. 39 (2007) 412–422, https://doi.org/10.1007/s00267-005-0317-9.
[31] H. Tanrivermis, Agricultural land use change and sustainable use of land resources in the mediterranean region of Turkey, J. Arid Environ. 54 Academic Press (2003) 553–564, https://doi.org/10.1006/jare.2002.1078.
[32] D.L. Johnson, Development trends and environmental deterioration in the agropastoral systems of the Central Middle Atlas, Morocco, in: W.D. Swearingen, A. Bencherifa (Eds.), North African Environment Risk, Westview, Boulder, CO, 1996, pp. 35–54.
[33] C.J. Barrow, H. Hicham, Two complimentary and integrated land uses of the western high Atlas Mountains, Morocco: the potential for sustainable rural livelihoods, Appl. Geogr. 20 (2000) 369–394, https://doi.org/10.1016/S0143-6228(00)00010-2.
[34] S.E. Zaimeche, The consequences of rapid deforestation: a North African example, Ambio 23 (1994) 136–140, https://doi.org/10.2307/4314180.
[35] FAO, Forest Assessment 1990. Non Tropical Countries. Mediterranean Region, Rome, 1994.
[36] M. Nash, D.J. Chaloud, W.G. Kepner, S. Sarri, Regional Assessment of Landscape and Land Use Change in the Mediterranean Region: Morocco Case Study (1981–2003), in: P.H. Liotta, et al., (Ed.), Environmental Change and Human Security, Springer Science, 2008, pp. 143–165.
[37] Y. Chebli, M. Chentouf, P. Ozer, J.L. Hornick, J.F. Cabaraux, Forest and silvopastoral cover changes and its drivers in northern Morocco, Appl. Geogr. 101 (2018) 23–35, https://doi.org/10.1016/j.apgeog.2018.10.006.
[38] G. del Barrio, M. Sanjuan, A. Hirche, M. Yassin, A. Ruiz, M. Ouessar, J. Martinez Valderrama, B. Essifi, J. Puigdefabregas, Land degradation states and trends in the northwestern Maghreb drylands, 1998–2008, Remote Sens. 8 (2016) 603, https://doi.org/10.3390/rs8070603.
[39] R. Nsibi, N. Souayah, L.M. Khouja, A. Khaldi, S. Bouzid, Impacts des facteurs biotiques et abiotiques sur la dégradation de suberaie tunisienne, Geo-Eco-Trop. 30 (1) (2006) 49–58.
[40] M.A. Belaid, Urban-rural use detection and analysis using GIS & RS technologies, in: 2nd FIG Reg. Conf. Marrakech, Morocco, December 2–5, 2003.
[41] B. Wright, C. Cafiero, Grain reserves and food security in the Middle East and North Africa, Food Secur. 3 (2011) 61–76.
[42] N.D. Mueller, J.S. Gerber, M. Johnston, D.K. Ray, N. Ramankutty, J.A. Foley, Closing yield gaps through nutrient and water management, Nature 490 (2012) 254–257, https://doi.org/10.1038/nature11420.
[43] UNWTO, Tourism statistics. In: UNWTO eLibrary. Madrid, World Tourism Organization. https://www.e-unwto.org/toc/unwtotfb/current, (2017).
[44] N.K. Sönmez, M. Sari, Monitoring land use change in the West Mediterranean region of Turkey: a case study on Antalya-Turkey coast, Fresen. Environ. Bull. 16 (2007) 1325–1330.
[45] A. Shalaby, R. Tateishi, Remote sensing and GIS for mapping and monitoring land cover and land-use changes in the Northwestern coastal zone of Egypt, Appl. Geogr. 27 (2007) 28–41, https://doi.org/10.1016/j.apgeog.2006.09.004.
[46] I. Jomaa, Y. Auda, B. Abi Saleh, M. Hamzé, S. Safi, Landscape spatial dynamics over 38 years under natural and anthropogenic pressures in Mount Lebanon, Landsc. Urban Plan. 87 (2008) 67–75, https://doi.org/10.1016/j.landurbplan.2008.04.007.
[47] S. Hepcan, C. Coskun Hepcan, C. Kilicaslan, M.B. Ozkan, N. Kocan, Analyzing landscape change and urban sprawl in a Mediterranean coastal landscape: a case study from Izmir, Turkey, J. Coast. Res. 29 (2012) 301, https://doi.org/10.2112/jcoastres-d-11-00064.1.
[48] D.M. Olson, E. Dinerstein, E.D. Wikramanayake, N.D. Burgess, G.V.N. Powell, E. C. Underwood, J.A. D'amico, I. Itoua, H.E. Strand, J.C. Morrison, C.J. Loucks, T.F. Allnutt, T. H. Ricketts, Y. Kura, J.F. Lamoreux, W.W. Wettengel, P. Hedao, K.R. Kassem, Terrestrial ecoregions of the world: a new map of life on earth, Bioscience 51 (2001) 933, https://doi.org/10.1641/0006-3568(2001)051[0933:teotwa]2.0.co;2.

[49] J.G. Pausas, M.M. Millán, Greening and browning in a climate change hotspot: the Mediterranean basin, Bioscience 69 (2019) 143–151, https://doi.org/10.1093/biosci/biy157.
[50] S.M. Vicente-Serrano, N. Martín-Hernández, F. Reig, C. Azorin-Molina, J. Zabalza, S. Beguería, F. Domínguez-Castro, A. El Kenawy, M. Peña-Gallardo, I. Noguera, M. García, Vegetation greening in Spain detected from long term data (1981–2015), Int. J. Remote Sens. (2019) 1–32, https://doi.org/10.1080/01431161.2019.1674460.
[51] A.D. Solomou, N.D. Proutsos, G. Karetsos, K. Tsagari, Effects of climate change on vegetation in Mediterranean forests: a review, Int. J. Environ. Agric. Biotechnol. 2 (2017) 240–247.https://doi.org/10.22161/ijeab/2.1.31.
[52] M. Merheb, R. Moussa, C. Abdallah, F. Colin, C. Perrin, N. Baghdadi, Hydrological response characteristics of Mediterranean catchments at different time scales: a meta-analysis, Hydrol. Sci. J. 61 (2016) 2520–2539, https://doi.org/10.1080/02626667.2016.1140174.
[53] P. Llorens, F. Domingo, Rainfall partitioning by vegetation under Mediterranean conditions. A review of studies in Europe, J. Hydrol. 335 (2007) 37–54, https://doi.org/10.1016/j.jhydrol.2006.10.032.
[54] H.R. Wallace, J.S. Oliver, Vegetation and hydroclimate, in: M.G. Anderson, T.P. Burt (Eds.), Process Studies in Hillslope Hydrology, John Wiley Sons Ltd, Chichester, 1990, pp. 9–41.
[55] L. Zhang, W.R. Dawes, G.R. Walker, Response of mean annual evapotranspiration to vegetation changes at catchment scale, Water Resour. Res. 37 (2001) 701–708, https://doi.org/10.1029/2000WR900325.
[56] J.M. Bosch, J.D. Hewlett, A review of catchment experiments to determine the effect of vegetation changes on water yield and evapotranspiration, J. Hydrol. 55 (1982) 3–23, https://doi.org/10.1016/0022-1694(82)90117-2.
[57] V. Andréassian, Waters and forests: from historical controversy to scientific debate, J. Hydrol. 291 (2004) 1–27, https://doi.org/10.1016/j.jhydrol.2003.12.015.
[58] J.D. Stednick, Monitoring the effects of timber harvest on annual water yield, J. Hydrol. 176 (1996) 79–95, https://doi.org/10.1016/0022-1694(95)02780-7.
[59] D. Lewis, M.J. Singer, R.A. Dahlgren, K.W. Tate, Hydrology in a California oak woodland watershed: a 17-year study, J. Hydrol. 240 (2000) 106–117, https://doi.org/10.1016/S0022-1694(00)00337-1.
[60] F. Gökbulak, K. Şengönül, Y. Serengil, S. Özhan, İ. Yurtseven, B. Uygur, M.S. Özçelik, Effect of forest thinning on water yield in a sub-humid Mediterranean oak-beech mixed forested watershed, Water Resour. Manag. 30 (2016) 5039–5049, https://doi.org/10.1007/s11269-016-1467-7.
[61] C. Cosandey, Forêt et écoulements: rôle de la forêt sur la formation des crues et le bilan de l'écoulement annuel, Impact d'une coupe forestière, Meudon 1993.
[62] V. Sahin, M.J. Hall, The effects of afforestation and deforestation on water yields, J. Hydrol. 178 (1996) 293–309, https://doi.org/10.1016/0022-1694(95)02825-0.
[63] E. Doblas-Miranda, R. Alonso, X. Arnan, V. Bermejo, L. Brotons, J. de las Heras, M. Estiarte, J.A. Hódar, P. Llorens, F. Lloret, F.R. López-Serrano, J. Martínez-Vilalta, D. Moya, J. Peñuelas, J. Pino, A. Rodrigo, N. Roura-Pascual, F. Valladares, M. Vilà, R. Zamora, J. Retana, A review of the combination among global change factors in forests, shrublands and pastures of the Mediterranean Region: beyond drought effects, Glob. Planet. Change 148 (2017) 42–54, https://doi.org/10.1016/j.gloplacha.2016.11.012.
[64] T.E. Huxman, B.P. Wilcox, D.D. Breshears, R.L. Scott, K.A. Snyder, E.E. Small, K. Hultine, W. T. Pockman, R.B. Jackson, Ecohydrological implications of woody plant encroachment, Ecology 86 (2005) 308–319, https://doi.org/10.1890/03-0583.
[65] B.P. Wilcox, M.K. Owens, W.A. Dugas, D.N. Ueckert, C.R. Hart, Shrubs, streamflow, and the paradox of scale, Hydrol. Process. 20 (2006) 3245–3259, https://doi.org/10.1002/hyp.6330.
[66] B. Wilcox, Y. Huang, J. Walker, Long-term trends in streamflow from semiarid rangelands: uncovering drivers of change, Glob. Chang. Biol. 14 (2008) 1676–1689, https://doi.org/10.1111/j.1365-2486.2008.01578.x.
[67] F. Gallart, P. Llorens, Catchment management under environmental change: impact of land cover change on water resources, Water Int. 28 (2003) 334–340, https://doi.org/10.1080/02508060308691707.

[68] S. Beguería, J.I. López-Moreno, A. Lorente, M. Seeger, J.M. García-Ruiz, Assessing the effect of climate oscillations and land-use changes on streamflow in the central Spanish Pyrenees, AMBIO A J. Hum. Environ. 32 (2003) 283–286, https://doi.org/10.1579/0044-7447-32.4.283.

[69] E. Morán-Tejeda, A. Ceballos-Barbancho, J.M. Llorente-Pinto, Hydrological response of Mediterranean headwaters to climate oscillations and land-cover changes: the mountains of Duero River basin (Central Spain), Glob. Planet. Change 72 (2010) 39–49, https://doi.org/10.1016/j.gloplacha.2010.03.003.

[70] F. Gallart, J. Delgado, S.J.V. Beatson, H. Posner, P. Llorens, R. Marcé, Analysing the effect of global change on the historical trends of water resources in the headwaters of the Llobregat and Ter river basins (Catalonia, Spain), Phys. Chem. Earth 36 (2011) 655–661, https://doi.org/10.1016/j.pce.2011.04.009.

[71] J.I. López-Moreno, S.M. Vicente-Serrano, E. Moran-Tejeda, J. Zabalza, J. Lorenzo-Lacruz, J. M. García-Ruiz, Impact of climate evolution and land use changes on water yield in the Ebro basin, Hydrol. Earth Syst. Sci. 15 (2011) 311–322, https://doi.org/10.5194/hess-15-311-2011.

[72] J.I. López-Moreno, S. Beguería, J.M. García-Ruiz, Trends in high flows in the central Spanish Pyrenees: response to climatic factors or to land-use change? Hydrol. Sci. J. 51 (2006) 1039–1050, https://doi.org/10.1623/hysj.51.6.1039.

[73] M. Albek, Ü.B. Öğütveren, E. Albek, Hydrological modeling of Seydi Suyu watershed (Turkey) with HSPF, J. Hydrol. 285 (2004) 260–271, https://doi.org/10.1016/j.jhydrol.2003.09.002.

[74] J. Delgado, P. Llorens, G. Nord, I.R. Calder, F. Gallart, Modelling the hydrological response of a Mediterranean medium-sized headwater basin subject to land cover change: The Cardener River basin (NE Spain), J. Hydrol. 383 (2010) 125–134, https://doi.org/10.1016/j.jhydrol.2009.07.024.

[75] R. Ranzi, M. Bochicchio, B. Bacchi, Effects on floods of recent afforestation and urbanisation in the Mella River (Italian Alps), Hydrol. Earth Syst. Sci. 6 (2002) 239–254, https://doi.org/10.5194/hess-6-239-2002.

[76] M. Antonellini, T. Dentinho, A. Khattabi, E. Masson, P.N. Mollema, V. Silva, P. Silveira, An integrated methodology to assess future water resources under land use and climate change: an application to the Tahadart drainage basin (Morocco), Environ. Earth Sci. 71 (2014) 1839–1853, https://doi.org/10.1007/s12665-013-2587-5.

[77] C. Cosandey, V. Andréassian, C. Martin, J.F. Didon-Lescot, J. Lavabre, N. Folton, N. Mathys, D. Richard, The hydrological impact of the Mediterranean forest: a review of French research, J. Hydrol. 301 (2005) 235–249, https://doi.org/10.1016/j.jhydrol.2004.06.040.

[78] F. Gallart, P. Llorens, J. Latron, D. Regüés, Hydrological processes and their seasonal controls in a small Mediterranean mountain catchment in the Pyrenees, Hydrol. Earth Syst. Sci. 6 (2002) 527–537, https://doi.org/10.5194/hess-6-527-2002.

[79] P. Llorens, F. Gallart, C. Cayuela, M. Roig-Planasdemunt, E. Casellas, A.J. Molina, M.M.D. L. Heras, G. Bertran, E. Sánchez-Costa, J. Latron, What have we learnt about Mediterranean catchment hydrology? 30 years observing hydrological processes in the Vallcebre research catchments, Geogr. Res. Lett. 44 (2018) 475–502, https://doi.org/10.18172/cig.3432.

[80] J.M. García-Ruiz, D. Regüés, B. Alvera, N. Lana-Renault, P. Serrano-Muela, E. Nadal-Romero, A. Navas, J. Latron, C. Martí-Bono, J. Arnáez, Flood generation and sediment transport in experimental catchments affected by land use changes in the Central Pyrenees, J. Hydrol. 356 (2008) 245–260, https://doi.org/10.1016/j.jhydrol.2008.04.013.

[81] N. Lana-Renault, E. Nadal-Romero, M.P. Serrano-Muela, B. Alvera, P. Sánchez-Navarrete, Y. Sanjuan, J.M. García-Ruiz, Comparative analysis of the response of various land covers to an exceptional rainfall event in the central Spanish Pyrenees, October 2012, Earth Surf. Process. Landf. 39 (2014) 581–592, https://doi.org/10.1002/esp.3465.

[82] N. Lana-Renault, J. Latron, D. Karssenberg, P. Serrano-Muela, D. Regüés, M.F. P. Bierkens, Differences in stream flow in relation to changes in land cover: a comparative study in two sub-Mediterranean mountain catchments, J. Hydrol. 411 (2011) 366–378, https://doi.org/10.1016/j.jhydrol.2011.10.020.

[83] M. Bonell, Progress in the understanding of runoff generation dynamics in forests, J. Hydrol. 150 (1993) 217–275, https://doi.org/10.1016/0022-1694(93)90112-M.

[84] E. Nadal-Romero, E. Cammeraat, M.P. Serrano-Muela, N. Lana-Renault, D. Regüés, Hydrological response of an afforested catchment in a Mediterranean humid mountain area: a comparative study with a natural forest, Hydrol. Process. 30 (2016) 2717–2733, https://doi.org/10.1002/hyp.10820.
[85] N. Lana-Renault, M. López-Vicente, E. Nadal-Romero, R. Ojanguren, J.A. Llorente, P. Errea, D. Regüés, P. Ruiz-Flaño, M. Khorchani, J. Arnáez, N. Pascual, Catchment based hydrology under post farmland abandonment scenarios, Geogr. Res. Lett. 44 (2018) 503–534, https://doi.org/10.18172/cig.3475.
[86] C. Cosandey, J.F. Dbdon-Lescot, Etude des crues Cévenoles: conditions d'apparition dans un petit bassin versant forestier sur le versant sud du Mont Lozère, 191, IAHS Publ, 1990, pp. 103–115.
[87] M. Robinson, B. Gannon, M. Schuch, A comparison of the hydrology of moorland under natural conditions, agricultural use and forestry, Hydrol. Sci. J. 36 (1991) 565–577, https://doi.org/10.1080/02626669109492544.
[88] L.H. Cammeraat, A.C. Imeson, The evolution and significance of soil-vegetation patterns following land abandonment and fire in Spain, Catena 37 (1999) 107–127, https://doi.org/10.1016/S0341-8162(98)00072-1.
[89] E. Rodríguez-Caballero, R. Lázaro, Y. Cantón, J. Puigdefábregas, A. Solé-Benet, Long-term hydrological monitoring in arid-semiarid almerÍa, SE Spain. What have we learned? Geogr. Res. Lett. 44 (2018) 581–600, https://doi.org/10.18172/cig.3462.
[90] E. Nadal-Romero, T. Lasanta, J.M. García-Ruiz, Runoff and sediment yield from land under various uses in a Mediterranean mountain area: long-term results from an experimental station, Earth Surf. Process. Landf. 38 (2013) 346–355, https://doi.org/10.1002/esp.3281.
[91] A.N. Nunes, C.O.A. Coelho, A.C. de Almeida, A. Figueiredo, Soil erosion and hydrological response to land abandonment in a central inland area of Portugal, Land Degrad. Dev. (2010), https://doi.org/10.1002/ldr.973.
[92] T. Lasanta, J.M. García-Ruiz, C. Pérez-Rontomé, C. Sancho-Marcén, Runoff and sediment yield in a semi-arid environment: the effect of land management after farmland abandonment, Catena 38 (2000) 265–278, https://doi.org/10.1016/S0341-8162(99)00079-X.
[93] J. Solé-Benet, A. Cantón, Y. Domingo, F. Del Barrio, G. Lázaro, R. Domene, M. A. Vidal, S. Puigdefábregas, Long term hydrological monitoring of two micro-catchments in semi-arid SE Spain, in: L. Holko, P. Miklánek (Eds.), Interdiscip, 2003, pp. 183–188 Approaches Small Catchment Hydrol. Monit. Res. IHP-VI, Tech. Doc. Hydrol. N° 67., UNESCO, Paris.
[94] J.B. Ries, U. Hirt, Permanence of soil surface crusts on abandoned farmland in the Central Ebro Basin/Spain, Catena 72 (2008) 282–296, https://doi.org/10.1016/j.catena.2007.06.001.
[95] J. Puigdefabregas, A. Sole, L. Gutierrez, G. Del Barrio, M. Boer, Scales and processes of water and sediment redistribution in drylands: results from the Rambla Honda field site in Southeast Spain, Earth Sci. Rev. 48 (1999) 39–70, https://doi.org/10.1016/S0012-8252(99)00046-X.
[96] L. Merino-Martín, M.M.-d.L. Heras, S. Pérez-Domingo, T. Espigares, J.M. Nicolau, Hydrological heterogeneity in Mediterranean reclaimed slopes: runoff and sediment yield at the patch and slope scales along a gradient of overland flow, Hydrol. Earth Syst. Sci. 16 (2012) 1305–1320, https://doi.org/10.5194/hess-16-1305-2012.
[97] G. Callegari, E. Ferrari, G. Garfi, F. Iovino, A. Veltri, Impact of thinning on the water balance of a catchment in a mediterranean environment, Forest Chron. (2003) 301–306, https://doi.org/10.5558/tfc79301-2 Canadian Institute of Forestry.
[98] F.T. Maestre, J. Cortina, Are Pinus halepensis plantations useful as a restoration tool in semiarid Mediterranean areas? For. Ecol. Manag. 198 (2004) 303–317, https://doi.org/10.1016/j.foreco.2004.05.040.
[99] G. Schiller, Y. Cohen, Water balance of *Pinus halepensis* Mill. afforestation in an arid region, For. Ecol. Manag. 105 (1998) 121–128, https://doi.org/10.1016/S0378-1127(97)00283-1.
[100] P. Llorens, R. Poyatos, J. Latron, J. Delgado, I. Oliveras, F. Gallart, A multi-year study of rainfall and soil water controls on scots pine transpiration under Mediterranean mountain conditions, Hydrol. Process. 24 (2010) 3053–3064, https://doi.org/10.1002/hyp.7720.
[101] J. Bellot, A. Bonet, J.R. Sanchez, E. Chirino, Likely effects of land use changes on the runoff and aquifer recharge in a semiarid landscape using a hydrological model, Landsc. Urban Plan. 55 (2001) 41–53, https://doi.org/10.1016/S0169-2046(01)00118-9.

[102] J.D. Lundquist, S.E. Dickerson-Lange, J.A. Lutz, N.C. Cristea, Lower forest density enhances snow retention in regions with warmer winters: a global framework developed from plot-scale observations and modeling, Water Resour. Res. 49 (2013) 6356–6370, https://doi.org/10.1002/wrcr.20504.

[103] J.I. López-Moreno, J. Latron, Influence of canopy density on snow distribution in a temperate mountain range, Hydrol. Process. 22 (2008) 117–126, https://doi.org/10.1002/hyp.6572.

[104] J. Revuelto, J.I. López-Moreno, C. Azorin-Molina, S.M. Vicente-Serrano, Canopy influence on snow depth distribution in a pine stand determined from terrestrial laser data, Water Resour. Res. 51 (2015) 3476–3489, https://doi.org/10.1002/2014WR016496.

[105] J. Revuelto, J.I. López-Moreno, C. Azorin-Molina, E. Alonso-González, A. Sanmiguel-Vallelado, Small-scale effect of pine stand pruning on snowpack distribution in the pyrenees observed with a terrestrial laser scanner, Forests 7 (2016) 1–15, https://doi.org/10.3390/f7080166.

[106] J.J. Camarero, J.M. García-Ruiz, G. Sangüesa-Barreda, J.D. Galván, A.Q. Alla, Y. Sanjuán, S. Beguería, E. Gutiérrez, Recent and intense dynamics in a formerly static Pyrenean Treeline, Arct. Antarct. Alp. Res. 47 (2015) 773–783, https://doi.org/10.1657/AAAR0015-001.

[107] Plan Bleu, L'Eau, in: G. Benoit, A. Comeau (Eds.), Méditerranée, les perspectives du Plan Bleu sur l'Environnement et le Développement, Editions de l'Aube, France, 2005, pp. 71–107.

[108] Y. Bekkar, M. Kuper, A. Hammani, M. Dionnet, A. Eliamani, Reconversion vers des systèmes d'irrigation localisée au Maroc. Quels enseignements pour l'agriculture familiale? Revue HTE (2007) 137.

[109] J.A. Rodríguez-Díaz, L. Pérez-Urrestarazu, E. Camacho-Poyato, P. Montesinos, La paradoja de la modernización de zonas regables: uso más eficiente del agua vinculado al aumento de la demanda energética, Spanish, J. Agric. Res. 9 (2011) 1000–1008, https://doi.org/10.5424/sjar/20110904-492-10.

[110] F. Molle, O. Tanouti, Squaring the circle: agricultural intensification vs. water conservation in Morocco, Agric. Water Manag. 192 (2017) 170–179, https://doi.org/10.1016/j.agwat.2017.07.009.

[111] L. Salvati, A. Sabbi, Exploring long-term land cover changes in an urban region of southern Europe, Int. J. Sustain. Dev. World Ecol. 18 (2011) 273–282, https://doi.org/10.1080/13504509.2011.560453.

[112] H. March, D. Saurí, The suburbanization of water scarcity in the Barcelona metropolitan region: sociodemographic and urban changes influencing domestic water consumption, Prof. Geogr. 62 (2010) 32–45, https://doi.org/10.1080/00330120903375860.

[113] A.M. Rico-Amoros, J. Olcina-Cantos, D. Sauri, Tourist land use patterns and water demand: evidence from the Western Mediterranean, Land Use Policy 26 (2009) 493–501, https://doi.org/10.1016/j.landusepol.2008.07.002.

[114] A. Hof, T. Schmitt, Urban and tourist land use patterns and water consumption: evidence from Mallorca, Balearic Islands, Land Use Policy 28 (2011) 792–804, https://doi.org/10.1016/j.landusepol.2011.01.007.

[115] A. Hof, E. Morán-Tejeda, J. Lorenzo-Lacruz, M. Blázquez-Salom, A. Hof, E. Morán-Tejeda, J. Lorenzo-Lacruz, M. Blázquez-Salom, Swimming pool evaporative water loss and water use in the Balearic Islands (Spain), Water 10 (2018) 1883, https://doi.org/10.3390/W10121883.

[116] R. Choukr-Allah, A. Nghira, A. Hirich, L. Bouchaou, Water resources master plan for sustainable development of the Souss-Massa river basin, in: R. Choukr-Allah et al., (Ed.), Souss-Massa River Basin, Morocco, Hdb Env Chem2017, pp. 1–26, https://doi.org/10.1007/698_2016_67.

[117] R.J. Batalla, C.M. Gómez, G.M. Kondolf, Reservoir-induced hydrological changes in the Ebro River basin (NE Spain), J. Hydrol. 290 (2004) 117–136, https://doi.org/10.1016/j.jhydrol.2003.12.002.

[118] M. Klein, The Change in Flood Regime Along the Lower Jordan River: Its Influence on Flood Plain Land Use, IAHS-AISH Publ.299 (2005) pp. 205–216.

[119] J.I. López-Moreno, S. Beguería, J.M. García-Ruiz, The management of a large Mediterranean reservoir: storage regimens of the Yesa reservoir, upper Aragon river basin, central Spanish Pyrenees, Environ. Manag. 34 (2004) 508–515, https://doi.org/10.1007/s00267-003-0249-1.

[120] O. Bonacci, T. Roje-Bonacci, The influence of hydroelectrical development on the flow regime of the karstic river Cetina, Hydrol. Process. 17 (2003) 1–15, https://doi.org/10.1002/hyp.1190.

[121] J. Zabalza-Martínez, S.M. Vicente-Serrano, J.I. López-Moreno, G. Borràs, R. Savé, D. Pascual, E. Plá, E. Morán-Tejeda, F. Domínguez-Castro, C.L. Tague, The influence of climate

and land-cover scenarios on dam management strategies in a high water pressure catchment in North-east Spain, Water 10 (2018) 1668.
[122] C. Leduc, S. Ben Ammar, G. Favreau, R. Beji, R. Virrion, G. Lacombe, J. Tarhouni, C. Aouadi, B. Zenati Chelli, N. Jebnoun, M. Oi, J.L. Michelot, K. Zouari, Impacts of hydrological changes in the Mediterranean zone: environmental modifications and rural development in the Merguellil catchment, Central Tunisia, Hydrol. Sci. J. 52 (2007) 1162–1178, https://doi.org/10.1623/hysj.52.6.1162.
[123] R.F. Warner, Gross channel changes along the Durance River, Southern France, over the last 100 years using cartographic data, Regul. Rivers Res. Manag. 16 (2000) 141–157, https://doi.org/10.1002/(SICI)1099-1646(200003/04)16:2<141::AID-RRR574>3.0.CO;2-2.
[124] J.I. López-Moreno, S. Beguería, J.M. García-Ruiz, Influence of the Yesa reservoir on floods of the Aragón River, central Spanish Pyrenees, Hydrol. Earth Syst. Sci. 6 (2002) 753–762, https://doi.org/10.5194/hess-6-753-2002.
[125] A. Shaban, Indicators and aspects of hydrological drought in Lebanon, Water Resour. Manag. 23 (2009) 1875–1891, https://doi.org/10.1007/s11269-008-9358-1.
[126] E. Custodio, J.M. Andreu-Rodes, R. Aragón, T. Estrela, J. Ferrer, J.L. García-Aróstegui, M. Manzano, L. Rodríguez-Hernández, A. Sahuquillo, A. del Villar, Groundwater intensive use and mining in south-eastern peninsular Spain: hydrogeological, economic and social aspects, Sci. Total Environ. 559 (2016) 302–316, https://doi.org/10.1016/j.scitotenv.2016.02.107.
[127] C. Rupérez-Moreno, J. Senent-Aparicio, D. Martinez-Vicente, J.L. García-Aróstegui, F.C. Calvo-Rubio, J. Pérez-Sánchez, Sustainability of irrigated agriculture with overexploited aquifers: the case of Segura basin (SE, Spain), Agric. Water Manag. 182 (2017) 67–76, https://doi.org/10.1016/j.agwat.2016.12.008.
[128] C. Garcia, J. Servera, Impacts of tourism development on water demand and beach degradation on the island of Mallorca (Spain), Geogr. Ann. Ser. A. Phys. Geogr. 85 (2003) 287–300, https://doi.org/10.1111/j.0435-3676.2003.00206.x.
[129] J. Martínez-Fernández, M.A. Esteve, A critical view of the desertification debate in southeastern Spain, Land Degrad. Dev. 16 (2005) 529–539, https://doi.org/10.1002/ldr.707.
[130] NTUA (National Technical University of Athens) National Programme for Water Resources Management and Preservation, Athens, Ministry of Environment, Planning and Public Works, 2007. http://www.itia.ntua.gr/e/docinfo/782/.
[131] A. Pulido-Bosch, P. Pulido-Leboeuf, L. Molina-Sánchez, A. Vallejos, W. Martin-Rosales, Intensive agriculture, wetlands, quarries and water management. A case study (Campo de Dalias, SE Spain), Environ. Geol. 40 (2000) 163–168, https://doi.org/10.1007/s002540000118.
[132] B. Bouya, M. Faouzi, M. Ben Abbou, A. Essahlaoui, B. Bahir, N. Youbi, M.A. Hessane, L'aquifère côtier des Mnasra (plaine du Gharb, Maroc): hydrogéologie et modélisation hydrodynamique, Comun. Geol. 98 (2011) 73–81.
[133] L. Lassaletta, H. García-Gómez, B.S. Gimeno, J.V. Rovira, Agriculture-induced increase in nitrate concentrations in stream waters of a large Mediterranean catchment over 25 years (1981-2005), Sci. Total Environ. 407 (2009) 6034–6043, https://doi.org/10.1016/j.scitotenv.2009.08.002.
[134] S.R. Lutz, S. Mallucci, E. Diamantini, B. Majone, A. Bellin, R. Merz, Hydroclimatic and water quality trends across three Mediterranean river basins, Sci. Total Environ. 571 (2016) 1392–1406, https://doi.org/10.1016/j.scitotenv.2016.07.102.
[135] D. Fatta, Z. Salem, M. Mountadar, O. Assobhei, M. Loizidou, Urban wastewater treatment and reclamation for agricultural irrigation: the situation in Morocco and Palestine, Environmentalist 24 (2004) 227–236, https://doi.org/10.1007/s10669-005-0998-x.
[136] M. Malki, L. Bouchaou, I. Mansir, H. Benlouali, A. Nghira, R. Choukr-Allah, Wastewater treatment and reuse for irrigation as alternative resource for water safeguarding in Souss-Massa region, Morocco, Eur. Water 59 2017.
[137] R. Aguilera, R. Marcé, S. Sabater, Detection and attribution of global change effects on river nutrient dynamics in a large Mediterranean basin, Biogeosciences 12 (2015) 4085–4098, https://doi.org/10.5194/bg-12-4085-2015.
[138] M. Blinda, G. Thivet, Ressources et demandes en eau en Méditerranée: situation et perspectives, Sécheresse 20 (2009) 9–16.

[139] P.H. Verburg, K.P. Overmars, Combining top-down and bottom-up dynamics in land use modeling: exploring the future of abandoned farmlands in Europe with the Dyna-CLUE model, Landsc. Ecol. 24 (2009) 1167–1181, https://doi.org/10.1007/s10980-009-9355-7.

[140] Ž. Malek, P.H. Verburg, I.R. Geijzendorffer, A. Bondeau, W. Cramer, Global change effects on land management in the Mediterranean region, Glob. Environ. Change 50 (2018) 238–254, https://doi.org/10.1016/j.gloenvcha.2018.04.007.

[141] A.D. del Campo, M. González-Sanchis, A. García-Prats, C.J. Ceacero, C. Lull, The impact of adaptive forest management on water fluxes and growth dynamics in a water-limited low-biomass oak coppice, Agric. For. Meteorol. 264 (2019) 266–282, https://doi.org/10.1016/j.agrformet.2018.10.016.

[142] M. González-Sanchis, G. Ruiz-Pérez, A.D. Del Campo, A. Garcia-Prats, F. Francés, C. Lull, Managing low productive forests at catchment scale: considering water, biomass and fire risk to achieve economic feasibility, J. Environ. Manag. 231 (2019) 653–665, https://doi.org/10.1016/j.jenvman.2018.10.078.

[143] A.D. del Campo, M. González-Sanchis, A. Lidón, A. García Prats, I. Lull, C. Bautista, G. Ruiz, F. Francés, Ecohydrological-based forest management in semi-arid climate, in: J. Křeček, M. Haigh, T. Hofer, E. Kubin, C. Promper (Eds.), Ecosyst. Serv. Headwater Catchments, Co-published by Springer Int. Publishing, Switzerland, with Capital Publishing Co., India, 2017, pp. 45–57.

[144] A.J. Molina, A.D. del Campo, The effects of experimental thinning on throughfall and stemflow: a contribution towards hydrology-oriented silviculture in Aleppo pine plantations, For. Ecol. Manag. 269 (2012) 206–213, https://doi.org/10.1016/j.foreco.2011.12.037.

[145] A.D. del Campo, T.J.G. Fernandes, A.J. Molina, Hydrology-oriented (adaptive) silviculture in a semi-arid pine plantation: how much can be modified the water cycle through forest management? Eur. J. For. Res. 133 (2014) 879–894, https://doi.org/10.1007/s10342-014-0805-7.

[146] J.E. Keeley, W.J. Bond, R.A. Bradstock, J.G. Pausas, P.W. Rundel, Fire in the Mediterranean basin, in: J.E. Keeley, W.J. Bond, R.A. Bradstock, J.G. Pausas, P.W. Rundel (Eds.), Fire in Mediterranean Ecosystems: Ecology, Evolution and Management, Cambridge University Press, 2012, pp. 83–112.

[147] N. Bauer, A. Wallner, M. Hunziker, The change of European landscapes: human-nature relationships, public attitudes towards rewilding, and the implications for landscape management in Switzerland, J. Environ. Manag. 90 (2009) 2910–2920, https://doi.org/10.1016/j.jenvman.2008.01.021.

[148] P. Tarolli, F. Preti, N. Romano, Terraced landscapes: from an old best practice to a potential hazard for soil degradation due to land abandonment, Anthropocene 6 (2014) 10–25, https://doi.org/10.1016/j.ancene.2014.03.002.

[149] G.-J. Nabuurs, M.-J. Schelhaas, M.J. Mohren, C.B. Field, Temporal evolution of the European forest sector carbon sink from 1950 to 1999, Glob. Chang. Biol. 9 (2003) 152–160, https://doi.org/10.1046/j.1365-2486.2003.00570.x.

[150] E. Nadal-Romero, E. Cammeraat, E. Pérez-Cardiel, T. Lasanta, How do soil organic carbon stocks change after cropland abandonment in Mediterranean humid mountain areas? Sci. Total Environ. 566–567 (2016) 741–752, https://doi.org/10.1016/j.scitotenv.2016.05.031.

[151] L.M. Navarro, H.M. Pereira, Rewilding abandoned landscapes in Europe, Ecosystems 15 2012) 900–912, https://doi.org/10.1007/s10021-012-9558-7.

[152] F. Robledano-Aymerich, A. Romero-Díaz, F. Belmonte-Serrato, V.M. Zapata-Pérez, C. Martínez-Hernández, V. Martínez-López, Ecogeomorphological consequences of land abandonment in semi-arid Mediterranean areas: integrated assessment of physical evolution and biodiversity, Agric. Ecosyst. Environ. 197 (2014) 222–242, https://doi.org/10.1016/j.agee.2014.08.006.

[153] T. Lasanta, M. Khorchani, F. Pérez-Cabello, P. Errea, R. Sáenz-Blanco, E. Nadal-Romero, Clearing shrubland and extensive livestock farming: active prevention to control wildfires in the Mediterranean mountains, J. Environ. Manag. 227 (2018) 256–266, https://doi.org/10.1016/j.jenvman.2018.08.104.

[153a] UNEP-MAP (United Nations Environment Programme – Mediterranean Action Plan), Mediterranean Strategy for Sustainable Development, UNEP-MAP, Athens, 2006.

[154] J. Alcamo, M. Flörke, M. Märker, Future long-term changes in global water resources driven by socio-economic and climatic changes, Hydrol. Sci. J. 52 (2007) 247–275, https://doi.org/10.1623/hysj.52.2.247.

[155] A. Giménez-Morera, J.D.R. Sinoga, A. Cerdà, The impact of cotton geotextiles on soil and water losses from Mediterranean rainfed agricultural land, Land Degrad. Dev. 21 (2010) 210–217, https://doi.org/10.1002/ldr.971.

[156] J.A. Gómez, C. Llewellyn, G. Basch, P.B. Sutton, J.S. Dyson, C.A. Jones, The effects of cover crops and conventional tillage on soil and runoff loss in vineyards and olive groves in several Mediterranean countries, Soil Use Manag. 27 (2011) 502–514, https://doi.org/10.1111/j.1475-2743.2011.00367.x.

[157] M. Prosdocimi, A. Jordán, P. Tarolli, S. Keesstra, A. Novara, A. Cerdà, The immediate effectiveness of barley straw mulch in reducing soil erodibility and surface runoff generation in Mediterranean vineyards, Sci. Total Environ. 547 (2016) 323–330, https://doi.org/10.1016/j.scitotenv.2015.12.076.

[158] T.P. Burt, J.J. McDonnell, Whither field hydrology? The need for discovery science and outrageous hydrological hypotheses, Water Resour. Res. 51 (2015) 5919–5928, https://doi.org/10.1002/2014WR016839.

[159] J. Latron, N. Lana-Renault, The relevance of hydrological research in small catchments—a perspective from long-term monitoring sites in Europe, Geogr. Res. Lett. 44 (2018) 387–395, https://doi.org/10.18172/cig.3499.

[160] C.G. Bates, A.J. Henry, Forest and streamflow experiment at Wagon Wheel Gap, Colorado, Mon. Weather Rev. Suppl. 30 (1928).

[160a] J.D. Hewlett, Principles of Forest Hydrology, The University of Georgia Press, Athens, 1982.

[161] J. Lavabre, D.S. Torres, F. Cernesson, Changes in the hydrological response of a small Mediterranean basin a year after a wildfire, J. Hydrol. 142 (1993) 273–299, https://doi.org/10.1016/0022-1694(93)90014-Z.

[162] D. Tetzlaff, S.K. Carey, J.P. McNamara, H. Laudon, C. Soulsby, The essential value of long-term experimental data for hydrology and water management, Water Resour. Res. 53 (2017) 2598–2604, https://doi.org/10.1002/2017WR020838.

[163] D. Boix, E. García-Berthou, S. Gascón, L. Benejam, E. Tornés, J. Sala, J. Benito, A. Munné, C. Solà, S. Sabater, Response of community structure to sustained drought in Mediterranean rivers, J. Hydrol. 383 (2010) 135–146, https://doi.org/10.1016/j.jhydrol.2010.01.014.

[164] J.M. García-Ruiz, N. Lana-Renault, S. Beguería, T. Lasanta, D. Regüés, E. Nadal-Romero, P. Serrano-Muela, J.I. López-Moreno, B. Alvera, C. Martí-Bono, L.C. Alatorre, From plot to regional scales: interactions of slope and catchment hydrological and geomorphic processes in the Spanish Pyrenees. Geomorphology 120 (2010) 248–257, https://doi.org/10.1016/j.geomorph.2010.03.038.

[165] J.D. Hewlett, H.W. Lull, K.G. Reinhart, In defense of experimental watersheds, Water Resour. Res. 5 (1969) 306–316, https://doi.org/10.1029/WR005i001p00306.

[166] J. Latron, P. Llorens, F. Gallart, The hydrology of Mediterranean mountain areas, Geogr. Compass 3 (2009) 2045–2064, https://doi.org/10.1111/j.1749-8198.2009.00287.x.

[167] C. Cudennec, C. Leduc, D. Koutsoyiannis, Dryland hydrology in Mediterranean regions—a review, Hydrol. Sci. J. 52 (2007) 1077–1087, https://doi.org/10.1623/hysj.52.6.1077.

[168] A. Bronstert, Rainfall-runoff modelling for assessing impacts of climate and land-use change, Hydrol. Process. 18 (2004) 567–570, https://doi.org/10.1002/hyp.5500.

[169] G.K. Devia, B.P. Ganasri, G.S. Dwarakish, A review on hydrological models, Aquat. Procedia 4 (2015) 1001–1007, https://doi.org/10.1016/j.aqpro.2015.02.126.

[170] M. Wegehenkel, Estimating of the impact of land use changes using the conceptual hydrological model THESEUS—a case study, Phys. Chem. Earth Parts A/B/C 27 (2002) 631–640, https://doi.org/10.1016/S1474-7065(02)00047-5.

[171] D. Niehoff, U. Fritsch, A. Bronstert, Land-use impacts on storm-runoff generation: scenarios of land-use change and simulation of hydrological response in a meso-scale catchment in SW-Germany, J. Hydrol. 267 (2002) 80–93, https://doi.org/10.1016/S0022-1694(02)00142-7.

[172] J.A. Vano, J.A. Foley, C.J. Kucharik, M.T. Coe, Controls of climatic variability and land cover on land surface hydrology of northern Wisconsin, USA, J. Geophys. Res. Biogeosci. 113 (2008) https://doi.org/10.1029/2007JG000681.

[173] N. Montaldo, R. Corona, J.D. Albertson, On the separate effects of soil and land cover on Mediterranean ecohydrology: two contrasting case studies in Sardinia, Italy, Water Resour. Res. 49 (2013) 1123–1136, https://doi.org/10.1029/2012WR012171.
[174] R. Alley, T. Berntsen, N.L. Bindoff, Z. Chen, A. Chidthaisong, P. Friedlingstein, J. Gregory, G. Hegerl, M. Heimann, B. Hewitson, B. Hoskins, F. Joos, J. Jouzel, V. Kattsov, U. Lohmann, M. Manning, T. Matsuno, M. Molina, N. Nicholls, J. Overpeck, D. Qin, G. Raga, V. Ramaswamy, J. Ren, M. Rusticucci, S. Solomon, R. Somerville, T.F. Stocker, P. Stott, R. J. Stouffer, P. Whetton, R.A. Wood, D. Wratt, J. Arblaster, G. Brasseur, J.H. Christensen, K. Denman, D.W. Fahey, P. Forster, E. Jansen, P.D. Jones, R. Knutti, H. Le Treut, P. Lemke, G. Meehl, P. Mote, D. Randall, D.A. Stone, K.E. Trenberth, J. Willebrand, F. Zwiers, Climate Change 2007: The Physical Science Basis, 2007.
[175] IPCC, in: Climate Change 2013: The Physical Science Basis, Contribution of Working Group I to the Fifth Assessment Report of the Intergovernmental Panel on Climate Change, Cambridge University Press, Cambridge, United Kingdom and New York, NY, USA, 2013.
[176] D. Serpa, J.P. Nunes, J. Santos, E. Sampaio, R. Jacinto, S. Veiga, J.C. Lima, M. Moreira, J. Corte-Real, J.J. Keizer, N. Abrantes, Impacts of climate and land use changes on the hydrological and erosion processes of two contrasting Mediterranean catchments, Sci. Total Environ. 538 (2015) 64–77, https://doi.org/10.1016/J.SCITOTENV.2015.08.033.
[177] E. Morán-Tejeda, J. Zabalza, K. Rahman, A. Gago-Silva, J.I. López-Moreno, S. Vicente-Serrano, A. Lehmann, C.L. Tague, M. Beniston, Hydrological impacts of climate and land-use changes in a mountain watershed: uncertainty estimation based on model comparison, Ecohydrology 8 (2015) 1396–1416, https://doi.org/10.1002/eco.1590.
[178] E. Mancosu, A. Gago-Silva, A. Barbosa, A. de Bono, E. Ivanov, A. Lehmann, J. Fons, Future land-use change scenarios for the Black Sea catchment, Environ. Sci. Pol. 46 (2015) 26–36, https://doi.org/10.1016/J.ENVSCI.2014.02.008.
[179] J.I. López-Moreno, J. Zabalza, S.M. Vicente-Serrano, J. Revuelto, M. Gilaberte, C. Azorin-Molina, E. Morán-Tejeda, J.M. García-Ruiz, C. Tague, Impact of climate and land use change on water availability and reservoir management: scenarios in the Upper Aragón River, Spanish Pyrenees, Sci. Total Environ. 493 (2014) 1222–1231, https://doi.org/10.1016/J.SCITOTENV.2013.09.031.
[180] G.J. Burch, R.K. Bath, I.D. Moore, E.M. O'Loughlin, Comparative hydrological behaviour of forested and cleared catchments in southeastern Australia, J. Hydrol. 90 (1987) 19–42, https://doi.org/10.1016/0022-1694(87)90171-5.
[181] N. Lana-Renault, J. Latron, D. Regüés, Streamflow response and water-table dynamics in a sub-Mediterranean research catchment (Central Pyrenees), J. Hydrol. 347 (2007) 497–507, https://doi.org/10.1016/j.jhydrol.2007.09.037.
[182] M.P. Serrano-Muela, N. Lana-Renault, E. Nadal-Romero, D. Regüés, J. Latron, C. Martí-Bono, J. M. García-Ruiz, Forests and their hydrological effects in Mediterranean Mountains, Mt. Res. Dev. 28 (2008) 279–285, https://doi.org/10.1659/mrd.0876.
[183] C. Carvalho-Santos, J.P. Nunes, A.T. Monteiro, L. Hein, J.P. Honrado, Assessing the effects of land cover and future climate conditions on the provision of hydrological services in a medium-sized watershed of Portugal, Hydrol. Process. 30 (2016) 720–738, https://doi.org/10.1002/hyp.10621.
[184] J.G. Arnold, R. Srinivasan, R.S. Muttiah, J.R. Williams, Large area hydrological modeling and assessment part I: model development, J. Am. Water Resour. Assoc. 34 (1998) 73–89, https://doi.org/10.1111/j.1752-1688.1998.tb05961.x.
[185] S. Neitsch, J. Arnold, J. Kiriny, J. Williams, Soil & Water Assessment Tool Theoretical Documentation Version 2009, College Station: Texas, USA, (2011).
[186] E. Morán-Tejeda, A. Ceballos-Barbancho, J.M. Llorente-Pinto, J.I. López-Moreno, Land-cover changes and recent hydrological evolution in the Duero Basin (Spain), Reg. Environ. Chang. 12 (2012) 17–33, https://doi.org/10.1007/s10113-011-0236-7.

CHAPTER 11

Evolution of wetlands in Mediterranean region

Christian Perennou, Elie Gaget, Thomas Galewski, Ilse Geijzendorffer, Anis Guelmami
Tour du Valat Research Institute for the Conservation of Mediterranean Wetlands, Arles, France

11.1 Introduction

The Mediterranean basin is a global biodiversity hotspot, and wetlands are an important, specific element in maintaining this status.

Wetlands, as per the Ramsar Convention on Wetlands (www.ramsar.org), include in the broad sense virtually all ecosystems, which are characterized by the presence of water during at least part of the year, except seas and oceans beyond 6m depth.

The Mediterranean region is famous for the prestigious delta wetlands such as Doñana in Spain, the Camargue in France, and the Nile delta in Egypt, as well as for its large inland salt lakes (chotts and sebkhas in North Africa), oases, temporary ponds, and marshes. These depend to a large extent on a variable climate, which leads to large inter-annual fluctuations in ecosystem extent and functions.

The status and trends of Mediterranean wetlands have been assessed by the Mediterranean Wetlands Observatory[a] since 2010, and two regional Outlooks published (Mediterranean Wetlands Outlooks 1 and 2; MWO1 and MWO2 respectively), making wetlands one of the very few ecosystems that have been assessed at this scale [1, 2]. Being placed under the aegis of the MedWet Initiative of the Ramsar Convention, the ongoing monitoring work of the Observatory is supported by all the countries in the Mediterranean Region, which are all members of MedWet[b].

The key specific scientific challenges are:

- Evaluating extent and trends of Mediterranean wetlands: despite recent progresses in methods based on Earth Observation, challenges remain for the monitoring of

[a] The definition of the "Mediterranean" as used by the Mediterranean Wetlands Observatory (e.g., MWO, 2012, 2018) in all its analysis presented here is broader than the geographical scope of the Barcelona Convention, as it also encompasses Portugal, Andorra, Serbia, Bulgaria, Jordan, Kosovo and North Macedonia.

[b] 28 countries from the Mediterranean and surroundings (parties of the Ramsar Wetland Convention); MedWet was founded in 1991

Water Resources in the Mediterranean Region
https://doi.org/10.1016/B978-0-12-818086-0.00011-X

wetlands [3], notably in terms of detecting spatial and temporal patterns and quantifying uncertainties.

- Pan-Mediterranean monitoring of changes in water availability for wetlands (quantities, flows, quality): temporal and spatial gaps in coverage remain, and making correct interpretations in a context of intra and interannual variability and political sensitivity in sharing data remains challenging.
- Correctly representing the complexity of changes in Mediterranean socio-ecosystems into indicators for ecosystem services [4].

11.2 State of the art, in the context of the Mediterranean region

11.2.1 Surface area of Mediterranean wetlands: current situation and trends

The Mediterranean basin holds 19–26 million hectares of wetlands [2, 5]. Although some wetland types are well covered in surface area inventories, others, such as the shallow marine waters are often overlooked—although in its definition of wetlands, the Ramsar Convention includes marine areas less than 6 m deep at low tide. No figures exist on the relative importance of coastal vs. inland wetlands in the Mediterranean, but at the global scale, natural inland wetlands cover a much larger extent (by a factor 7–8) than natural coastal wetlands [6].

The extent of Mediterranean wetlands has been greatly reduced for more than 2000 years due to multiple human pressures, for example to gain agricultural or urban land, or for sanitary reasons. While the losses were obvious, until recently there were few precise data available, and the trends of wetlands in the Mediterranean Basin remained poorly known.

Trends have recently been assessed using the WET Index [7], on a sample of 400 Mediterranean wetland sites. It showed that they have lost, on average, 48% of their natural wetland habitats between 1970 and 2013 (MWO, 2018), far worse than in Africa (−42%), Asia (−32%), and Europe (−35%), or than the world overall (−35%), as shown in the recent Global Wetlands Outlook [7, 8]. It should be noted that the WET Index applies to a sample of wetlands, and does not therefore necessarily reflect the total change in wetlands in the region. Also, it should not be taken as reflecting change in surface areas only, as this index aggregates other dimensions too (see Section 11.3).

Within a subset of 302 sites analyzed in more detail [2], south-eastern Mediterranean countries suffered the greatest losses of natural wetlands, but also the highest gains in artificial wetlands (Fig. 11.1). These latter figures, concerning emblematic major sites, probably underestimate the changes affecting all wetlands of the Mediterranean Basin as a whole.

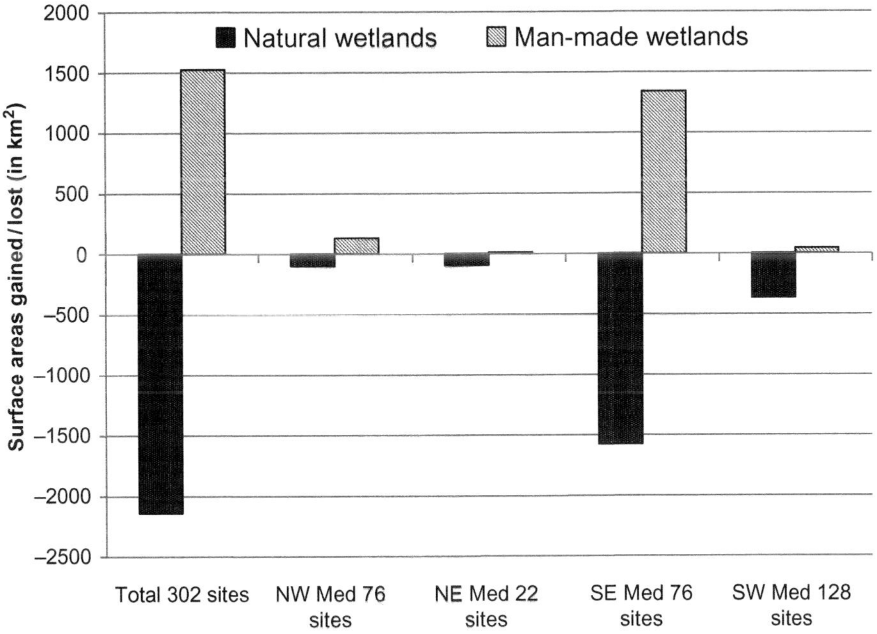

Fig. 11.1 Comparison of the losses/gains in the total area of natural and man-made wetlands for 302 Mediterranean sites (1975–2005) [2].

11.2.2 Biodiversity of Mediterranean wetlands: current situation and trends

The Mediterranean Basin is one of the planet's 34 biodiversity hotspots, characterized by exceptional rates of endemism. Preventing the extinction of endemic species is a crucial issue. Among Mediterranean ecosystems, wetlands are of capital importance for biodiversity: they cover just 2%–3% of the surface area of the Basin, but support more than 30% of its vertebrate species. Twice as many species threatened with extinction are found there than in all the other Mediterranean ecosystems. These figures illustrate the importance of preserving wetlands to prevent the erosion of biodiversity in the region [2].

Coastal wetlands in the Mediterranean basin are crucial for many species and ecological processes. For instance, coastal lagoons provide important feeding areas for many species of marine origin and therefore strongly contribute to the sustainability of fish stocks exploitation at sea (e.g., the sea bream *Sparus aurata*, the European bass *Dicentrarchus labrax*, or the common sole *Solea solea*). River deltas and their associated coastal lakes, lagoons, and marshlands host tens of millions of migratory, wintering, and breeding waterbirds traveling from as far away as the Arctic and Southern Africa. The emblematic Greater Flamingo *Phoenicopterus roseus* is one of the 150 waterbird species that shelter in

Mediterranean coastal wetlands. Among continental wetlands, the Mediterranean region is characterized by its seasonal flooded pools, marshes, or lakes, which support many species adapted to these difficult living conditions, which avoid competition from more widespread species inhabiting permanent flooded wetlands. Hundreds of plant or amphibian species are associated with these habitats. Several water courses and alluvial wetlands are also characterized by a high level of endemism with a number of highly range-restricted species (many freshwater mollusks, fish, and odonates).

Mediterranean wetland species face many anthropogenic pressures, including conversion of natural habitats, pollution, alteration of the natural functioning of watercourses and lagoons, hunting, intensive fishing and shellfish harvesting, rising temperatures, irregular precipitation, and invasive exotic species. As a result, the abundance of wetland-dependent species has been in long-term decline (−15% between 1990 and 2013; Fig. 11.2).

There are however contrasting trends between taxonomic groups. Waterbirds are on average on the increase (Fig. 11.3). This is very likely due to three reasons. First, many waterbirds have been heavily hunted in the past, which has greatly reduced

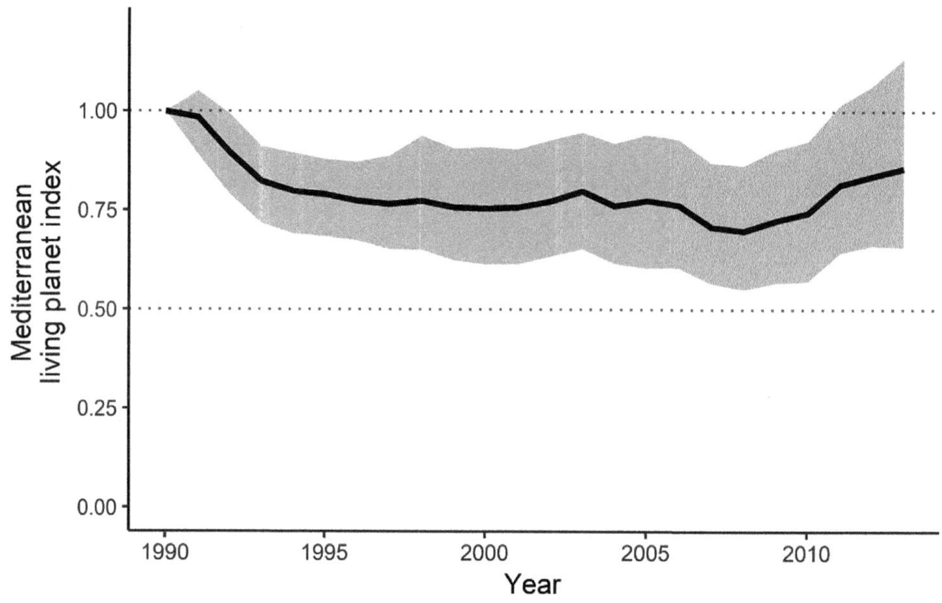

Fig. 11.2 The Living Planet Index calculated for the species dependent on Mediterranean wetlands. This index tracks the relative abundance of the populations of mammal, bird, reptile, amphibian, and fish species for which monitoring data are available. The results of more than 67,000 monitoring operations of populations belonging to 269 species were included in the calculation of the index. The value of the index is arbitrarily set at 1 in 1990. Gray areas represent the 95% confidence limits. *(Source: MWO (Mediterranean Wetlands Observatory), Mediterranean Wetlands Outlook 2: Solutions for sustainable Mediterranean wetlands, Tour du Valat/MedWet, Arles, France, 2018.)*

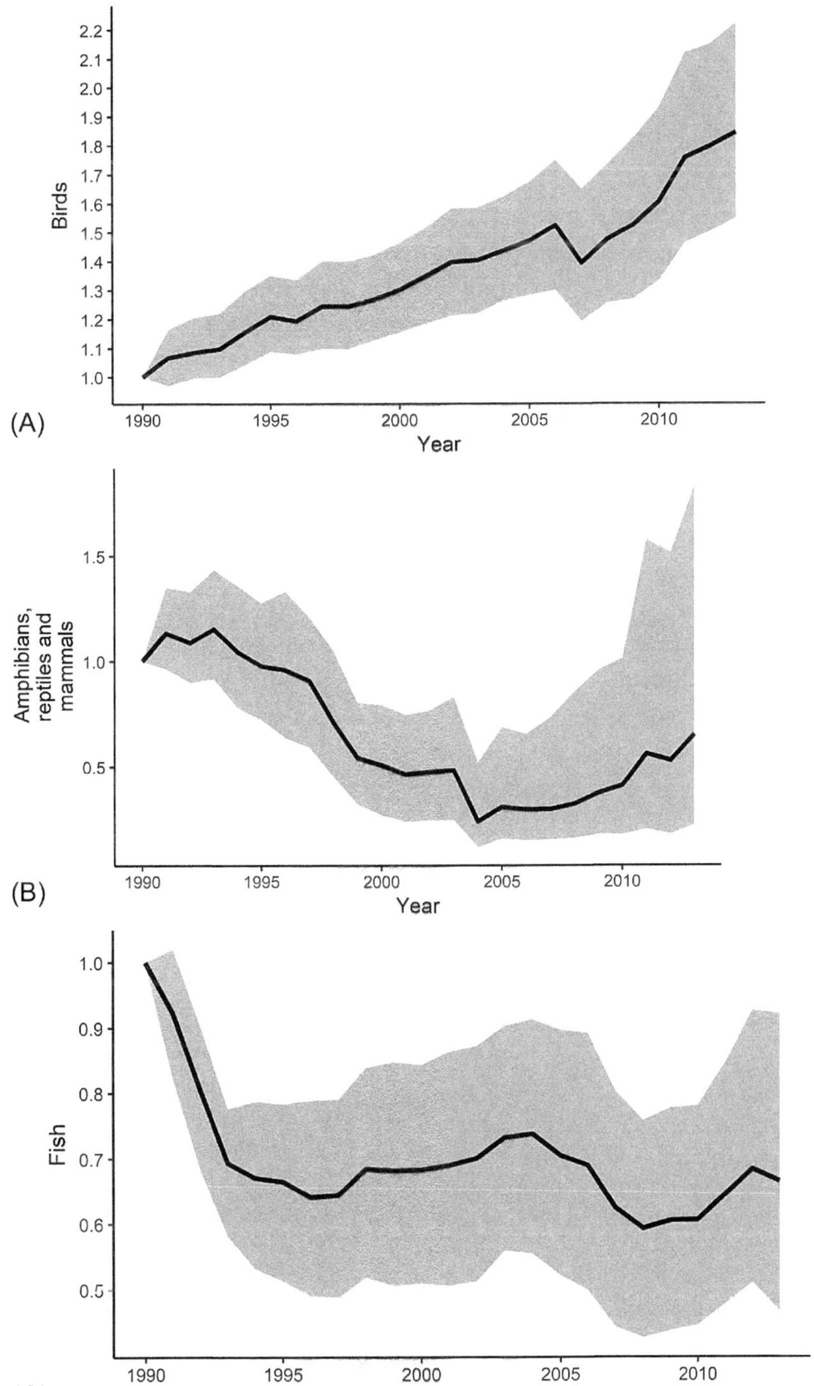

Fig. 11.3 Living Planet Index calculated for the species dependent on Mediterranean wetlands and by taxonomic groups. (A) Birds; (B) Amphibians, reptiles, and mammals; (C) Fish. *(Source: MWO (Mediterranean Wetlands Observatory), Mediterranean Wetlands Outlook 2: Solutions for sustainable Mediterranean wetlands, Tour du Valat/MedWet, Arles, France, 2018.)*

their numbers and range, resulting in a low baseline for 1990. Second, specific protection laws (e.g., the EU Birds Directive) associated with effective governance have led to a significant increase in the number of breeding populations in some countries, including the European Union [2, 9]. Third, artificial water bodies have also increased, which in contrast to many other wetland species, do provide additional habitat space for some water birds [2].

However, the Living Planet Index shows that the abundance of amphibians, reptiles, mammals, and fish in Mediterranean wetlands is far lower than in 1990 (Fig. 11.3). The species in decline are usually victims of the disappearance or degradation of their habitats. For example, small temporary wetlands are very important for amphibians, but are very often converted into agricultural or urban areas, or destroyed by transport infrastructure. The flow of watercourses is decreasing due to the creation of dams or water extraction, alluvial wetlands are drying out, and water quality is reduced by pollution, thus endangering fish populations. Only vertebrate species are included in the index, but IUCN Red List assessments also indicate a very worrying status for wetland plants and invertebrates, particularly freshwater mollusks. Overall, 36% of wetland-dependent species living in Mediterranean wetlands are threatened with extinction (Fig. 11.4), twice as much as in all other Mediterranean terrestrial and marine ecosystems [2]. One in ten species are expected to become extinct in the coming years if no conservation actions are taken in the near future. Climate change could further aggravate the situation in the next few decades, mainly through less frequent flooding of wetlands, reduction in the flow of watercourses, and the impacts of sea level rise on coastal wetlands.

Finally, decades of waterbird monitoring have shown that bird communities are better able to adapt to climate change in unmodified natural wetlands [10]. It is therefore logical to observe a better response of species to global warming in protected areas [11] because they limit the conversion of natural habitats into agricultural and urban environments. In this

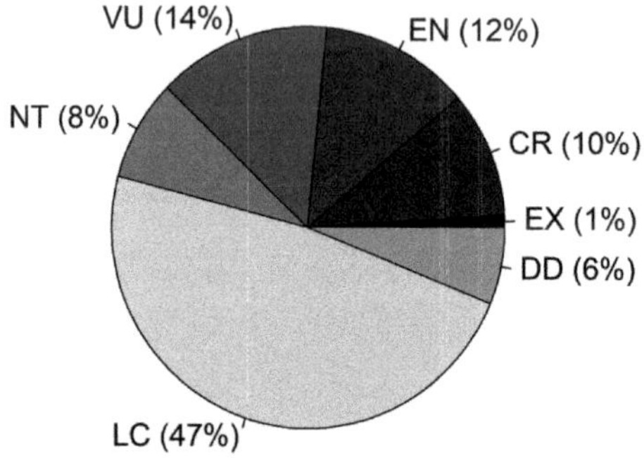

Fig. 11.4 Global conservation status of Mediterranean wetland species for eight taxonomic groups representing 1711 species. EX means "extinct," CR "critically endangered," EN "endangered," VU "vulnerable," NT "near threatened," LC "least concern." CR, EN, and VU species constitute together the pool of globally threatened species (36% of the total).

respect, the network of protected areas set up under the European Union's Nature Directives appears to be particularly effective [9]. Since global warming and the conversion of natural habitats are expected to accelerate in the Mediterranean basin in the coming decades [12], the protection of a large number of wetlands appears essential to mitigate the impact of rising temperatures on biodiversity.

11.2.3 Land use in and around Mediterranean wetlands: current situation and trends

The conversion of natural and seminatural habitats into land for farming or urban use is one of the main drivers that threaten Mediterranean wetlands. It leads to both the destruction and modification of natural habitats and species, while disturbing nearby habitats, e.g., by modifying sediment loads in nearby streams. In recent decades, urban development has accelerated in the Mediterranean Basin, notably in relation with demographic growth [13]. This process has been more rapid in southern and eastern Mediterranean countries, particularly in their coastal zones [2]. Agriculture development has been detrimental to wetlands, even though the total cultivated area did not increase in Mediterranean countries between 1961 and 2005 [14]. This apparent stability is caused by a continued conversion of farmland into cities and infrastructure, coupled with the fact that this "lost" farmland is then compensated by conversion of natural or seminatural habitats—including wetlands—into farmland [15].

Land-use changes have recently been assessed on a sample of 302 wetland sites throughout the Mediterranean Basin [2], which together cover 70,556 km^2 (Fig. 11.5). The study of their land cover on three dates (1975, 1990, and 2005) enabled us to quantify the trends. Urbanization increased by 1225 km^2 (+294%) and cultivated land by 4699 km^2 (+42%) between 1975 and 2005. These trends show geographical patterns, particularly in function of socio-economic conditions with less wetlands being converted in south-western Europe than in the rest of the Basin. In general, these transformations seem to be slowing down: they were less significant for 1990–2005 than for 1975–90.

The natural wetland habitats converted between 1975 and 2005 (in all 2150 km^2 for the 302 sites) were transformed into about equal parts of cropland and artificial wetlands (Fig. 11.6).[c] Urban development concerned smaller, but nonetheless significant areas.

Conversely, more than 1550 km^2 of artificial wetland habitats were created during 1975–2005. Nearly two-thirds of them replaced natural wetlands, for instance in areas such as alluvial plains in which gravel pits and sand pits have often replaced wet meadows and riverine woodlands (Fig. 11.7). Likewise, after the construction of dams, reservoirs

[c] It is worth noting that in these calculations, rice fields were included in the cropland, because the remote sensing methods could not separate them from other crops accurately, whereas in the Ramsar Convention, they are considered to be artificial wetlands.

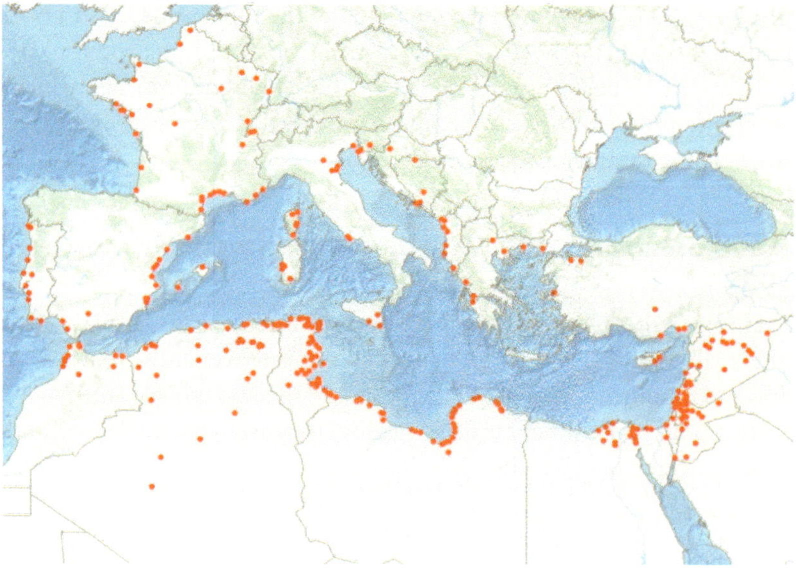

Fig. 11.5 Distribution of the 302 sites whose land cover was studied by the MWO [2] *(Source: MWO data.)*

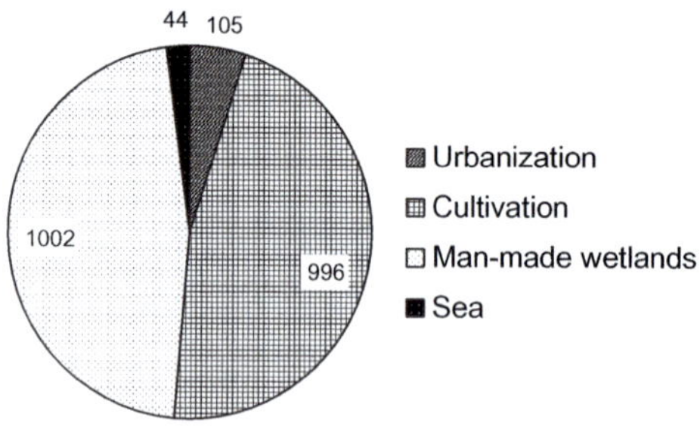

Fig. 11.6 Land cover (in km^2) conversion for 2150 km^2 of natural wetlands in 302 sites from 1975 to 2005 [2].

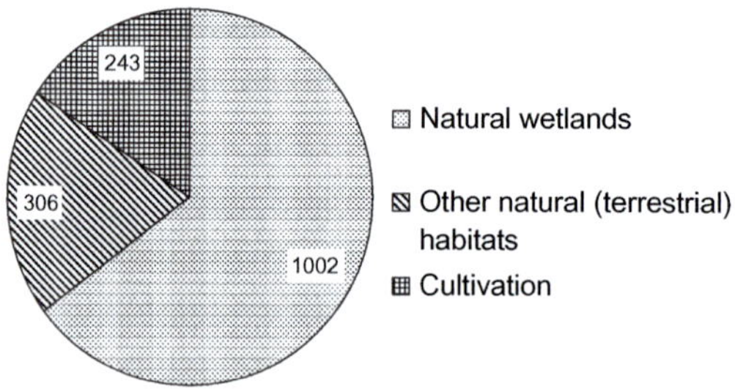

Fig. 11.7 Initial land cover (in km^2) on which 1550 km^2 of artificial wetland habitats were created (1975–2005) at 302 sites [2].

have replaced natural riparian habitats along watercourses. The other third corresponds essentially to the construction of water storage reservoirs in terrestrial or agricultural habitats, generally for the purposes of irrigation.

As part of their commitment to the Ramsar Convention on Wetlands, Mediterranean countries launched the MedWet Initiative in 1991, with the aim of preserving wetlands from unwanted developments. But our analyses above demonstrate to Mediterranean countries that their policies for the conservation and sustainable use of wetlands need to be consolidated or stepped up. This is because despite all of them signing the Ramsar Convention, natural wetlands continue to be lost throughout the Mediterranean basin, while man-made wetlands (reservoirs, rice fields, etc.) are increasing fast.

11.2.4 Water and Mediterranean wetlands: Current situation and trends

11.2.4.1 Open water surface extent

Despite the constant loss of natural wetlands, the area of inland surface water increased by almost 13% in the Mediterranean Basin during the last 30 years [2]. This trend is partly explained by the considerable rise in the number of dams and reservoirs constructed in the Mediterranean Basin during this period [16]. The regions with the highest increases of total flooded surface areas (North Africa, the Iberian Peninsula, part of the Balkans, Cyprus, and Turkey) include 1032 of the 1345 dams and reservoirs put into operation between 1984 and 2011 (Fig. 11.8).

The monitoring results, calculated for a sample of 1257 wetlands spread over all 28 MedWet countries, show that trends are rather the same as those previously observed for the whole Basin (increase in open water surface area between 1984 and 2015). However, there is a considerable contrast between natural and human-made wetlands. Fig. 11.9 suggests that this increase is above all linked to the major increase in open water surface areas in human-made wetlands (essentially dams and reservoirs in this case). This difference is even more significant in the region of the Mediterranean biodiversity hotspot, where sites dominated by artificial wetland habitats have had their open water surface area augmented by more than 60%, whereas there has been a drop of 0.17% in those dominated by natural habitats.

This situation could be due to the fact that in recent decades many Mediterranean countries have oriented their water management policies toward a major reinforcement of the storage capacities, in response to the needs of key socio-economic sectors (e.g., irrigation, aquaculture, demand for drinking water), to the detriment of natural ecosystems. Consequently, some natural wetlands have received less and less surface water, causing their permanent open water surface areas to progressively dry out, and to be transformed into temporarily flooded zones, or even to disappear completely (Fig. 11.10 gives the example of Lake Akşehir in Turkey).

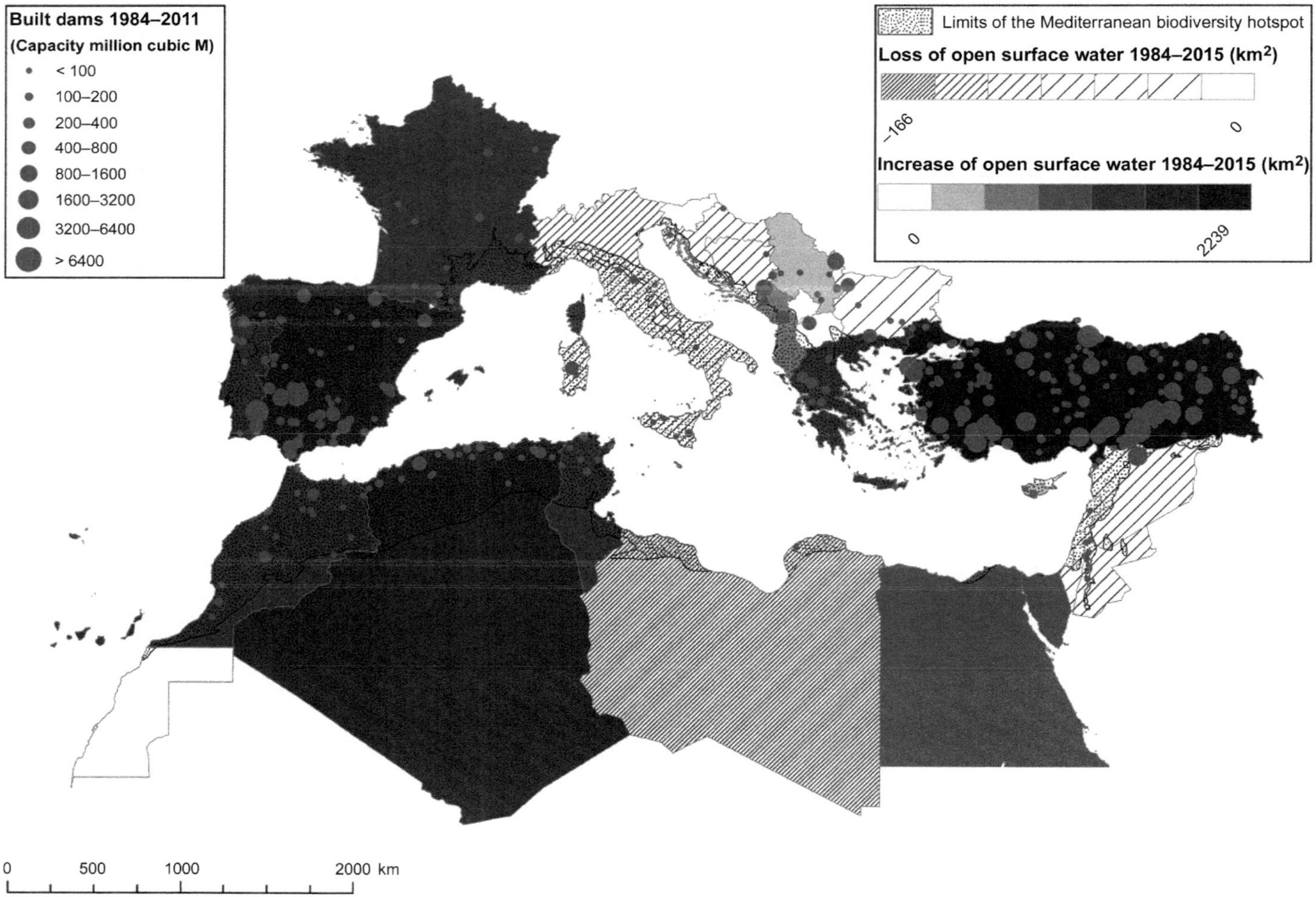

Fig. 11.8 Trends in total flooded surface areas in the MedWet countries and within the Mediterranean Biodiversity Hotspot [17] between 1984 and 2015 [18], with locations of the main dams and reservoirs (black circles proportional in size to their storage capacities) put into operation between 1984 and 2011 [16].

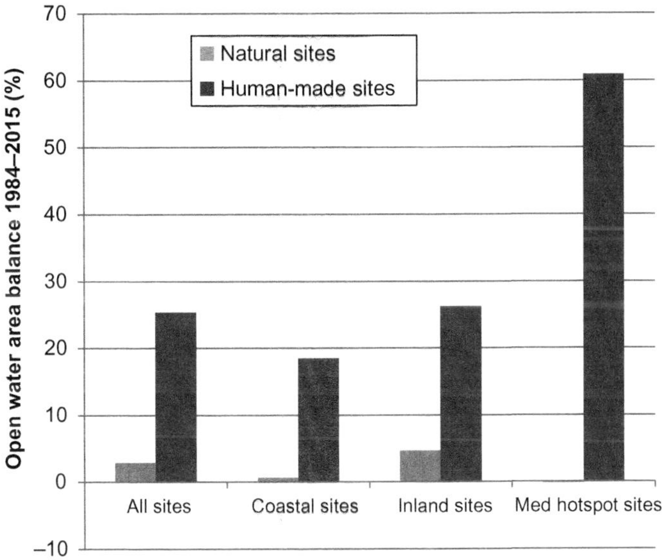

Fig. 11.9 Trends in open water surface areas (%) for a sample of 1257 Mediterranean wetlands (346 natural and 911 human-made with respective total surface areas of 75,929 km^2 and 31,913 km^2).

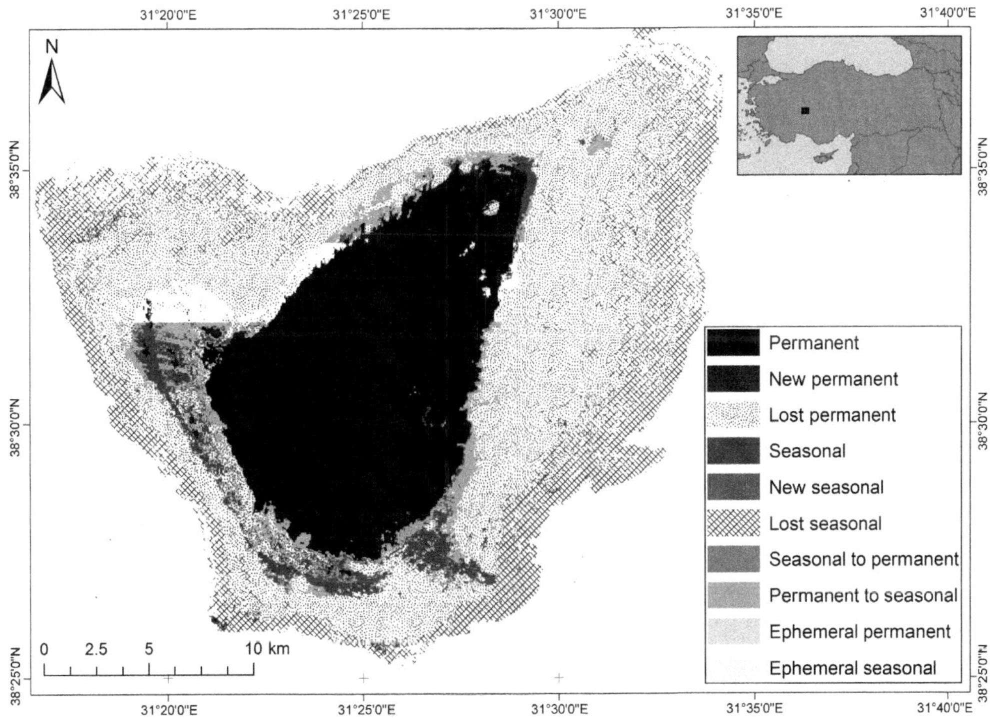

Fig. 11.10 Map of spatial-temporal surface water dynamics in Lake Akşehir (Turkey) between 1984 and 2015 [18]. Note the progressive drying out of the wetland, in particular with the regression of permanently flooded surface areas (transformed into temporarily flooded zones or totally dried out).

11.2.4.2 Pressures on wetlands: water quality, water abstraction

Freshwater is a crucial component of Mediterranean wetlands, and its characteristics (quantity, quality, and variation over time) are key ecological factors. This resource, which is unequally distributed in the region, is also used by human populations to meet their growing needs in terms of irrigation, manufacturing, drinking water, and tourism. In 2014, the annual withdrawals of water in the Mediterranean Basin were evaluated at between 313 km^3 and 323 km^3 (approximately 30% of its total renewable resources) [19]. Surface water abstraction from rivers, lakes, and dams accounts for 80% of this amount, and the rest is pumped from underground aquifers [19, 20]. Water consumption increased significantly during the 20th century, especially due to the demand for irrigation, which accounts for two-thirds of regional consumption [19]. The overexploitation of the underground water resources along the coast also often results in the intrusion of seawater and the salinization of water tables. In some inland areas, this has caused numerous wetlands which were fed by aquifers to dry up, such as the Azraq marshes in Jordan and the Tablas de Daimiel in Spain [21, 22]. Over-extraction for berry irrigation is also drying out temporary ponds in Doñana, Spain [23].

The flow of rivers largely determines the quantities of water which are effectively available for society, ecosystems, and various vital processes such as the transport of nutrients and sediments. Rivers are affected by ever increasing levels of water withdrawal for human activities, dams, and also climate changes, resulting in the degradation or disappearance of wetlands downstream. The drop in the flow of rivers between 1960 and 2000 was from 25% to 70%, reaching 80% for the River Cetina (Croatia) [24]. The Nile, previously a major source of freshwater for the Mediterranean, is an emblematic example of this decline (approximately −65%). All rivers considered, the total quantity of freshwater discharged into the Mediterranean each year (not including precipitation) declined by about 45% during the 20th century [24]. These global reductions in flow are often combined with other phenomena such as the reduction of minimum flows, and modified frequencies of flooding due to an increase in extreme weather events and river bed artificialization. A perceived increase in the need to ensure water security leads to more intensive hydraulic management, such as the construction of new dam–reservoirs or interbasin water transfers. The water resources available for wetlands thus become even scarcer throughout the Mediterranean region, and the water that does get to the wetlands is often more polluted due to reduced dilution levels.

The construction of dams in the Mediterranean Basin has increased massively, especially since 1950 (Fig. 11.11) [20, 25]. In 2013 their total storage capacity was estimated at 486 km^3, i.e., more than 1.5 times the current annual volume of freshwater discharged by rivers into the Mediterranean Sea. Dams cause multiple impacts, including the fragmentation of rivers and consequently of fish populations, a reduction in the duration and extent of flooding in many wetlands downstream, and a reduction in sediment transport leading to enhanced coastal erosion [26–28]. There is currently a major resurgence in

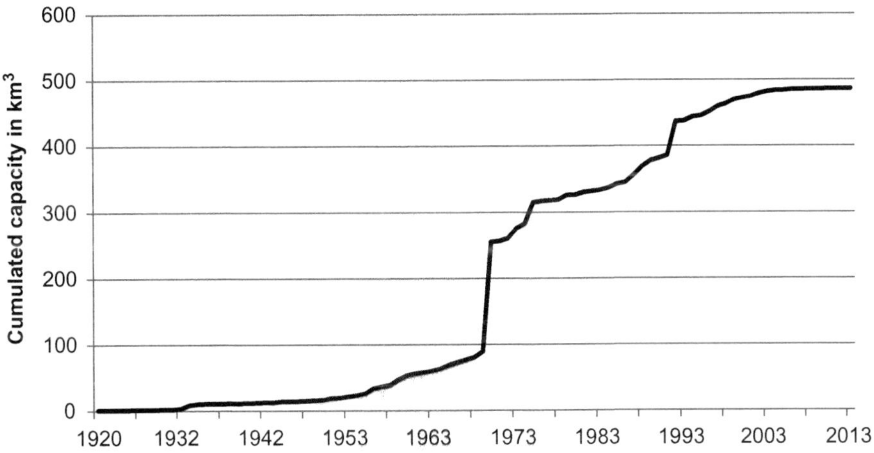

Fig. 11.11 Total storage capacity of dams on Mediterranean rivers, in km^3. *(Source: http://www.fao.org/nr/water/aquastat/dams/index.stm, accessed 04/01/2018.)*

dam construction in Anatolia and in the Balkans, a region that figures as one of the three world's "hotspots" for new dam projects [29]. On the other hand, the shallowest parts of reservoirs associated with dams can to some extent harbor important waterbird populations [30, 31].

Water quality is of primordial importance for the functioning of freshwater ecosystems and their biodiversity. It is threatened by the concentrations of numerous chemical components, largely attributable to human activity, including agriculture, industry, and domestic uses. Nutrients (nitrogen, phosphorus, and potassium), which cause the eutrophication of wetlands, are the most frequently monitored parameters, along with Biological Oxygen Demand (BOD), and heavy metals. But many other substances can affect water quality too, such as pesticides, polychlorobiphenyls (PCBs), drug residues (as shown clearly for Doñana in Spain: see [32–34]), nano-pollutants, and plastic waste. For coastal aquifers, salt levels are also a major issue. Wetlands play an important role in water purification, and therefore contribute to improving the quality of water in the environment. Water quality is mainly monitored in certain types of wetlands, such as artificial and natural lakes, rivers, coastal marine waters, and water tables (i.e., the water bodies covered by the European Union Water Framework Directive (WFD)). Other wetlands are relatively little monitored, and appropriate quality standards are less clear.

It is currently still impossible to obtain an overall assessment of water quality in Mediterranean wetlands, because too little monitoring of pollutants other than nutrients is carried out, and too little data shared [35]. Nonetheless, recent studies based on field data, socio-economic data, and modeling have enabled regional panoramas to be established for some criteria. For instance, the "Grey Water Footprint" maps the regions in which human activities cause the greatest pressures on water quality in general—and wetlands in

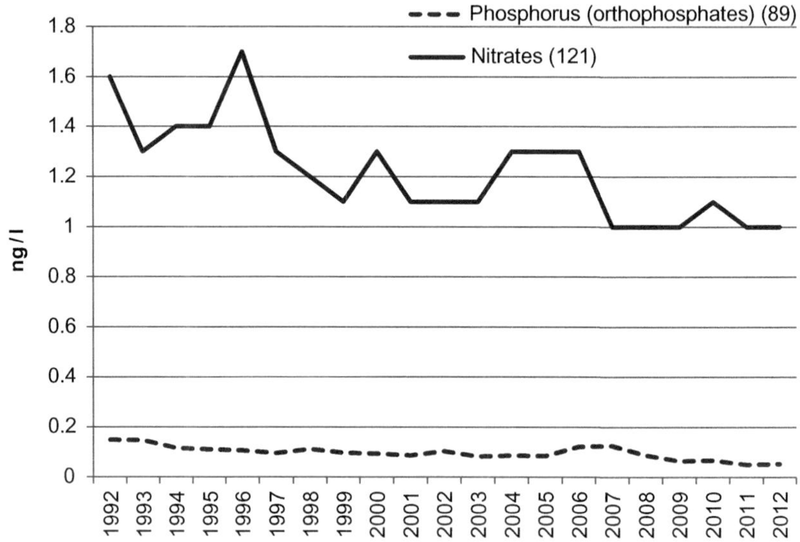

Fig. 11.12 Concentrations of nitrates and phosphorus (orthophosphates) in the European rivers that empty into the Mediterranean Sea. *Figures in brackets: number of measurement stations that provided usable data for each parameter. (Source: Online data from the European Environment Agency; EEA, 2015.)*

particular [36]. Despite the growing pressures, there are improvements to be noted. For Europe in general, water quality has improved in the main water masses with regard to BOD, nitrates, phosphates, and ammonium (cf. Fig. 11.12). Various factors, such as the development of wastewater treatment plants and phosphate-free detergents, have contributed to this progress. But in spite of these improvements for surface wetlands, the chemical status of a quarter of European water tables is still poor, mainly due to nitrates [37]. In North Africa and the Middle East, the trend is clearly toward a degradation of water quality, for example regarding BOD [38, 39]. Water quality has also declined recently in the watershed of Doñana, south-west Spain [40].

The degradation of water quality seems set to continue in the Mediterranean region [41]. Since there are no standardized data available for the whole Mediterranean Basin concerning most other pollutants, such as pesticides, heavy metals, drug residues, or various nano-pollutants, consolidated analysis is impossible. But significant quantities of sporadic data (e.g., Ref. [38]) suggest a general decline in water quality.

11.2.5 Ecosystem services provided by Med Wetlands

Apart from their importance for biodiversity globally, wetlands offer many ecosystem services that contribute to human well-being as well as nature-based solutions, for instance to buffer climate change impacts [2, 42, 43].

Human settlements have been using wetlands to ensure their well-being since a very long time. Among others, wetlands provide water for humans and cattle to drink as well as to produce food and construction materials. Many cultural traditions developed such as wetland adapted local breeds of horses and cattle in the Camargue (France), Tuscany (Maremma), Italy, Prespa (Greece), or Menorca (Spain) [44]. In North Africa, one out of three species of freshwater fish and aquatic plant species provide direct socio-economic advantages for local people [45, 46]. In Mediterranean coastal lagoons, the European eel (*Anguilla anguilla*) catch, which formed an important resource for local fishermen, has steadily declined in the last decades [47] as a result of a combination of various pressures on wetlands and rivers.

With the surface area and quality of Mediterranean wetlands decreasing, the capacity of Mediterranean wetlands to provide ecosystem services has been reducing rapidly in recent years.

Although this is clear for many of the actors depending on the services from Mediterranean wetlands, the lack of general baseline data is a real problem for quantifying the actual loss of services. Additionally, the complexity of the Mediterranean social-ecosystems limits the suitability of commonly used methods for ecosystem services assessments [4]. With Mediterranean wetlands being directly dependent on the quantity and quality of available water, the reducede quantity and quality of freshwater directly affect the potential supply of ecosystem services, such as clean drinking water, fisheries, rice production, and medicinal plants, whereas many other ecosystem services are indirectly affected through the impact on underlying ecological processes and biodiversity.

In addition to the potential reductions in the supply of ecosystem services, demand for ecosystem services has been increasing. With the current and projected impacts of climate change, the protection of people against hazards such as flooding or prolonged heat waves is more widely and more often required. Simultaneously, the number of people living in the Mediterranean basin has been steeply increasing, especially in coastal areas [2], and agriculture and industries have developed to cater for their consumption. This has caused an encroachment of both infrastructure and people into areas prone to flooding. Furthermore, the related artificialization has also reduced the capacity of ecosystems to provide regulating services. The combined result is a higher number of people being exposed to increasingly more frequent hazards, such as flooding or periods of droughts [44].

With increasing human population numbers and decreasing natural resources, increasing mismatches between the supply and the demand for ecosystem services are a prevailing trend for Mediterranean wetlands [44, 48]. For instance, the reduction of the surface of wetlands due to artificialization of riverbeds has greatly impacted the capacity of wetlands to mitigate floods, which increases the risk of flooding, while society has also become more vulnerable to the risk as more people live and more capital resides in sensitive areas [49].

For some services, the end of their supply has already been predicted, such as in the case of eel fishing. The combined impacts of a high demand for eels, pollution, introduced pathogens, and the impact of dams on river connectivity and dispersal capacities predict the disappearance of both this species and this traditional fishing practice, whereas other services are predicted to increase, such as bird watching tourism or education [2].

Given the dependence of the well-being of people on wetlands, investments in the conservation and sustainable use of wetlands directly contribute to the countries' objectives to meet the Paris agreement to combat climate change and adapt to its effects [50], to contribute to global biodiversity conservation targets [51], and to advance on the Sustainable Development Goals [52]. Although the indicators of the MWO2 unfortunately suggest a reduction of the contribution of Mediterranean wetlands to Sustainable Development Goals, examples of positive impacts that sustainable management, restoration, protection, and societal awareness can have on the conservation of Mediterranean wetlands do exist [2]. To achieve increases in the contributions of wetlands to human well-being and multiple policy objectives, the current challenge is to determine which interventions are really effective and provide this information to public, private, and societal actors to empower them in changes in current practices [2].

11.3 Description of the main methodologies to analyze the considered problems

11.3.1 Status and trends of Mediterranean wetland habitats

11.3.1.1 Land use/land cover in and around Mediterranean wetlands

The methodology used by the Mediterranean Wetland Observatory to assess and monitor Land Use Land Cover (LULC) status and trends in and around Mediterranean wetlands is based on Earth Observation (EO) data, i.e., satellite imagery. In the case of the MWO2, all results were derived from Landsat time series covering three epochs (1975, 1990, and 2005). The mapping approach (object-based supervised classification) was developed in the framework of the GlobWetland-II project (2010–2014 and funded by the European Space Agency), which had been further improved by the Mediterranean Wetland Observatory [3], especially by reducing the total number of identified classes in order to limit common confusions between some "spectrally similar" habitats (e.g., lakes and coastal lagoons). Thus, all LULC maps for the 302 sampled sites (mainly Ramsar sites, Pas, or Important Bird Areas, but other sites have also been taken into consideration) were ultimately represented using six broad classes: built-up areas, agricultural lands, natural dry lands, natural wetlands, human-made wetlands, and sea/oceans.

Once LULC maps were produced for all sites, many indicators were then computed and derived using GIS tools: areas of each of the listed classes above for each of the three time periods and some conversion indicators (e.g., conversion of the natural wetland habitats into agricultural lands, built-up areas, or artificial wetland classes). All these indicators were defined by the MWO, assuming that they represent the status and trends of wetland habitats identified within the 302 mapped sites.

11.3.1.2 The WET index

Based on a method developed by the UNEP's World Conservation Monitoring Centre (WCMC) and the Ramsar Convention Secretariat [7, 53], the WET index is the transposition in terms of wetland surface areas of the Living Planet Index (LPI) for biodiversity [54]. It represents the average of trends (gain or loss rates) in surface area of natural wetlands, calculated as the weighted aggregation of the trends for coastal/marine and inland wetlands relative to the reference year 1970. It requires a sufficient number (i.e., at least several hundreds of sites, S. Darrah/WCMC, personal communication) of diachronic data series (i.e., at least two measures of the wetland surface area at different dates), for the same sites. Data for a sample of 405 sites throughout the Mediterranean were obtained by the WCMC, through a literature search and a call for data launched by Ramsar. Out of these, data for 288 sites form a subset provided by the MWO.

The main advantage of the method is that it takes into account data series covering dates and durations that can vary considerably between sites. This configuration, the most frequent with real data, generally prevents the calculation of the loss rate between two dates by merely calculating the difference between the surface areas at those dates. Also, in order to avoid overrepresentation of certain subregions or types of wetlands that are likely to be better monitored, a weighting is applied. Each wetland site is first attributed to a major wetland type and to a Mediterranean subregion. The geometric mean of the loss rates is then calculated for all the sites for each possible pair (subregion x major wetland type), whatever the number of sites monitored for that subregion and wetland type (e.g., Adriatic x lagoons). Wetlands are not weighted in function of their size. Then the average for all these pairs is calculated to produce the WET Index for the total sample of sites.

It should be noted that the WET Index applies only to the sample of wetlands studied, and does not necessarily reflect the percentage change in the region they pertain to, overall. Also, the WET Index represents, in the Mediterranean, a weighted average of the individual loss rates (or gain rates) of natural wetland habitats within 405 sites, but not necessarily the change in total surface area of wetland habitats within this sample, let alone the whole region. These two metrics can actually be quite different. For instance, within the WCMC dataset, the MWO subset consists of 288 sites all monitored on the same dates, i.e., 1975–2005—a rare exception. This allows comparisons to be made on this subset, between the two possible approaches. The net loss in surface area of natural wetland habitats within these 288 sites (as calculated by subtraction) was 9.5%, whereas the concomitant WET index was 23% (i.e., the weighted average of the loss rates for each of the 288 sites). This 2.4-fold difference stems mainly from the fact that wetland loss rates are not equally distributed between small and large sites and between wetland types, and also from the WET Index weighting mechanism. The method can also be sensitive to the sample selected: the WET Index was −23% for the 288 MWO subset, but was −48% for the whole WCMC sample of 405 sites—including these 288. Thus, the precise wording for presenting the WET Index results is crucial for avoiding misinterpretations: on average, wetlands have declined by 23% in these 288 sites (weighted WET Index), but at the same time, these sites have lost 9.5% of the total surface area of their natural wetlands. In most cases of real data, however, only the WET Index can be calculated.

11.3.2 Biodiversity of Mediterranean wetlands: LPI, IUCN Red Lists, International Waterbird Census

As is commonly known, biodiversity data demonstrate geographical biases [55]. In the Mediterranean basin, the availability in biodiversity data is unbalanced between the northern and southern shores of the Mediterranean basin. For instance, the European Union has set up a reporting system providing detailed information on the status of species and habitats (http://ec.europa.eu/environment/nature/knowledge/rep_habitats/index_en.htm), for which there is no equivalent on the southern shore.

To obtain assessments on the status and trends of biodiversity that are as representative as possible of the entire Mediterranean basin, the Mediterranean Wetlands Observatory has given great importance to one of the oldest biodiversity monitoring programs in the world: the International Waterbird Census (http://ec.europa.eu/environment/nature/knowledge/rep_habitats/index_en.htm). These worldwide monitoring consists of counting all waterbirds visible on water bodies and wetlands around mid-January. When analyzed, these counts allow for an estimation of the population size of waterbird populations and the identification of sites of international importance (e.g., Ramsar sites). Waterbirds are also interesting indicators to inform us about the state of their main habitat, wetlands. These counts are carried out by professional and volunteer ornithologists who send their observations to a national coordinator who in turn sends them to the NGO Wetlands International. In recent years, the Mediterranean Waterbird Network coordinated by the Tour du Valat, the French national agency for wildlife, and several North African organizations (https://tourduvalat.org/en/actions/reseau-oiseaux-eau-mediterranee/) has helped in revitalizing counts in the Mediterranean.

One indicator used at the global level to inform on biodiversity trends, the Living Planet Index, was used to assess trends in vertebrates in Mediterranean wetlands (e.g., Ref. [56]). It has the key advantage of taking into account almost any type of abundance data regardless of the monitoring protocol used. Much of the raw data used to generate the index was obtained from studies published in scientific journals that are publicly accessible. Datasets were weighted to take into account possible biases in taxonomic or geographical representation.

Finally, the previous indicators and datasets only provide information on vertebrates, which represent a limited part of wetland biodiversity. In the Mediterranean basin, plants and invertebrates are particularly diverse and threatened. In addition, they can be more informative for different types of wetlands such as temporary pools, peat bogs, or ground waters. Fortunately, many competent experts exist in the Mediterranean and are regularly called upon to assess the conservation status of species according to the standard criteria of the IUCN Red List of Threatened Species (e.g., Ref. [57]). Following this methodology, serious assessments can be done even in the absence of abundance monitoring. Thanks to red lists, it is possible to identify regions and sites where threatened species are concentrated and, when these assessments are repeated over time, to produce an indicator that provides information on biodiversity trends.

11.3.3 Water in wetlands

11.3.3.1 Open water surface extent

Water is an essential component for Mediterranean wetlands: its quantity, quality, and temporality are key ecological determining factors. Yet hydrological functions are altered in various ways, particularly by withdrawal and storage (dams, retention basins, hydroelectric production) to respond to human and economic needs. These factors weigh heavily on all water resources, including wetlands.

Based on EO data and tools, it is possible to estimate the modifications of surface open water resources, whether increasing or decreasing, at several geographical scales (regional, national, catchment, and site). The causes of the trends are assumed to be essentially anthropogenic, with a minor influence of intra-annual climate changes and variations. However, surface water is different from measuring the dynamics of wetland habitats, because it essentially focuses on the flooded areas rather than habitats in the ecological sense. Moreover, the data used to calculate it are mainly derived from EO tools (optical satellites in this case), which have certain limitations, notably their inability to estimate the flooded surface area under dense vegetation covers, or with images featuring clouds.

The raw data are derived from the Global Surface Water Explorer [18]. They are available free of charge at worldwide scale, in the form of rasters (spatial resolution 30 m), enabling them to be cross-analyzed with other georeferenced databases (e.g., administrative boundaries within countries, or catchment watersheds obtained from the HydroSHEDS database), and therefore to be presented at different geographical scales (subregional, national, catchment, and site).

11.3.3.2 Other dimensions of water monitoring

Apart from remote sensing, water data can be gathered using a multitude of sources. For instance, the Mediterranean Wetlands Observatory uses official, national statistics produced by Mediterranean countries, which are secondarily gathered at a global level by the World Bank and FAO (e.g., [19, 25]), or at regional level by the Plan Bleu (2009) [13]. These allow the calculation of standard indicators that are now widely used, e.g., the Water Exploitation Index (WEI), which assesses the long-term utilization of freshwater. It is the relationship between the total volume of water used to cover human needs in a given area and the renewable water resources available in that area. This indicator is used in the framework of the Mediterranean Strategy for Sustainable Development, of Aichi Target 4 of the Convention on Biological Diversity (water stress), as well as in the monitoring of Target 6.4 of the Sustainable Development Goals, which gives countries a strong incentive to report on it on a regular basis.

International databases on hydrology, socio-economic statistics, climate, dams, etc., usually accessible online, are used to develop models for various aspects of water studies, e.g., water quality (e.g., [36]) and river flows [24].

11.3.4 Ecosystem services

One of the most important conceptual and methodological challenges in the monitoring of ecosystem service flows is that it requires multiple indicators and methods to obtain a full understanding of why an ecosystem services changes over time [2, 48]. Additionally, Mediterranean social-ecosystems have a high interannual variability and require the adaptation of commonly used indicators [4].

Data on how the potential supply is changing (e.g., a wetland is losing biodiversity) need to be combined with information on how the supply is managed (e.g., the implementation phase of a nature conservation management plan) to understand which changes are taking place in the ecosystem. Furthermore, the actual use of the service (i.e., nature tourism) can greatly depend on whether or not people have sufficient access (e.g., access roads, bird watching huts). Finally, whether people are willing to spend money or time to support the protection of the wetland and whether governance bodies actively seek to protect areas provide indices on the societal support for both the services and the protection of the wetland that provides the service. Generally, reporting of ecosystem services tends to focus on only a few elements of the ecosystem service flow and the selection process is strongly dependent on the availability of the existing data [58]. At larger scales, assessments using land cover data (e.g., mapping of ecosystem services) or national statistical data are often used, while at a local level, the perceptions and engagement of stakeholders are often included (e.g., participatory mapping). All these methods have their limitations (see for instance Refs. [59, 60] for participatory mapping and stakeholder biases, or Ref. [61] for mapping of ecosystem services), but the major challenge that currently still needs to be tackled is to obtain data on changes in all the major five elements of ecosystem services flows: potential supply, supply, use, demand, and governance [48, 58].

11.4 Main conclusions

The Mediterranean wetland Outlooks 1 and 2 demonstrate a continued trend of loss and degradation of Mediterranean wetlands [1, 2]. The loss of natural wetlands since 1970 amounts to 48%. The abundance of wetland-dependent species has declined by 15% between 1990 and 2013, with contrasting trends between taxonomic groups. The abundance of globally threatened, wetland-dependent species even declined by 46%. Overall, 36% of wetland-dependent species living in Mediterranean wetlands are threatened with extinction.

On a sample of wetlands, urbanization increased by almost 300% and cultivated land by 42% between 1975 and 2005. Conversely, many artificial wetland habitats were created, and nearly two-thirds of them replaced natural wetlands. But despite the constant loss of natural wetlands, the area of inland surface water increased by almost 13% in the Mediterranean Basin during the last 30 years, which is partly explained by the fast rise in dam construction in the Mediterranean Basin.

The pressure on freshwater resources continues to increase. Water consumption went up significantly during the 20th century, especially for irrigation, which today accounts for two-thirds of the regional consumption. The total quantity of freshwater discharged by all rivers into the Mediterranean each year has declined by about 45% during the 20th century, due both to climate change and water abstraction by people. Although water quality in Mediterranean wetlands continues to decrease, some improvements were noted in Europe with regard to BOD, nitrates, phosphates, and ammonium. In North Africa and the Middle East, the trend is clearly toward a degradation of water quality in most wetlands.

With the number, surface area, and quality of Mediterranean wetlands decreasing, their capacity to provide ecosystem services has been reducing rapidly in recent years. For instance, the region lost about 20% of the capacity of wetlands to regulate floods.

However, countries continue to commit to the protection of wetlands and the number of sites designated under the Ramsar Convention continues to increase—although only 44% of these Ramsar sites have a management plan being implemented. There remains therefore a pressing need to raise awareness at all levels in the society, from the general public to decision-makers, on the reasons why Mediterranean wetlands need to be protected and sustainably managed.

References

[1] MWO (Mediterranean Wetlands Observatory), Mediterranean Wetlands Outlook, Technical report, Tour du Valat, Arles, France, 2012.

[2] MWO (Mediterranean Wetlands Observatory), Mediterranean Wetlands Outlook 2: Solutions for sustainable Mediterranean wetlands, Tour du Valat/MedWet, Arles, France, 2018.

[3] C. Perennou, A. Guelmami, M. Paganini, P. Philipson, B. Poulin, A. Strauch, J. Truckenbrodt, C. Tottrup, I.R. Geijzendorffer, Mapping Mediterranean wetlands with remote sensing: a good looking map is not always a good map, Adv. Ecol. Res 58 (2018) 243–277.

[4] M.V. Balzan, A.M. Pinheiro, A. Mascarenhas, A. Morán-Ordóñez, A. Ruiz-Frau, C. Carvalho-Santos, I.N. Vogiatzakis, J. Arends, J. Santana-Garcon, J.V. Roces-Díaz, L. Brotons, C.S. Campagne, P.K. Roche, S. De Miguel, S. Targetti, E.G. Drakou, V. Vassiliki Vlami, F. Baro, I. Geizendorffer,

Improving ecosystem assessments in Mediterranean social-ecological systems: a DPSIR analysis, Ecosyst. People 15 (2019) 136–155. https://www.tandfonline.com/doi/full/10.1080/26395916.2019.1598499.

[5] C. Perennou, C. Beltrame, A. Guelmami, P. Tomas Vives, P. Caessteker, Existing areas and past changes of wetland extent in the Mediterranean region: an overview, Ecol. Medit. 38 (2012) 53–66.

[6] N.C. Davidson, C.M. Finlayson, Extent, regional distribution and changes in area of different classes of wetland, Mar. Freshw. Res. 69 (2018), https://doi.org/10.1071/MF17377.

[7] S.E. Darrah, Y. Shennan-Farpóna, J. Loh, N.C. Davidson, C.M. Finlayson, R.C. Gardner, M. J. Walpole, Improvements to the Wetland Extent Trends (WET) index as a tool for monitoring natural and human-made wetlands, Ecol. Indic. 99 (2019) 294–298.

[8] Ramsar Convention on Wetlands, Global Wetland Outlook: State of the World's wetlands and their services to people, Ramsar Convention Secretariat, Gland, Switzerland, 2018.

[9] E. Gaget, T. Galewski, F. Jiguet, I. Leviol, Waterbird communities adjust to climate warming according to conservation policy and species protection status, Biol. Conserv. 227 (2018) 205–212.

[10] E. Gaget, T. Galewski, F. Jiguet, A. Guelmami, C. Perennou, C. Beltrame, I. Leviol, Natural habitat conversion reduces wintering waterbird community adjustment to climate warming, Consulted on 09/03/2020.

[11] P. Gaüzere, F. Jiguet, V. Devictor, Can protected areas mitigate the impacts of climate change on bird's species and communities? Diversity Distrib. 22 (2016) 625–637.

[12] A. Mariotti, Y. Pan, N. Zeng, A. Alessandri, Long-term climate change in the Mediterranean region in the midst of decadal variability, Climate Dynam. 44 (2015) 1437–1456.

[13] Plan Bleu, État de l'environnement et du développement en Méditerranée, UNEP/MAP Plan Bleu, Athens, Greece, 2009.

[14] Mediterra, Repenser le développement rural en Méditerranée, Centre International des Hautes Etudes Agronomiques méditerranéennes, Presses de Sciences Po Paris, 2009.

[15] MWO (Mediterranean Wetlands Observatory), Land cover: Spatial dynamics in Mediterranean coastal wetlands from 1975 to 2005, Tour du Valat, Arles, France, 2014.

[16] B. Lehner, C. Reidy Liermann, C. Revenga, C. Vörösmarty, B. Fekete, P. Crouzet, P. Döll, M. Endejan, K. Frenken, J. Magome, C. Nilsson, J. Robertson, R. Rödel, N. Sindorf, D. Wisser, High resolution mapping of the world's reservoirs and dams for sustainable river flow management. Frontiers Ecol. Envir. (2011). https://doi.org/10.1890/100125.

[17] R.A. Mittermeier, W.R. Turner, F.W. Larsen, T.M. Brooks, C. Gascon, Global biodiversity conservation: the critical role of hotspots, in: F.E. Zachos, J.C. Habel (Eds.), Biodiversity Hotspots, Springer, Berlin/Heidelberg, Germany, 2011, pp. 3–22. https://doi.org/10.1007/978-3-642-20992-5_1.

[18] J.F. Pekel, A. Cottam, N. Gorelick, A.S. Belward, High-resolution mapping of global surface water and its long-term changes, Nature 540 (2016) 418–422, https://doi.org/10.1038/nature20584.

[19] World Bank. World Development Indicators: 3.5 Freshwater. http://wdi.worldbank.org/table/3.5#, consulted on 08/01/2018.

[20] J. Margat, S. Treyer, S., L'eau des Méditerranéens: situation et perspectives, MAP Technical Report Series No. 158, UNEP/MAP, Athens, 2004.

[21] Y.A. Al Zu'bi, The Azraq oasis, Jordan, in: C. Morillo, J.L. Gonzalez (Eds.), Management of Mediterranean Wetlands, Vol. II, MedWet/Ministerio de Medio Ambiente, Madrid, Spain, 1996, pp. 167–182.

[22] M. Peinado, The endorheic complex of la Mancha's lagoons, Spain, in: C. Morillo, J.L. Gonzalez (Eds.), Management of Mediterranean Wetlands, Vol. II, MedWet/Ministerio de Medio Ambiente, Madrid, Spain, 1996, , pp. 129–142.

[23] A.J. Green, P. Alcorlo, E.T.H.M. Peeters, E.P. Morris, J.L. Espinar, M.A. Bravo-Utrera, J. Bustamante, et al., Creating a safe operating space for wetlands in a changing climate, Frontiers Ecol. Envir. 15 (2017) 99–107.

[24] W. Ludwig, E. Dumont, M. Meybeck, S. Heussner, River discharges of water and nutrients to the Mediterranean and Black Sea: Major drivers for ecosystem changes during past and future decades, Prog. Oceanogr. 80 (2009) 199–217.

[25] FAO-AquaStat, www.fao.org/nr/water/aquastat/dams/index.stm, 2018. consulted on Jan. 4th 2018.

[26] C. Ibàñez, N. Prat, A. Canicio, Changes in the hydrology and sediment transport produced by large dams on the lower Ebro river and its estuary, River Res. Appl. 12 (1996) 51–62.

[27] M. Snoussi, S. Haida, S. Imassi, Effects of the construction of dams on the Moulouya and the Sebou rivers (Morocco), Reg. Environ. Change 3 (2002) 5–12.

[28] J. Day, C. Ibáñez, F. Scarton, D. Pont, P. Hensel, D. Jason, R. Lane, Sustainability of Mediterranean deltaic and lagoon wetlands with sea-Level rise: The importance of river input, Estuaries Coast 31 (2011) 183–193, https://doi.org/10.1007/s12237-011-9390-x.

[29] C. Zarfl, A.E. Lumsdon, J. Berlekamp, L. Tydecks, K. Tockner, A global boom in hydropower dam construction, Aquat. Sci. 77 (2014) 161–170.

[30] B. Afdhal, F. Charfi-Cheikhrouha, A. Moali, Tunisian man-made wetlands as alternative habitats for waterbirds and their role for conservation, Afr. J. Ecol. 51 (2012) 154–163.

[31] E. Sebastián-González, A.J. Green, Reduction of avian diversity in created versus natural and restored wetlands, Ecography 39 (2016) 1176–1184.

[32] D. Camacho-Munoz, J. Martin, J.L. Santos, I. Aparicio, E. Alonso, Occurrence, temporal evolution and risk assessment of pharmaceutically active compounds in Donana Park (Spain), J. Hazardous Mat. 183 (2010) 602–608.

[33] D. Camacho-Munoz, J. Martin, J.L. Santos, I. Aparicio, E. Alonso, Presence of pharmaceutically active compounds in Donana Park (Spain) main watersheds, J. Hazardous Mat. 177 (2010) 1159–1162.

[34] D. Camacho-Munoz, J. Martin, J.L. Santos, I. Aparicio, E. Alonso, Distribution and risk assessment of pharmaceutical compounds in river sediments from Donana Park (Spain), Water Air Soil Pollut. 224 (2013) 1665.

[35] P. Haener, Etude de faisabilité sur le développement d'un mécanisme d'observation régional sur l'eau en Méditerranée, Office International de l'Eau/SEMIDE, Sophia-Antipolis, 2008. http://www.emwis.net/documents/emwis-main-studies/medwip/MEDWIP2-final-EN.pdf.

[36] M.M. Mekonnen, A.Y. Hoekstra, Global gray water footprint and water pollution levels related to anthropogenic nitrogen loads to freshwater environments. Sci. Technol. 49 (2015) 12860–12868, https://doi.org/10.1021/acs.est.5b03191.

[37] EEA (European Environment Agency), SOER 2015: The European environment—state and outlook 2015. Cross-country comparisons > Freshwater quality > Nutrients in rivers, 2015. (https://www.eea.europa.eu/soer).

[38] MEDPOL, Indicator Fact Sheet 5. Nutrient concentrations in transitional, coastal and marine waters of the Mediterranean Sea, Indicator Specification Version 1, MEDPOL Monitoring database, 2015. http://enpi-seis.pbe.eea.europa.eu/data-and-indicators/enpi-seis-south-h2020-indicators/indicators-assessment/nutrient-concentration-ind5.

[39] UNEP, A snapshot of the World's water quality: towards a global assessment, United Nations Environment Programme, Nairobi, 2016.

[40] I. Paredes, F. Ramirez, M.G. Forero, A.J. Green, Stable isotopes in helophytes reflect anthropogenic nitrogen pollution in entry streams at the Doñana World Heritage Site, Ecol. Indic. 97 (2019) 130–140.

[41] Veolia, IFPRI (International Food Policy Research Institute), The murky future of global water quality. New global study projects rapid deterioration in water quality. A white paper, Washington DC, 2015.

[42] E. Cohen-Shacham, G. Walters, C. Janzen, S. Maginnis, Nature-based solutions to address global societal challenges. International Union for Conservation of Nature (IUCN), Gland, Switzerland, 2016.

[43] P. Bridgewater, A new context for the Ramsar convention: wetlands in a changing world, Rev. Eur. Community Int. Environ. Law 17 (2008) 100–106, https://doi.org/10.1111/j.1467-9388.2008.00582.x.

[44] I.R. Geijzendorffer, T. Galewski, A. Guelmami, C. Perennou, N. Popoff, P. Grillas, Mediterranean Wetlands: a Gradient from Natural Resilience to a Fragile Social-Ecosystem, in: M. Schröter, A. Bonn, S. Klotz, R. Seppelt, C. Baessler (Eds.), Atlas of Ecosystem Services: Drivers, Risks, and Societal Responses, Springer, 2019, pp. 83–90, https://doi.org/10.1007/978-3-319-96229-0.

[45] N. García, A. Cuttelod, D. Abdulmalak, The status and distribution of freshwater biodiversity in northern Africa, IUCN, Gland, Switzerland, 2010.

[46] D. Juffe-Bignoli, L. Rhazi, P. Grillas, The socio-economic value of aquatic plants in northern African region, D. Juffe-Bignoli, W.R.T. Darwall, (Eds.), in: Assessment of Socio-Economic Value of Freshwater Species in the North African Region, IUCN, Gland, Switzerland and Málaga, Spain, 2012, p. 84.
[47] E. Aalto, F. Capoccioni, J. Terradez Mas, M. Schiavina, C. Leone, G. De Leo, E. Ciccotti, Quantifying 60 years of declining European eel (*Anguilla anguilla* L., 1758) fishery yields in Mediterranean coastal lagoons. ICES J. Mar. Sci. J Cons. 73 (2016) 101–110.
[48] I.R. Geijzendorffer, B. Martin-Lopez, P.K. Roche, Improving the identification of mismatches in ecosystem services assessments, Ecol. Indic. 52 (2015) 320–331, https://doi.org/10.1016/j.ecolind.2014.12.016.
[49] W.J. Mitsch, J.G. Gosselink, The value of wetlands: importance of scale and landscape setting, Ecol. Econ. 35 (2000) 25–33.
[50] United Nations, Transforming Our World: The 2030 Agenda for Sustainable Development, United Nations, New York, 2015.
[51] CBD, Quick guides to the Aichi Biodiversity Targets, version 2. Decision document UNEP/CBD/COP/DEC/X/2, CBD Secretariat, Montreal, 2013.
[52] I.R. Geijzendorffer, C. Beltrame, L. Chazee, E. Gaget, T. Galewski, A. Guelmami, C. Perennou, N. Popoff, C. Guerra, R. Leberger, J. Jalbert, A more effective Ramsar Convention for the conservation of Mediterranean wetlands, Front. Ecol. Evol. 7 (2019) 21.https://www.frontiersin.org/articles/10.3389/fevo.2019.00021/full.
[53] M.J.R. Dixon, J. Loh, N.C. Davidson, et al., Tracking global change in ecosystem area: the Wetland Extent Trends index. Biol. Conserv. 193 (2016) 27–35, https://doi.org/10.1016/j.biocon.2015.10.023.
[54] J. Loh, R.E. Green, T. Ricketts, J. Lamoreux, M. Jenkins, V. Kapos, J. Randers, The living planet index: using species population time series to track trends in biodiversity, Phil. Trans. Roy. Soc. B 360 (2005) 289–295.
[55] C. Meyer, H. Kreft, R. Guralnick, W. Jetz, Global priorities for an effective information basis of biodiversity distributions, Nature Comm. 6 (2015) 8221.
[56] T. Galewski, B. Collen, L. McRae, et al., Long-term trends in the abundance of Mediterranean wetland vertebrates: from global recovery to localized declines, Biol. Conserv. 144 (2011) 1392–1399.
[57] E. Riservato, J.P. Boudot, S. Ferreira, M. Jovi, V.J. Kalkman, W. Schneider, B. Samraoui, A. Cuttelod, The Status and Distribution of Dragonflies of the Mediterranean Basin, IUCN, Gland, Switzerland and Malaga, Spain, 2009.
[58] I.R. Geijzendorffer, E. Cohen-Shacham, A.F. Cord, W. Cramer, C. Guerra, B. Martín-López, Ecosystem services in global sustainability policies, Environ. Sci. Policy 74 (2017) 40–48. https://www.sciencedirect.com/science/article/pii/S1462901117300047.
[59] B. Martín-López, I. Iniesta-Arandia, M. García-Llorente, I. Palomo, I. Casado-Arzuaga, D.G. Del Amo, E. Gómez-Baggethun, E. Oteros-Rozas, I. Palacios-Agundez, B. Willaarts, J.A. González, Uncovering ecosystem service bundles through social preferences, PLoS One 7 (2012) e38970.
[60] A.P. García-Nieto, C. Quintas-Soriano, M. García-Llorente, I. Palomo, C. Montes, B. Martín-López, Collaborative mapping of ecosystem services: the role of stakeholders′ profiles, Ecosyst. Serv. 13 (2015) 141–152.
[61] S. Jacobs, B. Burkhard, T. Van Daele, J. Staes, A. Schneiders, The Matrix Reloaded: a review of expert knowledge use for mapping ecosystem services, Ecol. Model. 295 (2015) 21–30.

Scientific committee

Clément Albergel, CNRS CNRM, France
Luca Brocca, CNR, Italy
Jean-Christophe Calvet, Météo France CNRM, France
Pierre Chevalier, IRD HSM, France
Père Quintana Segui, Observatori de l'Ebre, Spain
Valentina Lazarova, Water Globe Consultants, France
Michel Le Page, IRD CESBIO, France
Catherine Ottlé, CNRS LSCE, France
Olivier Merlin, CNRS CESBIO, France
François Molle IRD G-EAU, France
Damien Raclot, IRD LISAH, France
Yves Tramblay IRD HSM, France
Gerard Van der Schrier, Royal Netherlands Meteorological Institute, Netherlands
Mehrez Zribi CNRS CESBIO, France

Index

Note: Page numbers followed by *f* indicate figures, *t* indicate tables, and *b* indicate boxes.

N

O

P

Q

R

Z

9780128180860